T0281064

Lecture Notes in Computer Science 14833

Founding Editors

Gerhard Goos
Juris Hartmanis

The series Lecture Notes in Computer Science (LNCS), including its subseries Lecture Notes in Artificial Intelligence (LNAI) and Lecture Notes in Bioinformatics (LNBI), has established itself as a medium for the publication of new developments in computer science and information technology research, teaching, and education.

LNCS enjoys close cooperation with the computer science R & D community, the series counts many renowned academics among its volume editors and paper authors, and collaborates with prestigious societies. Its mission is to serve this international community by providing an invaluable service, mainly focused on the publication of conference and workshop proceedings and postproceedings. LNCS commenced publication in 1973.

Leonardo Franco · Clélia de Mulatier ·
Maciej Paszynski · Valeria V. Krzhizhanovskaya ·
Jack J. Dongarra · Peter M. A. Sloot
Editors

Computational Science – ICCS 2024

24th International Conference
Malaga, Spain, July 2–4, 2024
Proceedings, Part II

 Springer

Editors
Leonardo Franco (iD)
University of Malaga
Malaga, Spain

Clélia de Mulatier (iD)
University of Amsterdam
Amsterdam, The Netherlands

Maciej Paszynski (iD)
AGH University of Science and Technology
Krakow, Poland

Valeria V. Krzhizhanovskaya (iD)
University of Amsterdam
Amsterdam, The Netherlands

Jack J. Dongarra (iD)
University of Tennessee
Knoxville, TN, USA

Peter M. A. Sloot (iD)
University of Amsterdam
Amsterdam, The Netherlands

ISSN 0302-9743   ISSN 1611-3349 (electronic)
Lecture Notes in Computer Science
ISBN 978-3-031-63753-7   ISBN 978-3-031-63751-3 (eBook)
https://doi.org/10.1007/978-3-031-63751-3

This Springer imprint is published by the registered company Springer Nature Switzerland AG
The registered company address is: Gewerbestrasse 11, 6330 Cham, Switzerland

If disposing of this product, please recycle the paper.

Preface

Welcome to the proceedings of the 24th International Conference on Computational Science (https://www.iccs-meeting.org/iccs2024/), held on July 2–4, 2024 at the University of Málaga, Spain.

In keeping with the new normal of our times, ICCS featured both in-person and online sessions. Although the challenges of such a hybrid format are manifold, we have always tried our best to keep the ICCS community as dynamic, creative, and productive as possible. We are proud to present the proceedings you are reading as a result.

ICCS 2024 was jointly organized by the University of Málaga, the University of Amsterdam, and the University of Tennessee.

Facing the Mediterranean in Spain's Costa del Sol, Málaga is the country's sixth-largest city, and a major hub for finance, tourism, and technology in the region.

The University of Málaga (Universidad de Málaga, UMA) is a modern, public university, offering 63 degrees and 120 postgraduate degrees. Close to 40,000 students study at UMA, taught by 2500 lecturers, distributed over 81 departments and 19 centers. The UMA has 278 research groups, which are involved in 80 national projects and 30 European and international projects. ICCS took place at the Teatinos Campus, home to the School of Computer Science and Engineering (ETSI Informática), which is a pioneer in its field and offers the widest range of IT-related subjects in the region of Andalusia.

The International Conference on Computational Science is an annual conference that brings together researchers and scientists from mathematics and computer science as basic computing disciplines, as well as researchers from various application areas who are pioneering computational methods in sciences such as physics, chemistry, life sciences, engineering, arts, and the humanities, to discuss problems and solutions in the area, identify new issues, and shape future directions for research.

The ICCS proceedings series have become a primary intellectual resource for computational science researchers, defining and advancing the state of the art in this field.

We are proud to note that this 24th edition, with 17 tracks (16 thematic tracks and one main track) and close to 300 participants, has kept to the tradition and high standards of previous editions.

The theme for 2024, "Computational Science: Guiding the Way Towards a Sustainable Society", highlights the role of Computational Science in assisting multidisciplinary research on sustainable solutions. This conference was a unique event focusing on recent developments in scalable scientific algorithms; advanced software tools; computational grids; advanced numerical methods; and novel application areas. These innovative novel models, algorithms, and tools drive new science through efficient application in physical systems, computational and systems biology, environmental systems, finance, and others.

ICCS is well known for its excellent lineup of keynote speakers. The keynotes for 2024 were:

- David Abramson, University of Queensland, Australia
- Manuel Castro Díaz, University of Málaga, Spain
- Jiří Mikyška, Czech Technical University in Prague, Czechia
- Takemasa Miyoshi, RIKEN, Japan
- Coral Calero Muñoz, University of Castilla-La Mancha, Spain
- Petra Ritter, Berlin Institute of Health & Charité University Hospital Berlin, Germany

This year we had 430 submissions (152 to the main track and 278 to the thematic tracks). In the main track, 51 full papers were accepted (33.5%); in the thematic tracks, 104 full papers (37.4%). The higher acceptance rate in the thematic tracks is explained by their particular nature, whereby track organizers personally invite many experts in the field to participate. Each submission received at least 2 single-blind reviews (2.6 reviews per paper on average).

ICCS relies strongly on our thematic track organizers' vital contributions to attract high-quality papers in many subject areas. We would like to thank all committee members from the main and thematic tracks for their contribution to ensuring a high standard for the accepted papers. We would also like to thank Springer, Elsevier, and Intellegibilis for their support. Finally, we appreciate all the local organizing committee members for their hard work in preparing this conference.

We hope the attendees enjoyed the conference, whether virtually or in person.

July 2024

Leonardo Franco
Clélia de Mulatier
Maciej Paszynski
Valeria V. Krzhizhanovskaya
Jack J. Dongarra
Peter M. A. Sloot

Organization

Conference Chairs

General Chair

Valeria Krzhizhanovskaya   University of Amsterdam, The Netherlands

Main Track Chair

Clélia de Mulatier   University of Amsterdam, The Netherlands

Thematic Tracks Chair

Maciej Paszynski   AGH University of Krakow, Poland

Thematic Tracks Vice Chair

Michael Harold Lees   University of Amsterdam, The Netherlands

Scientific Chairs

Peter M. A. Sloot   University of Amsterdam, The Netherlands
Jack Dongarra   University of Tennessee, USA

Local Organizing Committee

Leonardo Franco (Chair)   University of Malaga, Spain
Francisco Ortega-Zamorano   University of Malaga, Spain
Francisco J. Moreno-Barea   University of Malaga, Spain
José L. Subirats-Contreras   University of Malaga, Spain

Thematic Tracks and Organizers

Advances in High-Performance Computational Earth Sciences: Numerical Methods, Frameworks & Applications (IHPCES)

Takashi Shimokawabe	University of Tokyo, Japan
Kohei Fujita	University of Tokyo, Japan
Dominik Bartuschat	FAU Erlangen-Nürnberg, Germany

Artificial Intelligence and High-Performance Computing for Advanced Simulations (AIHPC4AS)

Maciej Paszynski	AGH University of Krakow, Poland

Biomedical and Bioinformatics Challenges for Computer Science (BBC)

Mario Cannataro	University Magna Graecia of Catanzaro, Italy
Giuseppe Agapito	University Magna Graecia of Catanzaro, Italy
Mauro Castelli	Universidade Nova de Lisboa, Portugal
Riccardo Dondi	University of Bergamo, Italy
Rodrigo Weber dos Santos	Federal University of Juiz de Fora, Brazil
Italo Zoppis	University of Milano-Bicocca, Italy

Computational Diplomacy and Policy (CoDiP)

Roland Bouffanais	University of Geneva, Switzerland
Michael Lees	University of Amsterdam, The Netherlands
Brian Castellani	Durham University, UK

Computational Health (CompHealth)

Sergey Kovalchuk	Huawei, Russia
Georgiy Bobashev	RTI International, USA
Anastasia Angelopoulou	University of Westminster, UK
Jude Hemanth	Karunya University, India

Computational Optimization, Modelling, and Simulation (COMS)

Xin-She Yang	Middlesex University London, UK
Slawomir Koziel	Reykjavik University, Iceland
Leifur Leifsson	Purdue University, USA

Generative AI and Large Language Models (LLMs) in Advancing Computational Medicine (CMGAI)

Ahmed Abdeen Hamed	State University of New York at Binghamton, USA
Qiao Jin	National Institutes of Health, USA
Xindong Wu	Hefei University of Technology, China
Byung Lee	University of Vermont, USA
Zhiyong Lu	National Institutes of Health, USA
Karin Verspoor	RMIT University, Australia
Christopher Savoie	Zapata AI, USA

Machine Learning and Data Assimilation for Dynamical Systems (MLDADS)

Rossella Arcucci	Imperial College London, UK
Cesar Quilodran-Casas	Imperial College London, UK

Multiscale Modelling and Simulation (MMS)

Derek Groen	Brunel University London, UK
Diana Suleimenova	Brunel University London, UK

Network Models and Analysis: From Foundations to Artificial Intelligence (NMAI)

Marianna Milano	Università Magna Graecia of Catanzaro, Italy
Giuseppe Agapito	University Magna Graecia of Catanzaro, Italy
Pietro Cinaglia	University Magna Graecia of Catanzaro, Italy
Chiara Zucco	University Magna Graecia of Catanzaro, Italy

Numerical Algorithms and Computer Arithmetic for Computational Science (NACA)

Pawel Gepner	Warsaw Technical University, Poland
Ewa Deelman	University of Southern California, Marina del Rey, USA
Hatem Ltaief	KAUST, Saudi Arabia

Quantum Computing (QCW)

Katarzyna Rycerz	AGH University of Krakow, Poland
Marian Bubak	Sano and AGH University of Krakow, Poland

Simulations of Flow and Transport: Modeling, Algorithms, and Computation (SOFTMAC)

Shuyu Sun	King Abdullah University of Science and Technology, Saudi Arabia
Jingfa Li	Beijing Institute of Petrochemical Technology, China
James Liu	Colorado State University, USA

Smart Systems: Bringing Together Computer Vision, Sensor Networks and Artificial Intelligence (SmartSys)

Pedro Cardoso	University of Algarve, Portugal
João Rodrigues	University of Algarve, Portugal
Jânio Monteiro	University of Algarve, Portugal
Roberto Lam	University of Algarve, Portugal

Solving Problems with Uncertainties (SPU)

Vassil Alexandrov	Hartree Centre – STFC, UK
Aneta Karaivanova	IICT – Bulgarian Academy of Science, Bulgaria

Teaching Computational Science (WTCS)

Evguenia Alexandrova	Hartree Centre – STFC, UK
Tseden Taddese	UK Research and Innovation, UK

Reviewers

Ahmed Abdelgawad	Central Michigan University, USA
Samaneh Abolpour Mofrad	Imperial College London, UK
Tesfamariam Mulugeta Abuhay	Queen's University, Canada
Giuseppe Agapito	University of Catanzaro, Italy
Elisabete Alberdi	University of the Basque Country, Spain
Luis Alexandre	UBI and NOVA LINCS, Portugal
Vassil Alexandrov	Hartree Centre – STFC, UK
Evguenia Alexandrova	Hartree Centre – STFC, UK
Julen Alvarez-Aramberri	Basque Center for Applied Mathematics, Spain
Domingos Alves	Ribeirão Preto Medical School, University of São Paulo, Brazil
Sergey Alyaev	NORCE, Norway
Anastasia Anagnostou	Brunel University London, UK
Anastasia Angelopoulou	University of Westminster, UK
Rossella Arcucci	Imperial College London, UK
Emanouil Atanasov	IICT – Bulgarian Academy of Sciences, Bulgaria
Krzysztof Banaś	AGH University of Krakow, Poland
Luca Barillaro	Magna Graecia University of Catanzaro, Italy
Dominik Bartuschat	FAU Erlangen-Nürnberg, Germany
Pouria Behnodfaur	Curtin University, Australia
Jörn Behrens	University of Hamburg, Germany
Adrian Bekasiewicz	Gdansk University of Technology, Poland
Gebrail Bekdas	Istanbul University, Turkey
Mehmet Ali Belen	Iskenderun Technical University, Turkey
Stefano Beretta	San Raffaele Telethon Institute for Gene Therapy, Italy
Anabela Moreira Bernardino	Polytechnic Institute of Leiria, Portugal
Eugénia Bernardino	Polytechnic Institute of Leiria, Portugal
Daniel Berrar	Tokyo Institute of Technology, Japan
Piotr Biskupski	IBM, Poland
Georgiy Bobashev	RTI International, USA
Carlos Bordons	University of Seville, Spain
Bartosz Bosak	PSNC, Poland
Lorella Bottino	University Magna Graecia of Catanzaro, Italy

Bhaskar Dasgupta	University of Illinois at Chicago, USA
Clélia de Mulatier	University of Amsterdam, The Netherlands
Ewa Deelman	University of Southern California, Marina del Rey, USA
Quanling Deng	Australian National University, Australia
Eric Dignum	University of Amsterdam, The Netherlands
Riccardo Dondi	University of Bergamo, Italy
Rafal Drezewski	AGH University of Krakow, Poland
Simon Driscoll	University of Reading, UK
Hans du Buf	University of the Algarve, Portugal
Vitor Duarte	Universidade NOVA de Lisboa, Portugal
Jacek Długopolski	AGH University of Krakow, Poland
Wouter Edeling	Vrije Universiteit Amsterdam, The Netherlands
Nahid Emad	University of Paris Saclay, France
Christian Engelmann	ORNL, USA
August Ernstsson	Linköping University, Sweden
Aniello Esposito	Hewlett Packard Enterprise, Switzerland
Roberto R. Expósito	Universidade da Coruna, Spain
Hongwei Fan	Imperial College London, UK
Tamer Fandy	University of Charleston, USA
Giuseppe Fedele	University of Calabria, Italy
Christos Filelis-Papadopoulos	Democritus University of Thrace, Greece
Alberto Freitas	University of Porto, Portugal
Ruy Freitas Reis	Universidade Federal de Juiz de Fora, Brazil
Kohei Fujita	University of Tokyo, Japan
Takeshi Fukaya	Hokkaido University, Japan
Wlodzimierz Funika	AGH University of Krakow, Poland
Takashi Furumura	University of Tokyo, Japan
Teresa Galvão	University of Porto, Portugal
Luis Garcia-Castillo	Carlos III University of Madrid, Spain
Bartłomiej Gardas	Institute of Theoretical and Applied Informatics, Polish Academy of Sciences, Poland
Victoria Garibay	University of Amsterdam, The Netherlands
Frédéric Gava	Paris-East Créteil University, France
Piotr Gawron	Nicolaus Copernicus Astronomical Centre, Polish Academy of Sciences, Poland
Bernhard Geiger	Know-Center GmbH, Austria
Pawel Gepner	Warsaw Technical University, Poland
Alex Gerbessiotis	NJIT, USA
Maziar Ghorbani	Brunel University London, UK
Konstantinos Giannoutakis	University of Macedonia, Greece
Alfonso Gijón	University of Granada, Spain

Jorge González-Domínguez	Universidade da Coruña, Spain
Alexandrino Gonçalves	CIIC – ESTG – Polytechnic University of Leiria, Portugal
Yuriy Gorbachev	Soft-Impact LLC, Russia
Pawel Gorecki	University of Warsaw, Poland
Michael Gowanlock	Northern Arizona University, USA
George Gravvanis	Democritus University of Thrace, Greece
Derek Groen	Brunel University London, UK
Loïc Guégan	UiT the Arctic University of Norway, Norway
Tobias Guggemos	University of Vienna, Austria
Serge Guillas	University College London, UK
Manish Gupta	Harish-Chandra Research Institute, India
Piotr Gurgul	SnapChat, Switzerland
Oscar Gustafsson	Linköping University, Sweden
Ahmed Abdeen Hamed	State University of New York at Binghamton, USA
Laura Harbach	Brunel University London, UK
Agus Hartoyo	TU Kaiserslautern, Germany
Ali Hashemian	Basque Center for Applied Mathematics, Spain
Mohamed Hassan	Virginia Tech, USA
Alexander Heinecke	Intel Parallel Computing Lab, USA
Jude Hemanth	Karunya University, India
Aochi Hideo	BRGM, France
Alfons Hoekstra	University of Amsterdam, The Netherlands
George Holt	UK Research and Innovation, UK
Maximilian Höb	Leibniz-Rechenzentrum der Bayerischen Akademie der Wissenschaften, Germany
Huda Ibeid	Intel Corporation, USA
Alireza Jahani	Brunel University London, UK
Jiří Jaroš	Brno University of Technology, Czechia
Qiao Jin	National Institutes of Health, USA
Zhong Jin	Computer Network Information Center, Chinese Academy of Sciences, China
David Johnson	Uppsala University, Sweden
Eleda Johnson	Imperial College London, UK
Piotr Kalita	Jagiellonian University, Poland
Drona Kandhai	University of Amsterdam, The Netherlands
Aneta Karaivanova	IICT-Bulgarian Academy of Science, Bulgaria
Sven Karbach	University of Amsterdam, The Netherlands
Takahiro Katagiri	Nagoya University, Japan
Haruo Kobayashi	Gunma University, Japan
Marcel Koch	KIT, Germany

Harald Koestler	University of Erlangen-Nuremberg, Germany
Georgy Kopanitsa	Tomsk Polytechnic University, Russia
Sotiris Kotsiantis	University of Patras, Greece
Remous-Aris Koutsiamanis	IMT Atlantique/DAPI, STACK (LS2N/Inria), France
Sergey Kovalchuk	Huawei, Russia
Slawomir Koziel	Reykjavik University, Iceland
Ronald Kriemann	MPI MIS Leipzig, Germany
Valeria Krzhizhanovskaya	University of Amsterdam, The Netherlands
Sebastian Kuckuk	Friedrich-Alexander-Universität Erlangen-Nürnberg, Germany
Michael Kuhn	Otto von Guericke University Magdeburg, Germany
Ryszard Kukulski	Institute of Theoretical and Applied Informatics, Polish Academy of Sciences, Poland
Krzysztof Kurowski	PSNC, Poland
Marcin Kuta	AGH University of Krakow, Poland
Marcin Łoś	AGH University of Krakow, Poland
Roberto Lam	Universidade do Algarve, Portugal
Tomasz Lamża	ACK Cyfronet, Poland
Ilaria Lazzaro	Università degli studi Magna Graecia di Catanzaro, Italy
Paola Lecca	Free University of Bozen-Bolzano, Italy
Byung Lee	University of Vermont, USA
Mike Lees	University of Amsterdam, The Netherlands
Leifur Leifsson	Purdue University, USA
Kenneth Leiter	U.S. Army Research Laboratory, USA
Paulina Lewandowska	IT4Innovations National Supercomputing Center, Czechia
Jingfa Li	Beijing Institute of Petrochemical Technology, China
Siyi Li	Imperial College London, UK
Che Liu	Imperial College London, UK
James Liu	Colorado State University, USA
Zhao Liu	National Supercomputing Center in Wuxi, China
Marcelo Lobosco	UFJF, Brazil
Jay F. Lofstead	Sandia National Laboratories, USA
Chu Kiong Loo	University of Malaya, Malaysia
Stephane Louise	CEA, LIST, France
Frédéric Loulergue	University of Orléans, INSA CVL, LIFO EA 4022, France
Hatem Ltaief	KAUST, Saudi Arabia
Zhiyong Lu	National Institutes of Health, USA

Stefan Luding	University of Twente, The Netherlands
Lukasz Madej	AGH University of Krakow, Poland
Luca Magri	Imperial College London, UK
Anirban Mandal	Renaissance Computing Institute, USA
Soheil Mansouri	Technical University of Denmark, Denmark
Tomas Margalef	Universitat Autònoma de Barcelona, Spain
Arbitrio Mariamena	Consiglio Nazionale delle Ricerche, Italy
Osni Marques	Lawrence Berkeley National Laboratory, USA
Maria Chiara Martinis	Università Magna Graecia di Catanzaro, Italy
Jaime A. Martins	University of Algarve, Portugal
Paula Martins	CinTurs – Research Centre for Tourism Sustainability and Well-being; FCT-University of Algarve, Portugal
Michele Martone	Max-Planck-Institut für Plasmaphysik, Germany
Pawel Matuszyk	Baker-Hughes, USA
Francesca Mazzia	University di Bari, Italy
Jon McCullough	University College London, UK
Pedro Medeiros	Universidade Nova de Lisboa, Portugal
Wen Mei	National University of Defense Technology, China
Wagner Meira	Universidade Federal de Minas Gerais, Brazil
Roderick Melnik	Wilfrid Laurier University, Canada
Pedro Mendes Guerreiro	Universidade do Algarve, Portugal
Isaak Mengesha	University of Amsterdam, The Netherlands
Wout Merbis	University of Amsterdam, The Netherlands
Ivan Merelli	ITB-CNR, Italy
Marianna Milano	Università Magna Graecia di Catanzaro, Italy
Magdalena Misiak	Howard University College of Medicine, USA
Jaroslaw Miszczak	Institute of Theoretical and Applied Informatics, Polish Academy of Sciences, Poland
Dhruv Mittal	University of Amsterdam, The Netherlands
Fernando Monteiro	Polytechnic Institute of Bragança, Portugal
Jânio Monteiro	University of Algarve, Portugal
Andrew Moore	University of California Santa Cruz, USA
Francisco J. Moreno-Barea	Universidad de Málaga, Spain
Leonid Moroz	Warsaw University of Technology, Poland
Peter Mueller	IBM Zurich Research Laboratory, Switzerland
Judit Munoz-Matute	Basque Center for Applied Mathematics, Spain
Hiromichi Nagao	University of Tokyo, Japan
Kengo Nakajima	University of Tokyo, Japan
Philipp Neumann	Helmut-Schmidt-Universität, Germany
Sinan Melih Nigdeli	Istanbul University – Cerrahpasa, Turkey

Fernando Nobrega Santos	University of Amsterdam, The Netherlands
Joseph O'Connor	University of Edinburgh, UK
Frederike Oetker	University of Amsterdam, The Netherlands
Arianna Olivelli	Imperial College London, UK
Ángel Omella	Basque Center for Applied Mathematics, Spain
Kenji Ono	Kyushu University, Japan
Hiroyuki Ootomo	Tokyo Institute of Technology, Japan
Eneko Osaba	TECNALIA Research & Innovation, Spain
George Papadimitriou	University of Southern California, USA
Nikela Papadopoulou	University of Glasgow, UK
Marcin Paprzycki	IBS PAN and WSM, Poland
David Pardo	Basque Center for Applied Mathematics, Spain
Anna Paszynska	Jagiellonian University, Poland
Maciej Paszynski	AGH University of Krakow, Poland
Łukasz Pawela	Institute of Theoretical and Applied Informatics, Polish Academy of Sciences, Poland
Giulia Pederzani	Universiteit van Amsterdam, The Netherlands
Alberto Perez de Alba Ortiz	University of Amsterdam, The Netherlands
Dana Petcu	West University of Timisoara, Romania
Beáta Petrovski	University of Oslo, Norway
Frank Phillipson	TNO, The Netherlands
Eugenio Piasini	International School for Advanced Studies (SISSA), Italy
Juan C. Pichel	Universidade de Santiago de Compostela, Spain
Anna Pietrenko-Dabrowska	Gdansk University of Technology, Poland
Armando Pinho	University of Aveiro, Portugal
Pietro Pinoli	Politecnico di Milano, Italy
Yuri Pirola	Università degli Studi di Milano-Bicocca, Italy
Ollie Pitts	Imperial College London, UK
Robert Platt	Imperial College London, UK
Dirk Pleiter	KTH/Forschungszentrum Jülich, Germany
Paweł Poczekajło	Koszalin University of Technology, Poland
Cristina Portalés Ricart	Universidad de Valencia, Spain
Simon Portegies Zwart	Leiden University, The Netherlands
Anna Procopio	Università Magna Graecia di Catanzaro, Italy
Ela Pustulka-Hunt	FHNW Olten, Switzerland
Marcin Płodzień	ICFO, Spain
Ubaid Qadri	Hartree Centre – STFC, UK
Rick Quax	University of Amsterdam, The Netherlands
Cesar Quilodran Casas	Imperial College London, UK
Andrianirina Rakotoharisoa	Imperial College London, UK
Celia Ramos	University of the Algarve, Portugal

Robin Richardson	Netherlands eScience Center, The Netherlands
Sophie Robert	University of Orléans, France
João Rodrigues	Universidade do Algarve, Portugal
Daniel Rodriguez	University of Alcalá, Spain
Marcin Rogowski	Saudi Aramco, Saudi Arabia
Sergio Rojas	Pontifical Catholic University of Valparaiso, Chile
Diego Romano	ICAR-CNR, Italy
Albert Romkes	South Dakota School of Mines and Technology, USA
Juan Ruiz	University of Buenos Aires, Argentina
Tomasz Rybotycki	IBS PAN, CAMK PAN, AGH, Poland
Katarzyna Rycerz	AGH University of Krakow, Poland
Grażyna Ślusarczyk	Jagiellonian University, Poland
Emre Sahin	Science and Technology Facilities Council, UK
Ozlem Salehi	Özyeğin University, Turkey
Ayşin Sancı	Altinay, Turkey
Christopher Savoie	Zapata Computing, USA
Ileana Scarpino	University "Magna Graecia" of Catanzaro, Italy
Robert Schaefer	AGH University of Krakow, Poland
Ulf D. Schiller	University of Delaware, USA
Bertil Schmidt	University of Mainz, Germany
Karen Scholz	Fraunhofer MEVIS, Germany
Martin Schreiber	Université Grenoble Alpes, France
Paulina Sepúlveda-Salas	Pontifical Catholic University of Valparaiso, Chile
Marzia Settino	Università Magna Graecia di Catanzaro, Italy
Mostafa Shahriari	Basque Center for Applied Mathematics, Spain
Takashi Shimokawabe	University of Tokyo, Japan
Alexander Shukhman	Orenburg State University, Russia
Marcin Sieniek	Google, USA
Joaquim Silva	Nova School of Science and Technology – NOVA LINCS, Portugal
Mateusz Sitko	AGH University of Krakow, Poland
Haozhen Situ	South China Agricultural University, China
Leszek Siwik	AGH University of Krakow, Poland
Peter Sloot	University of Amsterdam, The Netherlands
Oskar Slowik	Center for Theoretical Physics PAS, Poland
Sucha Smanchat	King Mongkut's University of Technology North Bangkok, Thailand
Alexander Smirnovsky	SPbPU, Russia
Maciej Smołka	AGH University of Krakow, Poland
Isabel Sofia	Instituto Politécnico de Beja, Portugal
Robert Staszewski	University College Dublin, Ireland

Magdalena Stobińska	University of Warsaw, Poland
Tomasz Stopa	IBM, Poland
Achim Streit	KIT, Germany
Barbara Strug	Jagiellonian University, Poland
Diana Suleimenova	Brunel University London, UK
Shuyu Sun	King Abdullah University of Science and Technology, Saudi Arabia
Martin Swain	Aberystwyth University, UK
Renata G. Słota	AGH University of Krakow, Poland
Tseden Taddese	UK Research and Innovation, UK
Ryszard Tadeusiewicz	AGH University of Krakow, Poland
Claude Tadonki	Mines ParisTech/CRI – Centre de Recherche en Informatique, France
Daisuke Takahashi	University of Tsukuba, Japan
Osamu Tatebe	University of Tsukuba, Japan
Michela Taufer	University of Tennessee, USA
Andrei Tchernykh	CICESE, Mexico
Kasim Terzic	University of St Andrews, UK
Jannis Teunissen	KU Leuven, Belgium
Sue Thorne	Hartree Centre – STFC, UK
Ed Threlfall	United Kingdom Atomic Energy Authority, UK
Vinod Tipparaju	AMD, USA
Pawel Topa	AGH University of Krakow, Poland
Paolo Trunfio	University of Calabria, Italy
Ola Tørudbakken	Meta, Norway
Carlos Uriarte	University of the Basque Country, BCAM – Basque Center for Applied Mathematics, Spain
Eirik Valseth	University of Life Sciences & Simula, Norway
Rein van den Boomgaard	University of Amsterdam, The Netherlands
Vítor V. Vasconcelos	University of Amsterdam, The Netherlands
Aleksandra Vatian	ITMO University, Russia
Francesc Verdugo	Vrije Universiteit Amsterdam, The Netherlands
Karin Verspoor	RMIT University, Australia
Salvatore Vitabile	University of Palermo, Italy
Milana Vuckovic	European Centre for Medium-Range Weather Forecasts, UK
Kun Wang	Imperial College London, UK
Peng Wang	NVIDIA, China
Rodrigo Weber dos Santos	Federal University of Juiz de Fora, Brazil
Markus Wenzel	Fraunhofer Institute for Digital Medicine MEVIS, Germany

Lars Wienbrandt	Kiel University, Germany
Wendy Winnard	UKRI STFC, UK
Maciej Woźniak	AGH University of Krakow, Poland
Xindong Wu	Hefei University of Technology, China
Dunhui Xiao	Tongji University, China
Huilin Xing	University of Queensland, Australia
Yani Xue	Brunel University, UK
Abuzer Yakaryilmaz	University of Latvia, Latvia
Xin-She Yang	Middlesex University London, UK
Dongwei Ye	University of Amsterdam, The Netherlands
Karol Życzkowski	Jagiellonian University, Poland
Gabor Závodszky	University of Amsterdam, Hungary
Sebastian Zając	SGH Warsaw School of Economics, Poland
Małgorzata Zajęcka	AGH University of Krakow, Poland
Justyna Zawalska	ACC Cyfronet AGH, Poland
Wei Zhang	Huazhong University of Science and Technology, China
Yao Zhang	Google, USA
Jinghui Zhong	South China University of Technology, China
Sotirios Ziavras	New Jersey Institute of Technology, USA
Zoltan Zimboras	Wigner Research Center, Hungary
Italo Zoppis	University of Milano-Bicocca, Italy
Chiara Zucco	University Magna Graecia of Catanzaro, Italy
Pavel Zun	ITMO University, Russia

Contents – Part II

ICCS 2024 Main Track Full Papers

Stacking for Probabilistic Short-Term Load Forecasting

Grzegorz Dudek[(✉)] [iD]

Electrical Engineering Faculty, Czestochowa University of Technology, Czestochowa, Poland
grzegorz.dudek@pcz.pl

Abstract. In this study, we delve into the realm of meta-learning to combine point base forecasts for probabilistic short-term electricity demand forecasting. Our approach encompasses the utilization of quantile linear regression, quantile regression forest, and post-processing techniques involving residual simulation to generate quantile forecasts. Furthermore, we introduce both global and local variants of meta-learning. In the local-learning mode, the meta-model is trained using patterns most similar to the query pattern. Through extensive experimental studies across 35 forecasting scenarios and employing 16 base forecasting models, our findings underscored the superiority of quantile regression forest over its competitors.

Keywords: Ensemble forecasting · Meta-learning · Probabilistic forecasting · Quantile regression forest · Short-term load forecasting · Stacking

1 Introduction

Ensembling stands out as a highly effective strategy for enhancing the predictive power of forecasting models. By combining predictions from multiple models, the ensemble approach consistently yields heightened accuracy. It leverages the strengths of individual models while mitigating their weaknesses, thereby extending predictive capacity by capturing a broader range of patterns and insights within the data. Additionally, ensembling accommodates the inclusion of multiple influential factors in the data generation process, alleviating concerns regarding model structure and parameter specification [22]. This comprehensive approach minimizes the risk associated with relying solely on a single model's limitations or biases, providing a more holistic representation of the data generation process. Moreover, ensembling demonstrates remarkable robustness in handling outliers or extreme values and plays a crucial role in mitigating overfitting. It often achieves computational efficiency by leveraging parallel processing and optimization techniques.

Among the various methods for combining forecasts, the arithmetic average with equal weights emerges as a surprisingly robust and commonly used approach, often outperforming more complex weighting schemes [7]. Additionally,

L. Franco et al. (Eds.): ICCS 2024, LNCS 14833, pp. 3–18, 2024.
https://doi.org/10.1007/978-3-031-63751-3_1

alternative strategies such as median, mode, trimmed means, and winsorized means have been explored [14].

Linear regression serves as a valuable tool for assigning distinct weights to individual models, with weights estimated through ordinary least squares. These weights can effectively reflect the historical performance of base models [18]. They can be derived also from information criteria [12], diversity of individual learners [10], or time series characteristic features [17].

While averaging and linear regression are prevalent methods for combining forecasts, they may fall short in capturing nonlinear and complex relationships between base models' forecasts and the target value. In such cases, machine learning (ML) models offer an alternative through stacking procedures [1,6], which optimize forecast accuracy by learning the optimal combination of constituent forecasts in a data-driven manner. The literature underscores the advantages of stacking generalization, including forecasting tasks involving time series with intricate seasonality, such as short-term load forecasting (STLF). For instance, [20] proposed utilizing wavelet neural networks (NNs) both as base forecasters and meta-learners for handling series with multiple seasonal cycles. In [4], the authors employed regression trees, random forests (RFs), and NNs as base models for STLF, with gradient boosting as a meta-learner. For a similar forecasting challenge, [19] combined RFs, long short-term memory (LSTM), deep NNs, and evolutionary trees with gradient boosting models as meta-learners, achieving significant forecast error reduction. In [6], the authors compared forecast combination strategies, demonstrating the superior performance of stacking over methods like simple averaging and linear combination, showcasing its efficacy across diverse time series characteristics.

In this study, we propose a stacking approach to produce probabilistic STLF. STLF refers to the prediction of electricity demand over a relatively short period, typically ranging from a few hours to a few days ahead. Accurate STLF is crucial for efficient operation and planning of power systems, as it helps utility companies optimize generation, transmission, and distribution of electricity, leading to cost savings, improved reliability, and better utilization of resources. Probabilistic forecasting in power systems is essential for addressing uncertainties stemming from factors like weather variations, load fluctuations, and unexpected events [8]. Power system operators rely on probabilistic forecasts to inform decisions, improve operational efficiency, manage risks, and ensure a dependable power supply. Likewise, in energy trading, probabilistic forecasting assists traders in understanding potential outcomes and uncertainties related to electricity prices, load demand, and generation availability [2].

Several instances of stacking for probabilistic STLF exist in the literature. For instance, [9] employs a quantile regression LSTM meta-learner to combine point forecasts from tree-based models, including RF, gradient boosting decision tree, and light gradient boosting machine. It determines the probability density function (PDF) using kernel density estimation modified by Gaussian approximation of quantiles. In another study, [13], a Gaussian mixture distribution is utilized to combine probability density forecasts generated by various quantile regression models, such as Gaussian process regression, quantile regression NNs, and quantile regression gradient boosting. The results demonstrate the effectiveness of the proposed method in enhancing forecasting performance compared to methods employing simple averaging.

This study represents an extension of our previous conference paper [5], which primarily focused on deterministic STLF. Here, our primary objective is to introduce methodologies for generating probabilistic STLF, building upon the point base forecasts.

Our study makes the following contributions:

1. Exploration of three approaches for combining point forecasts to produce probabilistic forecasts: post-processing techniques involving residual simulation, quantile linear regression, and quantile regression forest.
2. Investigation of global and local meta-learning strategies, which comprehensively capture both overarching data patterns and subtle nuances crucial for addressing complex seasonality.
3. Validation through extensive experimentation across 35 STLF problems marked by triple seasonality, using 16 distinct base models.

The subsequent sections of this work are structured as follows. Section 2 defines the forecasting problem and discusses global and local meta-learning strategies. In Sect. 3, we provide a detailed description of the proposed meta-learners for probabilistic STLF. Section 4 presents application examples along with a thorough analysis of the achieved results. Finally, Sect. 5 outlines our key conclusions, summarizing the findings from this study.

2   Problem Statement

The challenge of forecast combination involves the goal of finding a regression function, denoted as f (meta-model). This function aggregates forecasts for time t generated by n forecasting models (base models) to create either point or probabilistic forecasts. A point forecast is a single-value prediction of a future outcome, represented by a single number. In contrast, probabilistic forecasts can take the form of predictive intervals (PI), a set of quantiles, or PDF, and this PDF can have a parametric or non-parametric form. In this work, we focus on generating probabilistic forecasts in the form of a set of quantiles. For a continuous cumulative distribution function (CDF), $F(y|X = x) = P(Y \leq y|X = x)$, the α-quantile is defined as:

$$Q_\alpha(x) = \inf\{y : F(y|X = x) \geq \alpha\} \tag{1}$$

where $\alpha \in [0, 1]$ denotes a nominal probability level.

Function f can make use of all available information up to time $t - h$, where h signifies the forecast horizon. However, in this study, we limit this information to the base forecasts, represented by vector $\hat{\mathbf{y}}_t = [\hat{y}_{1,t}, ..., \hat{y}_{n,t}]$. The combined quantile forecast is given as $\tilde{\mathbf{q}}_t = f(\hat{\mathbf{y}}_t; \boldsymbol{\theta}_t)$, where $\tilde{\mathbf{q}}_t = [\tilde{q}_t(\alpha)]_{\alpha \in \Pi}$, Π represents the assumed set of probabilities α, and $\boldsymbol{\theta}_t$ denotes model parameters.

The class of regression functions f encompasses a wide range of mappings, including both linear and nonlinear ones. The meta-model parameters can either remain static or vary over time. To enhance the performance of the meta-model, we employ an approach where the parameters are learned individually for each forecasting task, using a specific training set tailored for that task, represented as $\Phi = \{(\hat{\mathbf{y}}_\tau, y_\tau)\}_{\tau \in \Xi}$. Here, y_τ denotes the target value, and Ξ is a set of selected time indices from the interval $T = 1, ..., t - h$.

The base forecasting models generate forecasts for successive time points $T = 1, ..., t$. To obtain an ensemble forecast for time t, a meta-model can be trained using all available historical data from period $\Xi = \{1, ..., t - h\}$, referred to as the global approach. This method allows the model to leverage all past information to generate a forecast for time point t.

In the alternative local mode, the goal is to train the meta-model locally around query pattern $\hat{\mathbf{y}}_t$. To achieve this, k most similar input vectors to $\hat{\mathbf{y}}_t$ are selected and included into the local training set. The Euclidean metric is used to determine the nearest neighbors. The rationale behind adopting local training mode is rooted in the assumption that by focusing on a more narrowly defined segment of the target function — as opposed to the broader scope of the global function – there will be an enhancement in forecasting accuracy for the given query pattern. However, a crucial aspect of this approach is the determination of the optimal size for the local area, i.e. the choice of the number of neighbors, k.

3   Meta-models for Probabilistic STLF

Quantile regression is concerned with estimating the conditional quantiles of a response variable. Unlike traditional regression that models the conditional mean of the response, quantile regression provides a more comprehensive view by estimating various quantiles. This approach facilitates a thorough statistical analysis of the stochastic relationships among random variables, capturing a broader range of information about the distribution of the response variable.

In our research, we present three approaches for probabilistic STLF, each centered around the concept of quantile modeling.

3.1   Quantile Estimation Through Residual Simulation (QRS)

This method operates under the assumption that the distribution of residuals observed for historical data remains consistent with the distribution for future data. The process involves the following steps. Firstly, a meta-model for deterministic forecasting is trained, as outlined in [5]. Then residuals are computed for historical data, which could encompass all training data, selected training patterns, or validation patterns. In our approach, we compute residuals for all training patterns, but note that in the local training mode, training patterns are chosen based on their similarity to the query pattern.

In the subsequent step, these residuals are added to the point forecasts, and a distribution function is fitted to the resulting values. To accomplish this, we employ a nonparametric kernel method, known for its flexibility compared to parametric alternatives. Once the distribution is estimated, quantiles are calculated using the inverse CDF.

Drawing from the findings reported in [5], the RF model was chosen for deterministic forecasting. The formulation of this model is outlined as follows:

$$f(\hat{\mathbf{y}}) = \sum_{\tau \in \Xi} w_\tau(\hat{\mathbf{y}}) y_\tau \tag{2}$$

$$w_\tau(\hat{\mathbf{y}}) = \frac{1}{p} \sum_{j=1}^{p} \frac{\mathbb{1}\{\hat{\mathbf{y}}_\tau \in \ell_j(\hat{\mathbf{y}})\}}{\sum_{\kappa \in \Xi} \mathbb{1}\{\hat{\mathbf{y}}_\kappa \in \ell_j(\hat{\mathbf{y}})\}} \tag{3}$$

where p is the number of trees in the forest, ℓ_j denotes the leaf that is obtained when dropping $\hat{\mathbf{y}}$ down the j-th tree, and $\mathbb{1}$ denotes the indicator function.

The RF response is the average of all training patterns $\hat{\mathbf{y}}_\tau$ that reached the same leaves (across all trees) as query pattern $\hat{\mathbf{y}}$. It approximates the conditional mean $E(Y|X = \hat{\mathbf{y}})$ by a weighted mean over the observations of the response variable Y.

3.2 Quantile Linear Regression (QLR)

Classical linear regression, focused on minimizing sums of squared residuals, is adept at estimating models for conditional means. In contrast, quantile regression provides a powerful framework for estimating models for the entire spectrum of conditional quantile functions.

As observed by Koenker in [11], quantiles can be conceptualized as the solution to a straightforward optimization problem using the pinball loss function:

$$L_\alpha(y, q) = \begin{cases} (y - q)\alpha & \text{if } y \geq q \\ (y - q)(\alpha - 1) & \text{if } y < q \end{cases} \tag{4}$$

where y represents the true value, and q is its predicted α-quantile.

Thus, the linear model in the form of

$$f(\hat{\mathbf{y}}) = \sum_{i=1}^{n} a_i \hat{y}_i + a_0 \tag{5}$$

where $a_0, ..., a_n$ are coefficients, can effectively generate quantiles when the loss function is (4). The optimization problem is linear and can be efficiently solved using the interior point (Frisch-Newton) algorithm.

3.3 Quantile Regression Forest (QRF)

QRF extends the principles of RF to accommodate quantile regression [16]. Similar to traditional RF, QRF builds an ensemble of regression trees, with each tree constructed using a bootstrap sample of the training data. Notably, at each split, a random subset of predictor variables is considered, introducing an element of randomness and promoting diversity among the trees.

In contrast to standard RF, which estimates the conditional mean, QRF focuses on estimating the full conditional distribution, $E(\mathbb{1}\{Y \leq y\}|X = \hat{\mathbf{y}})$. This distribution is approximated in QRF through the weighted mean across observations of $\mathbb{1}\{Y \leq y\}$ [16]:

$$\hat{F}(y|X = \hat{\mathbf{y}}) = \sum_{\tau \in \Xi} w_\tau(\hat{\mathbf{y}})\mathbb{1}\{y_\tau \leq y\} \tag{6}$$

To obtain estimates of the conditional quantiles $Q_\alpha(\hat{\mathbf{y}})$, the empirical CDF determined by QRF in (6) is incorporated into (1).

It is worth noting that the weights in (6) align with those used in RF, as defined in (3). The construction process of QRF mirrors that of RF, and QRF shares the same set of hyperparameters for tuning. The primary distinction between RF and QRF lies in how they treat each leaf: while RF retains only the mean of the observations falling into a leaf, QRF retains the values of all observations in the leaf, not just their mean. Subsequently, QRF leverages this information to assess the conditional distribution.

4 Experimental Study

In this section, we assess the effectiveness of our proposed stacking approaches for combining forecasts in the context of probabilistic STLF for 35 European countries. The time series under consideration exhibit triple seasonality, encompassing daily, weekly, and yearly patterns. The ensemble of base models comprises 16 forecasting models of diverse types, elaborated upon in Subsect. 4.5.

4.1 Dataset

We gathered real-world data from the ENTSO-E repository (www.entsoe.eu/data/power-stats) for our study. The dataset comprises hourly electricity loads recorded from 2006 to 2018, spanning 35 European countries. This dataset provides a rich array of time series, each showcasing distinctive features such as levels and trends, variance, patterns of seasonal fluctuations across different periods (annual, weekly, and daily), and random fluctuations.

4.2 Training and Evaluation Setup

The forecasting base models underwent optimization and training on data spanning from 2006 to 2017. These models were then employed to generate hourly forecasts for the entire year of 2018, on a daily basis (for a detailed methodology, refer to [21]). To assess the effectiveness of the meta-models, we selected 100 specific hours (test hours) for each country from the latter half of 2018, evenly distributed across this period. The forecasts for these chosen hours were aggregated by the meta-models.

Each meta-model was trained for every test hour to generate a vector of quantiles of probabilities $\alpha \in \Pi = 0.01, 0.02, ..., 0.99$. The training utilized data from January 1, 2018, up to the hour just before the forecasted hour ($h = 1$). The meta-models were trained in both global and local modes. The local modes involved training on the k nearest training patterns to the query pattern, with $k \in K = \{20, 40, ..., 200, 250, 300\}$.

4.3 Hyperparameters

We conducted a comprehensive evaluation of the meta-models, considering different hyperparameter values:

- QRS: RF was utilized for generating point forecasts. After conducting preliminary simulations, we opted for default values for the RF hyperparameters: number of predictors to select at random for each decision split $r = n/3$ (as recommended by the RF inventors), minimum number of observations per tree leaf $q = 1$ (leading to overtrained trees, mitigated by combining them in the forest), and number of trees in the forest $p = 100$. Additionally, for estimating the forecast distribution, a nonparametric kernel method leveraging a normal kernel with the bandwidth optimized for normal densities was employed.
- QLR: Implemented using Matlab code provided by Roger Koenker, solving a linear program via the interior point method (www.econ.uiuc.edu/~roger/research/rq/rq).
- QRF: The minimum number of observations per leaf, q, was searched over the set $\{1, 5, 10, 15, 20, 30, 40, 50, 60\}$, while other hyperparameters remained the same as for RF used in QRS.

The proposed meta-models were implemented in Matlab 2022b, and experiments were conducted on a Microsoft Windows 10 Pro operating system, with an Intel(R) Core(TM) i7-6950x CPU @3.0 GHz processor, and 48 GB RAM.

4.4 Evaluation Metrics

To assess the performance and effectiveness of the base models, we use the following metrics for point STLF:

- MAPE: mean absolute percentage error,
- MdAPE: median of the absolute percentage error (more robust to outlier forecasts than MAPE),
- MSE: mean squared error (more sensitive to outlier forecasts than MAPE),

- MPE: mean percentage error (a measure of the forecast bias),
- StdPE: standard deviation of the percentage error (a measure of the forecast dispersion).

For probabilistic STLF we apply:

- MPQRE: mean percentage quantile regression error, which is defined based on pinball loss function (4) as follows:

$$MPQRE = \frac{100}{N|\Pi|} \sum_{i=1}^{N} \sum_{\alpha \in \Pi} \frac{L_\alpha(y_i, \tilde{q}_i(\alpha))}{y_i} \qquad (7)$$

where N is the number of forecasts.

This metric is expressed in percentage of true value y_i to facilitate comparison across countries.

To provide a more comprehensive view of the results, we introduce two additional metrics: MdPQRE (median of PQRE), which is robust to outlier errors, and StdPQRE (standard deviation of PQRE), which measures the dispersion of errors.

- ReFr: relative frequency, also known as calibration or coverage, defined as follows [15]:

$$ReFr(\alpha) = \frac{1}{N} \sum_{i=1}^{N} \mathbb{1}\{y_i \le \tilde{q}_i(\alpha)\} \qquad (8)$$

The desired value of $ReFr(\alpha)$ is nominal probability level α. In other words, the predicted α-quantiles should exceed the realized values in $100\alpha\%$ of cases, ensuring an ReFr of α. To assess the average deviation of $ReFr(\alpha)$ from the desired α across all $\alpha \in \Pi$, we define the mean absolute ReFr error:

$$MARFE = \frac{1}{|\Pi|} \sum_{\alpha \in \Pi} |ReFr(\alpha) - \alpha| \qquad (9)$$

We calculate also the median and standard deviation of ARFE: MdARFE and StdARFE, respectively.

- MPWS: mean percentage Winkler score defined as:

$$MPWS = \frac{100}{N} \sum_{i=1}^{N} \frac{WS(y_i, \tilde{q}_{l,i}, \tilde{q}_{u,i})}{y_i} \qquad (10)$$

$$WS(y, \tilde{q}_l, \tilde{q}_u) = \begin{cases} (\tilde{q}_u - \tilde{q}_l) + \frac{2}{\alpha}(\tilde{q}_l - y) & \text{if } y < \tilde{q}_l \\ (\tilde{q}_u - \tilde{q}_l) & \text{if } \tilde{q}_l \le y \le \tilde{q}_u \\ (\tilde{q}_u - \tilde{q}_l) + \frac{2}{\alpha}(y - \tilde{q}_u) & \text{if } y > \tilde{q}_u \end{cases} \qquad (11)$$

where y represents the true value, $\tilde{q}_l$ and $\tilde{q}_u$ represent the predicted lower and upper quantiles defining the $100(1-\alpha)\%$ PI, $\alpha = \alpha_u - \alpha_l$, α_u and α_l define $\tilde{q}_l$ and $\tilde{q}_u$, respectively.

This measure evaluates PI, not a forecast distribution expressed by a set of quantiles as MPQRE and MARFE do. In this study, we assumed 90% PI, $\alpha_l = 0.05$ and $\alpha_u = 0.95$. MPWS, like MPQRE, is a normalized measure to enable comparison results for different countries.

As in the case of MPQRE and MARFE, we define two additional metrics: MdPWS – median of PWS, and StdPWS – standard deviation of PWS.

- inPI, belowPI, abovePI: percentage of observed values in PI, below PI and above PI. This simple measure allows us to compare the results with the desired results: 90/5/5 in our case.

- QMAPE, QMdAPE: quality metrics specifically devised to evaluate point forecasts derived from predicted quantiles. These metrics operate under the assumption that the point forecast aligns with the 0.5-quantile (median). By contrasting these median-based metrics with conventional point forecasting metrics like MAPE and MdAPE, we can gauge the effectiveness of probabilistic forecasting meta-models in producing accurate point forecasts.

4.5   Base Forecasting Models

We utilized a diverse array of forecasting models as the base learners, encompassing statistical models, ML models, and various recurrent, deep, and hybrid NN architectures retrieved from [21]. This broad selection spans a spectrum of modeling techniques, each equipped with distinct mechanisms for capturing temporal patterns in data. The incorporation of this varied ensemble of models was intentional, aiming to ensure ample diversity among the base learners to enhance the overall forecasting performance.

Table 1 presents the results of the base models averaged over 100 selected test hours across 35 countries. The models exhibit variations in MAPE from 1.70 to 3.83 and in MSE from 224,265 to 1,641,288. Among these, cES-adRNN emerges as the most accurate model based on MAPE, MdAPE, and MSE, while Prophet ranks as the least accurate.

4.6   Results

Figure 1 showcases the performance metrics of QRF across different training approaches and varying values of q (the minimum number of observations per leaf). This visualization clearly illustrates that QRF attains its lowest error rates under global training conditions, especially when the q parameter is set to 10 (resulting in the lowest values of MPQRE, MPWS, and QMAPE). However, it is important to note that MARFE draws slightly different conclusions. The MARFE metric for differnt k exhibits less variability compared to other metrics, achieving its minimum value for $k = 250$ and $q = 5$ ($MARFE = 0.0301$).

Figure 2 provides a comparative evaluation of QRS, QRE, and QRF, focusing on their accuracy across global and local training methodologies. Notably, the QRE model displayed significant errors when trained on smaller datasets. As the size of the training dataset increased, all models demonstrated a gradual improvement in accuracy, with the most substantial enhancements observed in

Table 1. Point STLF quality metrics for the base models.

Base model	MAPE	MdAPE	MSE	MPE	StdPE
ARIMA	2.86	1.82	777012	0.0556	4.60
ES	2.83	1.79	710773	0.1639	4.64
Prophet	3.83	2.53	1641288	−0.5195	6.24
N-WE	2.12	1.34	357253	0.0048	3.47
GRNN	2.10	1.36	372446	0.0098	3.42
MLP	2.55	1.66	488826	0.2390	3.93
SVM	2.16	1.33	356393	0.0293	3.55
LSTM	2.37	1.54	477008	0.0385	3.68
ANFIS	3.08	1.65	801710	−0.0575	5.59
MTGNN	2.54	1.71	434405	0.0952	3.87
DeepAR	2.93	2.00	891663	−0.3321	4.62
WaveNet	2.47	1.69	523273	−0.8804	3.77
N-BEATS	2.14	1.34	430732	−0.0060	3.57
LGBM	2.43	1.70	409062	0.0528	3.55
XGB	2.32	1.61	376376	0.0529	3.37
cES-adRNN	1.70	1.10	224265	−0.1860	2.57

ARIMA: autoregressive integrated moving average model,
ES: exponential smoothing model,
Prophet: modular additive regression model with nonlinear trend and seasonal components,
N-WE: Nadaraya-Watson estimator,
GRNN: general regression NN,
MLP: perceptron with a single hidden layer and sigmoid nonlinearities,
SVM: linear epsilon-insensitive support vector machine,
LSTM: long short-term memory,
ANFIS: adaptive neuro-fuzzy inference system,
MTGNN: graph NN for multivariate time series forecasting,
DeepAR: autoregressive recurrent NN model for probabilistic forecasting,
WaveNet: autoregressive deep NN model combining causal filters with dilated convolutions,
N-BEATS: deep NN with hierarchical doubly residual topology,
LGBM: light gradient-boosting machine,
XGB: extreme gradient boosting algorithm,
cES-adRNN – contextually enhanced hybrid and hierarchical model combining exponential smoothing and dilated recurrent NN with attention mechanism.

the global learning mode. It is worth highlighting that QRF stands out by achieving commendable results even with a limited training dataset size, as indicated by MARFE. Among the three models, QRF distinguishes itself by delivering superior accuracy and exhibiting minimal sensitivity to variations in the size of the training dataset. The subsequent results for QRF presented below are based on the global training mode with a q-value of 10.

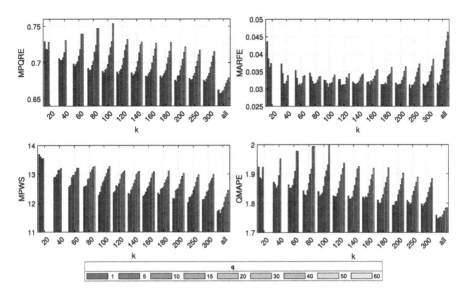

Fig. 1. Quality metrics for QRF.

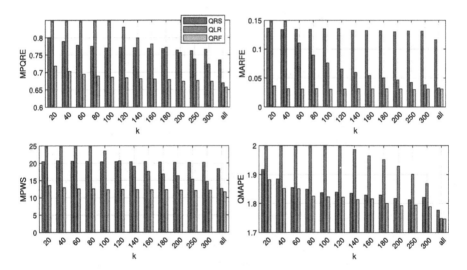

Fig. 2. Comparison of quality metrics for probabilistic STLF.

Table 2 provides a concise summary of the quality metrics. Here, MARFE, MdARFE and StdARFE were determined based on 35·99 ReFr values calculated individually for each country. A closer analysis of the metrics highlights QRF as the top-performing meta-model, exhibiting superior performance across most metrics. Both QRF and QLR produce comparable PIs that closely resemble the ideal distribution, with a 90% PI width and 5% coverage on each side. In contrast, the PIs generated by QRS are notably narrower, providing coverage of less than 60% on average, which falls short of the desired 90% coverage. Furthermore, when examining metrics like QMAPE and MdAPE, which are derived from the medians of the predicted distributions, it is evident that these values are larger compared to those generated by meta-models tailored for point forecasting (as observed in Table II in [5]).

Table 2. Probabilistic STLF quality metrics for meta-models.

Metric	QRS	QLR	QRF
MPQRE	0.735	0.669	**0.657**
MdPQRE	**0.386**	0.406	0.399
StdPQRE	1.066	0.913	**0.877**
MARFE	0.1163	0.0327	**0.0321**
MdARFE	0.1100	0.0300	**0.0200**
StdARFE	0.0696	**0.0268**	0.0295
MPWS	18.38	12.62	**11.64**
MdPWS	**4.09**	7.9	7.24
StdPWS	38.44	23.69	**20.99**
inPI	59.26	**90.34**	90.77
belowPI	18.4	**4.74**	4.23
abovePI	22.34	4.91	**5.00**
QMAPE	1.78	**1.75**	**1.75**
QMdAPE	1.12	1.12	**1.09**

Table 3 provides a comprehensive breakdown of the MPQRE, MARFE and MPWS for each country, offering a detailed view of the model performance on a country-specific level. Notably, the QRF stands out as the most accurate model in terms of all metrics for the majority of countries included in the analysis. In contrast, QRS emerges as the most accurate model for just one country in this evaluation.

For a more robust assessment of model performance, a Diebold-Mariano test [3] was executed to evaluate the statistical significance of differences between the probabilistic forecasts generated by each pair of models, considering individual country errors. The results based on MPQRE are visually depicted in Fig. 3, where the diagram illustrates the instances in which the model represented on the y-axis is statistically more accurate than the model represented on the x-axis. It is worth highlighting that the statistical analysis reveals QRF's superior accuracy over QRS in a substantial majority, specifically 31 out of 35, of the considered countries. Additionally, QLR outperforms QRS in terms of accuracy in 19 countries. Interestingly, QRF surpasses QLR in accuracy for just one country in this comparison.

The relative frequency (ReFr) charts illustrated in Fig. 4 provide a comprehensive view of the probabilistic forecasting outcomes. The desired value of $ReFr(\alpha)$ is nominal probability level α, shown with the dashed line. ReFrs calculated for each country and each $\alpha \in \Pi$ are represented by dots, while the red

Table 3. Probabilistic STLF quality metrics for each country.

Country	MPQRE			MARFE			MPWS		
	QRS	QRE	QRF	QRS	QRE	QRF	QRS	QRE	QRF
AL	0.85	**0.75**	0.76	0.108	0.035	**0.032**	20.3	13.8	**13.1**
AT	0.67	**0.57**	0.58	0.124	**0.012**	0.030	16.7	10.9	**9.2**
BA	0.56	0.54	**0.53**	0.081	**0.034**	0.044	13.7	10.0	**9.8**
BE	1.09	0.98	**0.96**	0.128	0.021	**0.019**	27.9	17.7	**17.6**
BG	0.66	**0.58**	**0.58**	0.112	0.029	**0.015**	14.9	**10.2**	10.3
CH	1.23	**1.06**	**1.06**	0.149	0.032	**0.018**	30.4	**19.1**	19.4
CZ	0.54	**0.51**	**0.51**	0.125	**0.021**	0.024	14.3	10.0	**9.1**
DE	0.51	**0.44**	0.46	0.108	**0.028**	0.036	12.3	8.1	**7.8**
DK	0.86	**0.73**	0.75	0.135	**0.040**	0.045	20.2	**13.6**	13.7
EE	0.81	0.72	**0.70**	0.091	**0.048**	0.057	18.9	**13.4**	13.4
ES	0.45	**0.37**	0.39	0.127	0.026	**0.020**	10.2	**6.6**	6.8
FI	0.51	**0.44**	**0.44**	0.113	**0.016**	0.026	12.6	7.5	**7.1**
FR	0.57	**0.52**	**0.52**	0.088	**0.019**	0.030	14.4	10.8	**9.9**
GB	1.40	1.32	**1.26**	0.106	0.022	**0.016**	35.9	22.3	**21.8**
GR	0.84	**0.71**	0.77	0.111	0.047	**0.037**	22.5	**13.5**	13.8
HR	0.94	**0.81**	**0.81**	0.122	0.029	**0.024**	24.9	**13.9**	14.5
HU	0.74	0.65	**0.62**	0.146	0.029	**0.027**	20.4	13.5	**10.8**
IE	0.52	0.56	**0.51**	0.116	0.022	**0.020**	12.1	10.9	**8.7**
IS	0.56	**0.46**	**0.46**	0.198	**0.027**	0.059	15.0	**10.0**	10.5
IT	0.81	0.74	**0.69**	0.126	0.032	**0.030**	21.7	14.4	**10.8**
LT	0.58	0.55	**0.53**	0.089	0.037	**0.033**	15.8	10.3	**8.7**
LU	**0.67**	0.90	0.69	**0.037**	0.068	0.066	**12.1**	20.6	14.7
LV	0.62	**0.57**	**0.57**	0.126	0.048	**0.026**	13.9	9.2	**8.2**
ME	0.98	0.91	**0.90**	0.122	0.022	**0.018**	23.4	17.1	**15.6**
MK	1.50	**1.30**	**1.30**	0.125	**0.031**	0.033	40.0	23.7	**21.1**
NL	0.65	**0.55**	0.59	0.108	0.050	**0.046**	16.0	11.3	**11.0**
NO	0.73	**0.62**	0.65	0.125	**0.029**	0.032	19.0	**10.1**	10.4
PL	0.58	0.51	**0.50**	0.119	0.035	**0.024**	15.8	11.0	**9.1**
PT	0.51	**0.46**	0.47	0.117	**0.044**	0.049	11.8	9.0	**8.7**
RO	0.54	**0.44**	0.45	0.109	0.049	**0.037**	13.6	8.0	**7.7**
RS	0.53	0.51	**0.50**	0.106	**0.032**	0.037	10.8	9.1	**8.6**
SE	0.70	0.67	**0.63**	0.112	**0.036**	0.049	16.9	12.4	**10.5**
SI	0.69	0.73	**0.64**	0.112	0.037	**0.028**	16.4	12.8	**9.9**
SK	0.59	0.54	**0.52**	0.131	0.034	**0.012**	15.9	11.3	**9.8**
TR	0.78	0.72	**0.71**	0.114	0.024	**0.023**	22.8	16.2	**15.4**

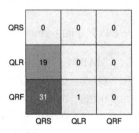

Fig. 3. Results of the Diebold-Mariano tests for MPQRE.

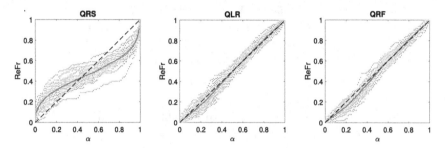

Fig. 4. Relative frequency.

line signifies the mean ReFr value across all countries. Observing this figure, it is evident that QRS exhibits the most distorted ReFr distribution, indicating excessively narrow PIs. In contrast, ReFrs for QLR and QRF are less dispersed and closely aligned with the desired values, with the average line almost coinciding with the dashed line. It is worth noting that MARFE presented in Table 2 serves as an indicator of the average deviation of ReFr values for each $\alpha \in \Pi$ and each country (dots) from the desired value (dashed line).

5   Conclusion

The practice of forecast combination has gained widespread recognition as an effective strategy for enhancing forecast accuracy and reliability. In this study, we focused on combining point forecasts using various stacking approaches to produce probabilistic forecasts. We introduced three approaches capable of generating quantile forecasts based on the point forecasts of base models. Notably, in the domain of short-term load forecasting, the quantile regression forest outperformed quantile linear regression slightly and demonstrated significant improvement over a method reliant on quantile estimation through residual simulation.

Our future research endeavors will be concentrated on the development of advanced machine learning models and approaches specifically tailored for forecast combination. This pursuit aims to further enhance predictive capabilities within bagging and boosting scenarios, ultimately advancing the field of forecasting.

References

1. Babikir, A., Mwambi, H.: Evaluating the combined forecasts of the dynamic factor model and the artificial neural network model using linear and nonlinear combining methods. Empirical Economics **51**(4), 1541–1556 (2016)
2. Beykirch, M., Janke, T., Steinke, F.: Bidding and scheduling in energy markets: which probabilistic forecast do we need? In: 2022 17th International Conference on Probabilistic Methods Applied to Power Systems (PMAPS), pp. 1–6 (2022). https://doi.org/10.1109/PMAPS53380.2022.9810632
3. Diebold, F., Mariano, R.: Comparing predictive accuracy. J. Bus. Economic Stat. **13**, 253–263 (1995)
4. Divina, F., Gilson, A., Goméz-Vela, F., Torres, M.G., Torres, J.: Stacking ensemble learning for short-term electricity consumption forecasting. Energies **11**(4), 949 (2018)
5. Dudek, G.: Combining forecasts using meta-learning: a comparative study for complex seasonality. In: IEEE 10th International Conference on Data Science and Advanced Analytics (DSAA'23), pp. 1–10. IEEE (2023). https://doi.org/10.1109/DSAA60987.2023.10302585
6. Gastinger, J., Nicolas, S., Stepić, D., Schmidt, M., Schülke, A.: A study on ensemble learning for time series forecasting and the need for meta-learning. In: Proc. 2021 International Joint Conference on Neural Networks (IJCNN), pp. 1–8 (2021)
7. Genre, V., Kenny, G., Meyler, A., Timmermann, A.: Combining expert forecasts: can anything beat the simple average? Int. J. Forecast. **29**(1), 108–121 (2013)
8. Haupt, S.E., et al.: The use of probabilistic forecasts: applying them in theory and practice. IEEE Power Energ. Mag. **17**(6), 46–57 (2019)
9. He, Y., Xiao, J., An, X., Cao, C., Xiao, J.: Short-term power load probability density forecasting based on GLRQ-Stacking ensemble learning method. Int. J. Electr. Power Energy Syst. **142**, 108243 (2022). https://doi.org/10.1016/j.ijepes.2022.108243
10. Kang, Y., Cao, W., Petropoulos, F., Li, F.: Forecast with forecasts: diversity matters. Eur. J. Oper. Res. **301**(1), 180–190 (2022)
11. Koenker, R., Hallock, K.F.: Quantile regression. J. Econ. Perspect. **15**(4), 143–156 (2001)
12. Kolassa, S.: Combining exponential smoothing forecasts using akaike weights. Int. J. Forecast. **27**(2), 238–251 (2011)
13. Li, S., et al.: Enhancing the locality and breaking the memory bottleneck of transformer on time series forecasting. In: Advance Neural Information Processing System. 32, pp. 5243–5253 (2019)
14. Lichtendahl, K., Winkler, R.: Why do some combinations perform better than others? Int. J. Forecast. **36**, 142–149 (2020)
15. Makridakis, S., et al.: The M5 uncertainty competition: results, findings and conclusions. Int. J. Forecast. **38**(4), 1365–1385 (2022)
16. Meinshausen, N.: Quantile regression forests. J. Mach. Learn. Res. **7**, 983–999 (2006)
17. Montero-Manso, P., Athanasopoulos, G., Hyndman, R., Talagala, T.S.: FFORMA: feature-based forecast model averaging. Int. J. Forecast. **36**(1), 86–92 (2020)
18. Pawlikowski, M., Chorowska, A.: Weighted ensemble of statistical models. Int. J. Forecast. **36**(1), 93–97 (2020)
19. Reddy, A.S., Akashdeep, S., Harshvardhan, R., Kamath, S.S.: Stacking deep learning and machine learning models for short-term energy consumption forecasting. Adv. Eng. Inform. **52**, 101542 (2022)

20. Ribeiro, G., Mariani, V., Coelho, L.: Enhanced ensemble structures using wavelet neural networks applied to short-term load forecasting. Eng. Appl. Artif. Intell. **82**, 272–281 (2019)
21. Smyl, S., Dudek, G., Pełka, P.: ES-dRNN: A hybrid exponential smoothing and dilated recurrent neural network model for short-term load forecasting. IEEE Transactions on Neural Networks and Learning Systems, pp. 1–13 (2023). https://doi.org/10.1109/TNNLS.2023.3259149
22. Wang, X., Hyndman, R., Li, F., Kang, Y.: Forecast combinations: An over 50-year review. International Journal of Forecasting (2022), in print

Trends in Computational Science: Natural Language Processing and Network Analysis of 23 Years of ICCS Publications

Lijing Luo[1]([✉]), Sergey Kovalchuk[4], Valeria Krzhizhanovskaya[1], Maciej Paszynski[2], Clélia de Mulatier[1], Jack Dongarra[3], and Peter M. A. Sloot[1]

[1] University of Amsterdam, Amsterdam, The Netherlands
lijing.luo@student.uva.nl,
{v.krzhizhanovskaya,c.m.c.demulatier,p.m.a.sloot}@uva.nl
[2] AGH University of Krakow, Krakow, Poland
paszynsk@agh.edu.pl
[3] University of Tennessee, Knoxville, USA
dongarra@icl.utk.edu
[4] Amsterdam, The Netherlands

Abstract. We analyze 7826 publications from the International Conference on Computational Science (ICCS) between 2001 and 2023 using natural language processing and network analysis. We categorize computer science into 13 main disciplines and 102 sub-disciplines sourced from Wikipedia. After lemmatizing full texts of these papers, we calculate the similarity scores between the papers and each sub-discipline using vectors built with TF-IDF evaluation. Among the 13 main disciplines, machine learning & AI have become the most popular topics since 2019, surpassing parallel & distributed computing, which peaked in the early 2010 s. Modeling & simulation, and algorithms & data structure have always been popular disciplines in ICCS over the past 23 years. The most frequently researched sub-disciplines, on average, are algorithms, numerical analysis, and machine learning. Deep learning shows the most rapid growth, while parallel computing has declined over the past 23 years in ICCS publications. The network of sub-disciplines exhibits a scale-free distribution, indicating certain disciplines are more connected than others. We also present correlation analysis of sub-disciplines, both within the same main disciplines and between different main disciplines.

Keywords: natural language processing · topic modelling · computational science · graph theory · network analysis · scientometrics · ICCS

1 Introduction

The continuous growth and digitization of scientific publications offer extensive research opportunities in scientometrics. As an integral part in science of

International Conference Computational Science—http://www.iccs-meeting.org/.

science, scientometrics plays a crucial role in guiding policies related to scientific development [1]. Additionally, it enables exploration of the progress within current scientific research fields [2]. The necessity to utilize quantitative methods for modeling and analyzing the progress of science has emerged as a key area of research [3]. The International Conference on Computational Science (ICCS) is an annual conference in the field that provides a prestigious platform for researchers, scientists, and engineers to explore computational disciplines encompassing mathematics and computer science [4–6]. Computational science is inherently interdisciplinary, offering advanced computing methodologies for addressing problems, identifying new issues, and shaping future directions in physics, chemistry, social sciences, and other fields. Since its inception in 2001, ICCS has consistently attracted an average of 340 highly cited papers per year [4–6]. This remarkable achievement establishes it as one of the most influential events within the field of computational science.

As a noteworthy asset in the field of computational science, the rapidly expanding proceedings series serve as a valuable corpus for quantifying scientific advancements. In this study, we apply a topic modeling technique to model and analyze the content of research papers. Building upon the concept that documents consist of various topics corresponding to specific disciplines, we apply text classification techniques and natural language processing methods to discover and analyze these topics. Utilizing a standardized corpus categorized by discipline, we conduct an annual analysis to explore changes in the distribution of research fields over 23 years. Simultaneously, we investigate emerging and declining research fields based on their popularity. Additionally, static and dynamic network analyses are conducted to examine how correlations between disciplines evolve and how network structures change over time. We answer the following questions: Which disciplines are gaining prominence or diminishing in computational science research? How have popular disciplines emerged or disappeared over the past 23 years? What is the structure of disciplinary networks and how does it evolve? How do correlations between disciplines change? The general methodology and tools we present can be applied to other fields of science.

This paper will be divided into seven sections presenting our studies: Sect. 2 summarizes related work; Sect. 3 describes data collection, pre-processing, and relevant methodologies; Sect. 4 demonstrates results about first-level disciplines; Sect. 5 provides an analysis from the perspective of second-level disciplines; Sect. 6 analyzes disciplinary networks; finally, Sect. 7 presents conclusions and future work.

2   Related Work

As the two most commonly used topic modelling methods, Latent Dirichlet Allocation (LDA) and Non-Negative Matrix Factorization (NMF) are widely used [7,8,10]. Blei et al. first introduced the Latent Dirichlet Allocation (LDA) as a generative probabilistic model to collect discrete tokens to provide an explicit representation of a document [7]. Similar to the LDA method, Greene et al.

proposed that the NMF method can be used to model topics in documents [8]. However, some studies pointed out the disadvantages of NMF. In particular, Wang et al. demonstrated that NMF based topic modelling may suffer from optimization and high computational complexity issues [9]. Pan also indicate that the Non-Uniqueness of NMF would cause multiple different factorization for a given input [10]. This may lead to the interpretation and comparison of results being more challenging. The feature selection and the result of tokens in topics would also cause differences in results, which causes ambiguity and mislabeling.

In the analysis of ICCS publication activity in 2017, Abuhay et al. used the NMF topic modelling method and classified the corpus into 13 high-level topics [12,13]. The authors found that modelling, HPC and e-science were the most popular topics between 2001 and 2017 [13]. However, as the disadvantage stated from previous research: NMF topic modelling is not unique and requires manually labelling the extracted keywords in the topics. It may cause ambiguity, non-exclusive, and cannot be extended to other fields or subjects. On the other hand, the research focus on computer science has changed since 2017. The Council of Europe and the European Union reported that machine learning and artificial intelligence experienced a rapid increase after 2016 and had a profound impact on society [14,15]. The application and research on machine learning increased rapidly after 2018 [15]. It is necessary to re-evaluate the most popular topics in computational science after 2017 and see if new topics have emerged.

In terms of text similarity comparison, past research indicates the method of comparing the cosine similarity of TF-IDF vector to measure similarity between papers [16–19]. Gunawan et al. demonstrated that a measure of cosine Similarity could be implemented to classify papers into subject types from text keywords [16]. The study gives a method to classify the research fields of the document from a series of keywords. When comparing highly specialized terminologies or disciplines, a corpus that is standardized into a unique document-term matrix shows its advantages in labeling documents. Wang et al. pointed out the shortcomings of the traditional "Bag of Words" (BoW) representation and introduced a method of using Wikipedia to apply content-based measure to compare the similarity between two texts [19]. Although Wang et al. indicated that Wikipedia's category structure does not form a tree taxonomy, but a directed acyclic graph in which multiple classification schemes coexist simultaneously [19]. This suggests that the method can be improved by building tree structures of domains with parents and subclasses. By applying the TF-IDF vector and the cosine similarity, terminologies within the disciplines of computer science can be compared. Nastase et al. [20] also pointed out that the knowledge base of Wikipedia could be transformed into a large-scale multilingual concept network. On the other side, the Association for Computing Machinery (ACM) classified the entire underlying disciplines of computer and computational science into 17 bodies of knowledge and dozens of sub-disciplines [21]. Curlie also provides related classification libraries in the field of computer science [22].

3   Data Collection and Preprocessing

In this study, the data is divided into two corpora: ICCS corpus and classification corpus. We collected all the papers published in the ICCS proceedings by Springer Lecture Notes in Computer Science (LNCS) (2001 - 2009 & 2018 - 2023), as well as Elsevier Procedia Computer Science from 2010 to 2017 [4,12,13]. The links to published volumes can be found in the conference webpage. The ICCS corpus encompasses 7826 papers over the twenty-three years, which is 340 papers on average each year with an average length of ten pages each. For the text classification corpus, we referred to the computer science curricula for 2023 provided by the Association for Computing Machinery (ACM) [21] and Curlie's outline of computer science [22] to classify computer science into thirteen first-level disciplines and one hundred and two second-level disciplines (see Table 1). We utilized Wikipedia's public API known as English Wikipedia API to extract the textual content as the second-level discipline standard classification library [19,20].

   We perform the following pre-processes to the text content:

1) We removed the HTML tags and the unrelated content from the classification corpus to create a standardized classification corpus for each second-level discipline.
2) In the ICCS corpus, we first standardized all the documents into text. We only keep the main content of the papers (from abstract to conclusion & discussion).
3) We removed all the English stop words, punctuation marks, and numbers, which contain no topical information. We extracted and excluded information such as place name, names, organization, etc. which may interfere with the classification process.
4) We chose lemmatisation instead of Porter stemmization to standardise different forms of words used by authors for grammatical reasons without changing the information the word contains [13].

4   The First-Level Disciplines of ICCS Papers

After applying the TF-IDF vectorizer to generate the document-term matrices, we computed the cosine similarity between ICCS papers and the corpus of each second-level discipline. Initially, we examined the topical structure of first-level disciplines. Similar to most topic modelling problems encountered, it is crucial to determine the number of disciplines (K) in each paper. The exclusive classification between the disciplines prevents the disadvantage of numerous highly similar topics. On the other hand, a large number of chosen disciplines will lead to irrelevant topics being included. Consequently, we conducted five preliminary experiments with K = 10, 20, 30, 40, 50 respectively. Based on our experiment results, we decided to adopt K = 20 as it represents a significant number of strongly relevant disciplines (see Table 2). Comparing this result with that from 2017 revealed a high degree of similarity between both methods which further validated our approach [13].

Table 1. Discipline classification structure

First-level disciplines	Number of second-level disciplines
Mathematical foundations	8
Modelling and simulation	6
Algorithms and data structures	7
Machine learning & AI	10
Network and security	7
Computer architecture	5
Computer graphics	5
Parallel, and distributed systems	8
Database	5
Programming languages and compilers	10
Scientific computing & Interdisciplinary	14
Software engineering	11
Theory of computation	6

Table 2. Comparison of new results with the previous analysis [12,13]: Pearson correlation coefficient R for different number of topics K. The p-values are indicated by stars: $^*p < 0.1$, $^{**}p < 0.05$, $^{***}p < 0.01$

Old & new disciplines	K = 10	K = 20	K = 30	K = 40	K = 50
Machine Learning	0.252	0.578***	0.161	0.086	-0.310
Network & Security	0.919***	0.919***	0.817***	0.886***	0.809***
HPC	0.803***	0.820***	0.829***	0.832***	0.822***
Programming	0.610***	0.638***	0.636***	0.607***	0.760***

We then calculated the popularity score for each year by aggregating the similarity scores across all first-level disciplines. To ensure comparability among different years' popularity scores, we standardized them into average popularity scores per every 100 papers. After smoothing the data to observe long-term trend by calculating the rolling mean using a window size of 2, thirteen first-level disciplines displays different trends during the past 23 years (see Fig. 1). Simultaneously, we calculated and visualized the proportion distribution among these thirteen first-level disciplines (see Fig. 2).

We look into some key time points to study the change of proportions of the first-level discipline (See Fig. 1 & 2). Combining the two sets of figures, we can draw some results:

1 . Machine learning & artificial intelligence did not garner significant attention prior to 2016. It was only in 2017 that topics related to machine learning & artificial intelligence began exhibiting a growth trajectory. In 2019, it surpassed modelling and simulation to emerge as the most prominent topics

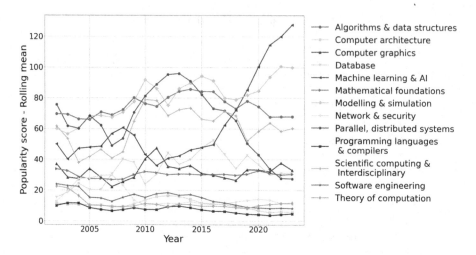

Fig. 1. The trend of popularity scores for the first-level disciplines

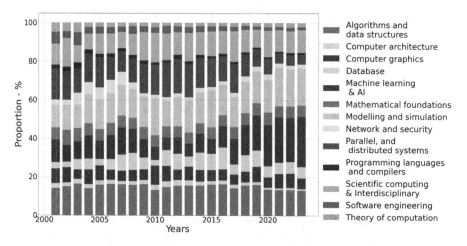

Fig. 2. The percentage of each discipline

in ICCS. Figure 2 illustrates an escalating proportion of machine learning & artificial intelligence across the entire corpus, encompassing a substantial share of total topics at approximately 26.2% by 2023.

2. Parallel and distributed computing has undergone significant transformations over the past 23 years. It exhibited a fluctuating and declining trend from 2001 to 2007, followed by rapid growth after 2008 that culminated in its peak in 2012 at 18%. During the period between 2011 and 2014, it remained the most sought-after topic. However, its proportion gradually decreased after 2015, accounting for only 5.4% of the total research popularity in 2023. This observation aligns closely with the intense competition of supercomputers between 2009 and 2016 when there was a two-order-of-magnitude improvement in the

performance of the fastest supercomputer (Jaguar:1.759 PFLOPS - Sunway TaihuLight:93.01 PFLOPS) [23]. In other periods, high-performance computing has never experienced such remarkable progress.

3. Under the 23-year time frame, modelling & simulation and algorithms & data structures have always been popular among all the ICCS papers. The proportion illustrates that the two disciplines cover more than 30% of the topic (see Fig. 2). It implies the essential contribution of these two disciplines toward computer science and computational science. Among other theoretical disciplines, such as the theory of computation and mathematical foundation, they show a stable trend with no significant increase or decrease.

4. Notably, we notice the Network & Security experienced a rise from 2003 to 2008 and reached its peak in 2007 at 6.9%. This finding aligns with the results of the 2017 study which shows the same sudden increase [12,13]. The explanation is the growing interest in early IPv6 deployment within universities [24]. These academic institutions provided a testing platform for evaluating and pre-commercializing IPv6 products and networks.

5   The Second-Level Disciplines

After analysing the topical trends of the first-level disciplines, we then study the evolution of the second-level disciplines. We first look into the average rank of the second-level disciplines (see Table 3). Throughout the entire span of 23 years, algorithms, numerical analysis, machine learning, mathematical models, and computer simulation emerge as consistently popular research topics that align closely with computational science. Subsequently, we explore the most prevalent secondary disciplines in 2023 (see Table 3). Notably, machine learning and artificial intelligence-related disciplines dominate six positions. This finding corroborates our earlier observations from first-level discipline trends: Machine learning and artificial intelligence-related topics are progressively gaining popularity.

We then analyze the evolution of the second-level disciplines. The disciplines are sorted based on changes in popularity scores, and we rank the ten most increased and decreased second-level disciplines (see Fig. 3). According to Fig. 3, deep learning shares the highest contribution in research popularity growth within machine learning & artificial intelligence. Deep learning is the most important driving force behind the rapid growth of artificial intelligence. It is followed by cross-validation, machine learning, reinforcement learning, and artificial intelligence. In terms of data science categories, data mining undergoes rapid development from 2005 to 2012. Agent-based modeling (ABM) stabilizes after a period of rapid growth between 2003 and 2010. Concerning scientific computing & interdisciplinary applications, computational social science contributes significantly to the growth of its first-level disciplines. Additionally, mathematical modeling exhibits a strong increase after 2017.

Table 3. Top 20 s-level disciplines for 23 years and in 2023

Top 20 in 23 years	Average rank	Top 20 in 2023	Rank
Algorithm	2.83	Machine learning	1.00
Numerical analysis	4.13	Deep learning	2.00
Machine learning	4.61	Mathematical model	3.00
Mathematical model	5.83	Artificial intelligence	4.00
Computer simulation	7.96	Algorithm	5.00
Data & information visualization	8.35	Numerical analysis	6.00
Parallel computing	10	Computational sociology	7.00
Computational sociology	10.78	ABM	8.00
Mathematical optimization	10.82	Data & information visualization	9.00
Genetic algorithm	11.26	Data mining	10.00
GPGPU	11.52	Stochastic simulation	11.00
Stochastic simulation	11.95	Mathematical optimization	12.00
ABM	13.48	Cross-validation	13.00
Data mining	13.96	Computer simulation	14.00
Distributed computing	14.57	Scientific modelling	15.00
Computational science	15.70	Genetic algorithm	16.00
Data structure	17.96	Reinforcement learning	17.00
Artificial intelligence	18	Computational chemistry	18.00
Deep learning	20.04	GPGPU	19.00
Database	20.52	Natural language processing	20.00

We then look into Fig. 3 bottom to explore the top ten decreasing disciplines. The second-level disciplines associated with parallel computing, distributed computing, and supercomputers have witnessed a significant decrease in popularity. It is noteworthy that these three disciplines were once the most popular and crucial topics in computer and computational science between 2010 and 2015. However, starting from 2016, this group of research areas has experienced a substantial downturn. In conjunction with Table 3, it becomes evident that the current research focus within parallel and distributed computing lies in general-purpose graphics processing units (GPGPU). Furthermore, we observe a diminishing presence of programming (Computer programming) and operating systems (software engineering), which have declined by 81.06% and 77.88%, respectively. Additionally, topics such as computational complexity and computational geometry enjoyed popularity prior until 2004 but gradually faded away after 2005.

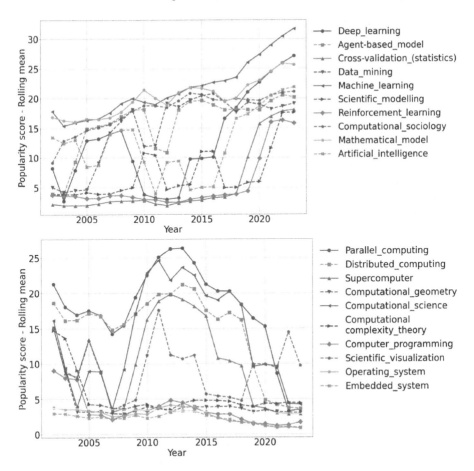

Fig. 3. The top 10 s-level disciplines. Top: rising disciplines, bottom: falling disciplines.

6   The Dynamic Analysis of Networks

We then proceed to investigate the interrelationships between topics. To construct the network of disciplines, we utilize a methodology akin to previous studies involving co-occurrence: if two disciplines are mentioned in the same paper, it signifies a correlation between them [12,13]. Subsequently, we assign weights to edges based on the similarity scores obtained for each discipline and generate twenty-three undirected networks corresponding to twenty-three years. In these networks, nodes represent disciplines, edges signify correlations among disciplines, and edge weights indicate the strength of co-occurrence [25]. Simultaneously, we conduct three experiments by pruning edges below 0.01, 0.05, and 0.09 respectively. We set the threshold at 0.05 for constructing networks based on the distribution of edge weights (see Fig. 4). Using Gephi software, we visualize the network matrix of 2022 as an example (see Fig. 5). The larger the network node, the higher the research popularity in this disciplines. And a wider edge means a stronger correlation between the two topics.

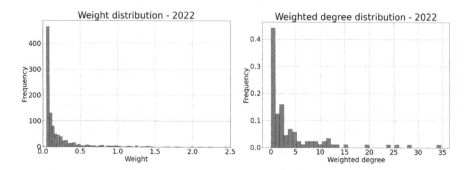

Fig. 4. Network analysis. Left: node degree distribution, right: edge weight distribution for year 2022.

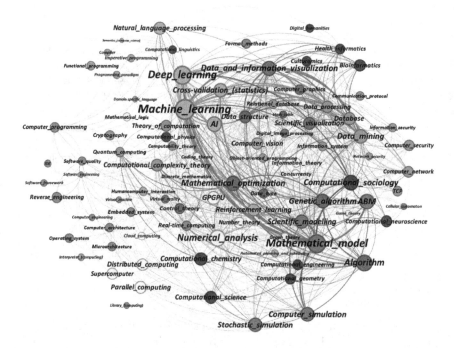

Fig. 5. The visualisation of the network - 2022.

After analyzing the network structure, we observed that the fraction of nodes with degree k follows a power-law distribution ($f(x) = cx^{-\alpha-1}$) where the exponent parameter $\alpha > 1$, and $p > 0.1$ (see Table 4 & Fig. 4). This indicates that the networks are scale-free [26,27]. Additionally, we notice a decreasing trend in both the average clustering coefficient and the total number of edges (see Table 4). Coupled with an increase in α (The exponent of the power), this suggests a decline in network density and decentralization of the network.

The structure of the network is undergoing changes, and throughout the timeline, there have been shifts in the central points and essential edges within

Table 4. The network structure & node degree distribution(2013 - 2023)

Parameters	2013	2014	2015	2016	2017	2018	2019	2020	2021	2022	2023
Total edges (N)	1296	1279	1328	1058	1004	823	1025	1095	794	1006	809
Cluster coefficient	0.65	0.67	0.66	0.60	0.62	0.58	0.63	0.62	0.56	0.60	0.56
The exponent α	1.83	1.96	1.77	2.60	1.68	2.24	1.85	3.68	1.80	3.00	3.20
KS statistics	0.124	0.117	0.132	0.117	0.137	0.095	0.103	0.131	0.131	0.103	0.097
p-value	0.417	0.219	0.631	0.762	0.297	0.655	0.854	0.665	0.744	0.982	0.221

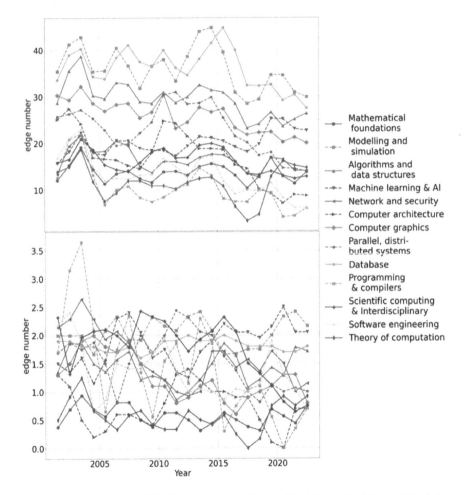

Fig. 6. The standardised N of community edges by first-level disciplines. Top:inter-community Bottom: intra-community

the network. To explore the correlation between various disciplines in the networks, we first look into the trend between first-level disciplines. By regarding the division of first-level disciplines as network communities, Fig. 6 shows

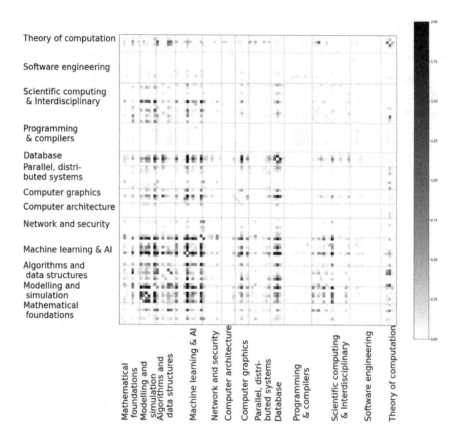

Fig. 7. The heatmap of edge correlations - 2023

standardized average intra-community and inter-community edges of each first-level disciplines [29]. The same result as metrics in Table 4 indicates, the number of both intra- and inter-community edges are decreasing. Each first-level disciplines presents different characteristics. Disciplines such as algorithms & data structure, modelling & simulation, and database show strong external correlations. Figure 6 shows that these three first-level disciplines are the highly connected hubs of multiple disciplines and provide collaborative bridges for multidisciplinary interactions. On the other side, disciplines such as programming languages & compilers, software engineering, and scientific computing & interdisciplinary application present a strong internal correlation. It suggests that these communities are more independent than other communities. Papers in these communities have begun to form a tight internal structure which connect outwards through few key nodes. For example, the only junction between programming languages community and software engineering community is computer programming (see Fig. 5).

We then examine the evolution of weighted edges from the first-level disciplines' perspectives. We select 2023 as an example to illustrate the heat map of the network for both first and second-level disciplines (see Fig. 7). We found that machine learning and artificial intelligence is highly correlated with other disciplines. Machine learning & AI is actively collaborated with computer graphics and scientific computing & interdisciplinary applications. This indicates that machine learning has significant impact on computer graphics and interdisciplinary application research in 2023. Additionally, the heat maps reveal high activity levels for database, algorithms & simulation, modeling & simulation and theory of computation. From the perspective of computer and computational science, data, algorithms, modelling and computational theory are the foundations of research and the bridge between disciplines. They play a very important role among the research of computer and computational science.

7 Conclusion and Future Work

In this paper, we have discussed the changes in popular disciplines within ICCS research over the past 23 years. The most prominent disciplines are machine learning & AI, modelling & simulation, and algorithms. From 2005 to 2008, there was a minor peak in computer network due to the rising interest in application of IPv6. Between 2009 and 2016, research focused on parallel and distributed computing reached its zenith, which correlates with the rapid growth of supercomputer computing power. Simultaneously, the popularity of parallel and distributed computing on ICCS has declined after 2017. As academia and institutions recognizes 2016 as the rise of artificial intelligence [14,15], we provide evidence that machine learning and artificial intelligence-related research experienced rapid growth thereafter. By 2019, it surpassed modelling and simulation as the most popular discipline.

Our network analysis revealed that the degree distribution of second-level cipline networks are scale-free networks. This also suggests a preferential attachment phenomenon where new research disciplines tend to align with existing popular disciplines. Regarding the network structure itself, the decrease of clustering coefficient and the increase of α imply that the ICCS research network has become more decentralized over the past decade and a few nodes will have more connections. This indicates that a few popular subjects are widely utilised in more research, while the overall cluster structure of the research network is gradually weaken.

At the same time, the network analysis also suggests that machine learning & AI are interacting with multiple disciplines: we see strong correlations with modelling & simulation, algorithms & data structures, computer graphics, and interdisciplinary applications. The communities formed by each first-level discipline also demonstrate different intra- and inter-community characteristics: Algorithms, modelling & simulation and database show strong external and internal correlations, while programming language & compilers and software engineering present strong internal but weak external correlation.

However, we only discussed the correlation rather than the causal relationship between disciplines due to the methodology. Future research can apply more advanced natural language processing techniques to explore the causal relationship for more accurate directed correlations.

Finally, it is important to note that although ICCS holds prominence within the computational science conferences, the results may be biased due to its limited coverage. At the same time, our classification of disciplines can be further refined. The accuracy of the results can be improved by more detailed division of disciplines. A larger corpus with wider coverage could depict the full picture of the entire computational science research. Towards this end, we will analyze the corpus of publications of the Journal of Computational Science (JoCS).

References

1. Fortunato, S., et al.: Science of science. Science **359**(6349), (2018). https://doi.org/10.1126/science.aao0185
2. Mingers, J., Leydesdorff, L.: A review of theory and practice in scientometrics. Eur. J. Oper. Res. **246**(1), 1–19 (2015). https://doi.org/10.1016/j.ejor.2015.04.002
3. Wang, D., Barabási, A.-L.: The science of science. Sci. Sci. (2021). https://doi.org/10.1017/9781108610834
4. Derek Groen, Clélia De Mulatier, Maciej Paszynski, Valeria V Krzhizhanovskaya, Jack J Dongarra, Peter M A Sloot: The Computational Planet. Preface to ICCS (2022). https://www.iccs-meeting.org/iccs2022/
5. Paszynski, M., Kranzlmüller, D., Krzhizhanovskaya, V.V., Dongarra, J.J., Sloot, P.M.A. (eds.): Computational Science – ICCS 2021: 21st International Conference, Krakow, Poland, June 16–18, 2021, Proceedings, Part III. Springer International Publishing, Cham (2021)
6. Krzhizhanovskaya, V.V., et al. (eds.): Computational Science – ICCS 2020: 20th International Conference, Amsterdam, The Netherlands, June 3–5, 2020, Proceedings, Part VII. Springer International Publishing, Cham (2020)
7. Blei, D.M., Andrew, Y. Ng., Jordan, M.I.: Latent Dirichlet Allocation. J. Mach. Learn. Res. **3**, 993–1022 (2003)
8. Derek Greene, James P. Cross: Exploring the Political Agenda of the European Parliament Using a Dynamic Topic Modeling Approach. **25**(1), pp .77–94 (2016) http://arxiv.org/abs/1607.03055
9. Wang, J.: Xiao Lei Zhang: Deep NMF topic modeling. Neurocomputing **515**, 157–173 (2023). https://doi.org/10.1016/j.neucom.2022.10.002
10. W. Pan, F. Doshi-Velez: A Characterization of the Non-Uniqueness of nonnegative matrix factorizations (2016). http://arxiv.org/abs/1604.00653
11. Kim, S.W., Gil, J.M.: Research paper classification systems based on TF-IDF and LDA schemes. Hum-centric Comput. Inf. Sci. **9**(1), (2019) . https://doi.org/10.1186/s13673-019-0192-7
12. Abuhay, T.M., Kovalchuk, S.V., Bochenina, K.O., Kampis, G., Krzhizhanovskaya, V.V., Lees, M.H.: Analysis of computational science papers from ICCS 2001–2016 using topic modeling and graph theory. Procedia Comput. Sci. **108**, 7–17 (2017). https://doi.org/10.1016/j.procs.2017.05.183
13. Abuhay, T.M., et al.: Analysis of publication activity of computational science society in 2001–2017 using topic modelling and graph theory. J. Comput. Sci. **26**, 193–204 (2018). https://doi.org/10.1016/j.jocs.2018.04.004

14. Verbeek, A., Lundqvist, M.: European Union: artificial intelligence, blockchain and the future of Europe: How disruptive technologies create opportunities for a green and digital economy (2021). https://www.eib.org/attachments/thematic/artificial_intelligence_blockchain_and_the_future_of_europe_report_en.pdf

15. The council of Europe: History of Artificial Intelligence (2024) https://www.coe.int/en/web/artificial-intelligence/history-of-ai

16. Gunawan, D., Sembiring, C.A., Budiman, M.A. : The Implementation of cosine similarity to calculate text relevance between two documents. J. Phys. Conf. Ser. **978**(1), (2018) . https://doi.org/10.1088/1742-6596/978/1/012120

17. Li, B., Han, L.: LNCS 8206 - Distance weighted cosine similarity measure for text classification. LNCS **8206**(1), 611–618 (2013)

18. Singh, R., Singh, S.: Text Similarity measures in news articles by vector space model using NLP. J. Inst. Eng. (India): Series B **102**(2), 329–338 (2021). https://doi.org/10.1007/s40031-020-00501-5

19. Wang, P., et al.: Using wikipedia knowledge to improve text classification. Knowledge and Information Systems **19**(3), 265–281.Springer London (2009). https://doi.org/10.1007/s10115-008-0152-4

20. Nastase, V., Strube, M.: Transforming Wikipedia into a large scale multilingual concept network. Artif. Intell. **194**, 62–85 (2013). https://doi.org/10.1016/j.artint.2012.06.008

21. Association for computing machinery (ACM), IEEE-computer society (IEEE-CS), Association for Advancement of Artificial Intelligence (AAAI): Computer Science Curricula 2023. (2023) https://csed.acm.org/wp-content/uploads/2023/09/Version-Gamma.pdf

22. Curlie: Computers (2024). https://curlie.org/en/Computers/

23. Top500: NOVEMBER 2017. (2018). https://www.top500.org/lists/top500/2017/11/

24. Colitti, L., Gunderson, S.H., Kline, E., Refice, T. : Evaluating IPv6 Adoption in the Internet. PAM 2010 (2010).http://www.pam2010.ethz.ch/papers/full-length/15.pdf

25. Jiming, H., Zhang, Y.: Discovering the interdisciplinary nature of Big Data research through social network analysis and visualization. Scientometrics **112**(1), 91–109 (2017). https://doi.org/10.1007/s11192-017-2383-1

26. Broido, A.D., Clauset, A.: Scale-free networks are rare. Nat. Commun. **10**(1), (2019). https://doi.org/10.1038/s41467-019-08746-5

27. Choi, J., Yi, S.: Kun Chang Lee: analysis of keyword networks in MIS research and implications for predicting knowledge evolution. Inf. Manage. **48**(8), 371–381 (2011). https://doi.org/10.1016/j.im.2011.09.004

28. Intel: Building Optimized High Performance Computing (HPC) Architectures and Applications. https://www.intel.com/content/www/us/en/high-performance-computing/hpc-architecture.html

29. Kang, Y., Lee, J.-S., Shin, W.-Y., Kim, S.-W.: Community reinforcement: an effective and efficient preprocessing method for accurate community detection. Knowl.-Based Syst. **237**(8), 107741 (2022). https://doi.org/10.1016/j.knosys.2021.107741

Target-Phrase Zero-Shot Stance Detection: Where Do We Stand?

Dawid Motyka$^{(\boxtimes)}$ and Maciej Piasecki

Department of Artificial Intelligence, Wrocław University of Science and Technology,
Wrocław, Poland
dawid.motyka@pwr.edu.pl

Abstract. Stance detection, i.e. recognition of utterances in favor, against or neutral in relation to some targets is important for text analysis. However, different approaches were tested on different datasets, often interpreted in different ways. We propose a unified overview of the state-of-the-art stance detection methods in which targets are expressed by short phrases. Special attention is given to zero-shot learning settings. An overview of the available multiple target datasets is presented that reveals several problems with the sets and their proper interpretation. Wherever possible, methods were re-run or even re-implemented to facilitate reliable comparison. A novel modification of a prompt-based approach to training encoder transformers for stance detection is proposed. It showed comparable results to those obtained with large language models, but at the cost of an order of magnitude fewer parameters. Our work tries to reliably show where do we stand in stance detection and where should we go, especially in terms of datasets and experimental settings.

Keywords: stance detection · zero-shot learning · prompt based learning for transformers

1 Introduction

People not only communicate some information or opinions, but also often express their stance against or in favor of the topic.

Stance is orthogonal to sentiment and emotions to a very large extent. While writing in favor is naturally more likely to express positive sentiment, this is not guaranteed. Mohhamad et al. [27] characterise *stance detection* as "the task of automatically determining from text whether the author of the text is in favor of, against, or neutral towards a proposition or target.". Such broad definition results in several variants, e.g. rumour stance [7]. First of all, different target types are considered, e.g. [12]: headline, claim, topic, person. Dealing with different target types, across different domains [12] is challenging. Some datasets use only two labels: *in-favor* and *against*, e.g. p-stance [17]. It may work if targets are clearly identifiable as proper names, but in the case of multiple targets two-label annotation blurs the difference between neutral utterances towards some

L. Franco et al. (Eds.): ICCS 2024, LNCS 14833, pp. 34–49, 2024.
https://doi.org/10.1007/978-3-031-63751-3_3

target, and non-related ones. Some authors consider inter-connected targets, like multi-target [30] (stance towards two targets, e.g. Clinton-Sanders), but this is the question of target definition.

Building data sets for supervised learning gets more laborious with the increasing number of targets, while the generalisation becomes harder. Thus, zero-shot-learning approach to stance detection is a good direction. [27] already discussed it, and [2] characterise it as stance detection for targets absent in the training data. [3] considers also different zero-shot perspectives, like language, genre, label set in relation to stance-detection. However, such perspective have not received limited attention so far.

We focus on stance detection tasks with targets expressed by short phrases. This case may be called *target-phrase stance detection* – contrary to tasks with targets represented by sentences or short texts. We focus on problems described with three labels: *in favor*, *against* and *neutral*, as the lack of the neutral category or implicit assumption that all non-labelled samples are neutral, significantly simplifies the problem. Target-phrase stance detection is represented by the most influential stance detection datasets, see Sect 2. Our main motivation was observation that different approaches in different papers were tested with slightly different experimental settings, also in combination with different interpretations of the annotation of datasets. All this causes problems with comparison and reproducibility. Focusing only on target-phrase stance detection may seem to be limitation, but target phrase stance detection is important for applications and enables to confine the overview within a limited size of the paper.

Due to the unclear picture of zero-shot stance detection, we aim at investigating where do we stand in the stance detection, at least in the target phrase subtype. What are the real results of approaches when compared in fairly comparable conditions?

The main novelty of our work is in clarifying the picture of zero-shot target phrase stance detection in relation to:

- available datasets for multiple target stance detection suitable for zero-shot learning together with their limitations,
- comparison of the SOTA approaches to multiple target stance detection performed on a carefully set-up common ground,
- re-running and in some cases even re-implementing several approaches[1],
- significant problems observed in an important benchmark, namely VAST [2], causing issues in comparison,
- supplementing the map of approaches with methods based on prompting transformers that were not tested in this task so far.

Our prompting-based methods express very good results in a zero-shot setting, better than the current SOTA. They are also in many cases better than a prompt-based learning stance detection with gpt-3.5-turbo or FLAN-UL2 models, while much more efficient, that makes them especially interesting.

---

[1] The source code: https://github.com/dawidm/zssd-iccs-2024.

2   Stance Detection Datasets

For zero-shot target phrase stance detection, it is of primary importance that a dataset includes multiple and diversified targets. In the cases of small number of targets, e.g. six [27], zero-shot methods struggle to achieve good generalisation. Datasets with a small number of targets are not suitable for evaluation and development of such methods, while they dominate in stance detection. There are only two commonly used datasets, namely *SemEval 2016 Task 6* [27] and *VAried Stance Topics* [2] that are useful for target phrase zero-shot stance detection, but not free of some shortcomings, characterised below.

SemEval 2016 Task 6 [27] (henceforth Sem2016T6) is a commonly used stance detection dataset based on Twitter posts. It consists of 4 870 samples with six stance targets labelled by *favor*, *against* and *neither*. The last one is used to mark texts that are not related to a given target, but not necessarily including its mention. It may be used in a zero-shot configuration with the data for five targets used for training and the remaining one reserved for testing. During annotation authors labelled a few samples as *neutral*, i.e. referring to the target, but where stance cannot be deduced. However, finally all of them were assigned to the *neither* class.

VAried Stance Topics (VAST) [2], also popular in research, differs from Sem2016T6. Texts come from the *New York Times Room for Debate* forum and are on average longer and of a less casual character. VAST includes 18 515 samples divided into train, validation and test parts with 5 634 stance targets (called *topics* by the authors). Three labels are used: *pro*, *con* and *neutral*. A subset of the test samples includes mentions of targets not occurring in the training and validation parts, so this gives an opportunity for zero-shot experiment setting. It is worth to notice that only about 9% of samples can be considered to be *ingeniously neutral* stance: a given target occurs, but the sample has neutral stance in respect to it. Other 'neutral' samples have been created synthetically by assigning a random target to a sample in which it does not occur (so in a trivial way the stance cannot be other than neutral). We will call them *synthetic neutral* and others *true neutral*. More than half of the sample targets (counted with the exclusion of synthetic neutrals) are automatically extracted noun phrases, which is a distinctive feature of VAST. They tend to be relatively specific and, in some cases, even arguably suitable for a task e.g. *comment sections, healthier, a smart investment*. In this work we use the same label *neutral* for both *neither* samples from Sem2016T6 and *neutral* samples from VAST.

In addition to the two main datasets, several other were proposed, but all of them lack properties required for target phrase zero-shot stance detection.

P-Stance [17] is a large dataset of more than 20 000 samples but only 3 stance targets, which are also highly related (Trump, Biden, Sanders).

The encryption debate [1] consists of 3 000 samples but for only 1 target. Only tweets IDs are available, so there could be a problem with obtaining texts.

A Dataset for Multi-Target Stance Detection [30] incorporates 4 455 samples and 4 stance targets, which are highly related (e.g. Hillary Clinton, Bernie Sanders), but it enables cross-target or multi-target stance detection.

Stance Detection in COVID-19 Tweets [11] with 6 133 samples and 4 highly related stances targets (e.g. *stay at home orders, school closures*). It uses same annotation instructions as Sem2016T6. Only tweets IDs are available.

BASIL [9] includes only 300 news articles with sentence-level annotations, but only article-level annotations can be considered suitable for detecting the author's stance. The results on article-level annotations are very low even in non-zero-shot settings [24], probably due to the small number of samples.

3 Existing Approaches

Existing zero-shot stance detection methods aim at improving language model generalisation to unseen stance targets. For a dataset like VAST, particularly rich in stance targets, only transformer models achieve high results, leaving behind other models based on recurrent (LSTM) neutral networks [2]. This is also true for Sem2016T6 in zero-shot setting, though the performance gap is smaller [4]. BERT-base [8] is most commonly used transformer encoder, followed by RoBERTa [23]. Recently, also the encoder-decoder transformer BART [16] was utilised [18,32]. Various techniques described below are applied along with the mentioned models to improve the results.

Latent Target Representations from training datasets are incorporated to relate unseen targets to known ones to obtain better target-aware representations. Allaway et al. [2] used clustering of input representations, while [24] proposed a method with learned topic clusters. Another approach [20] used graph neural networks (GNNs) to link latent representations with the new samples.

External Knowledge Sources. Conditioning the stance detection model on an external knowledge source may be beneficial for generalisation to new targets, if the source contains relevant information. Also, updating knowledge should be easier with such methods compared to retraining the underlying language model. A general knowledge graph with graph neural networks to obtain graph embeddings was used in [22,26]. A different approach is to use knowledge in a form of plain text, which could be used directly as an input to a language model. Wikipedia's texts were used in [13,32] for creating stance target definitions. In [36] detailed information were obtained using keyword matching.

Contrastive Learning. Contrastive learning was applied in stance detection for different purposes. Liang et al. [20] applied supervised contrastive learning in order to improve generalisation ability. In [35] contrastive approach were combined with word masking to capture target-invariant stance features. [14] proposed a solution able to leverage unlabelled data to acquire better target representations.

Pre-training on an Auxiliary Task. Liu et al. [24] investigated pretraining RoBERTa language model using a large collection of political texts to improve stance detection in this domain. [33] utilised similarities in textual entailment and stance detection to pre-train RoBERTa with textual entailment task.

Dataset Augmentations. In zero-shot stance detection, data augmentation could be used to generate new pairs: target and text to improve generalisation. Such an approach with the help of a large language model (LLM), namely GPT3 [6] was proposed in [33]. A method with a smaller generative model used to extract keywords as potential new stance targets was also developed [18].

Using Prompts for transformer encoder models has been shown to increase performance, especially in a few-shot settings for many tasks [10,28,29]. [24] apply the RoBERTa-base model and test it on VAST and Sem2016T6 datasets, but does not focus specifically on zero-shot performance.

Opportunities of Generative Models. Using generative models brings new possibilities in improving learning for stance detection, e.g. [32]. They proposed predicting not only the stance label but also the target and using unlikelihood to leverage samples with assigned opposite labels as an auxiliary tasks.

Large Language Models (LLM), exhibit an ability to solve various tasks in zero-shot settings [6]. Kocoń et al. [15] tested ChatGPT (a GPT-family model) on numerous tasks, including stance detection, showing results lower compared to fine-tuned transformer encoder models. [34] also provided results for 3 targets from Sem2016T6. Both approaches were conducted using ChatGPT web interface before official OpenAI API was available. More tests are needed to examine the performance and problems of LLMs in stance detection.

4   Experiments

4.1   Datasets and Metrics

We chose the Sem2016T6 and VAST datasets. For Sem2016T6 we use six configurations created by leaving out samples with a given target as a test set and splitting the rest of the samples into 85% and 15% for a train and validation set, respectively. For VAST, we use the default splits.

We apply metrics commonly used for the selected datasets. For Sem2016T6 the average of F_1 for *favor* and *against* classes (F_{1mfa}) [27]. It should be noted that *neutral* samples are included in these calculations. For VAST, the average of F_1 score of all three classes is used, calculated both for the whole test dataset F_{1m} and its zero-shot part F_{1mZS} [2]. We also report the average of F_{1m} for Sem2016T6 in some experiments, for comparison with [33]. All experiments were run 10 times and the average result is reported.

Most neutral samples in VAST are *synthetic* ones. We verified performance on *true neutral* (i.e. related to a target but with no clear stance): test set containing only *true neutral* samples left in the test set and train two models, one with also *synthetic neutrals* and the other without. The limitation of this approach is that there are only 114 and 45 *true neutral* samples in the training and test set respectively.

4.2   Baseline Models

Establishing good, reliable baseline results for models of sizes comparable to BERT-base for both considered datasets on the basis of literature appeared to be challenging. Our results for **Sem2016T6** were not consistent with the commonly used baseline from [4]. We re-implemented their setup with BERT-base to recreate the results. We also trained RoBERTa-base with our setup for comparison, as it was examined so far only for the non-zero-shot variant [5].

For the **VAST** dataset, due to its rich original data structure, it may be unclear which values to use. We distinguished and tested four possible variants using BERT-base and the best configuration also with RoBERTa-base:

Using Unprocessed Texts. Lemmatised and with stop words removed texts are stored in `text_s` along with unprocessed ones in `post`. We found it ambiguous which values are used in the works from literature, and we test both variants.

Using Unprocessed Targets. The similar situation is with stance targets, but there is no specific column where all unprocessed values are valid, and it requires additional effort to extract them. We test both target sets with our baselines.

Discarding type 3 (*list*) Samples. It can be concluded from the annotation task that samples of a certain type of *list* samples often have the wrong stance label. Such samples can be removed, not affecting the validation and test sets.

Discarding Ambiguous Samples. there are some samples in the training set with different labels for the same text and stance target, that can be removed. The training set sizes after different modifications are shown in a Table. 1.

4.3   Reproduced Methods

For fair comparison we tried to reproduce results for all the methods using models of the size comparable to BERT-base running the published source codes. In the **JoinCL** [20] implementation we find out that for Sem2016T6 in zero-shot configuration, the test part is also used as a *validation* set during development (sic!), while it should be a subset of the training set, e.g. [4]. In our reproduction we did not introduce any changes to the method itself and used the provided parameters. We only added an evaluation metric for the full VAST dataset.

In **WS-BERT** [13] target definitions for VAST are automatically retrieved from Wikipedia, but the targets come from `new_topic` column and are not correct for *synthetic neutral* samples, i.e. they have not been changed to random ones in order to make the stance neutral. This makes the obtained definitions highly related to texts, which potentially adds positive bias to classification. For reproduction, we fixed this problem by correcting stance targets in `new_topic` column in VAST, but has not introduced any other changes, and used parameters provided by authors. For Sem2016T6 we used the WS-BERT-Single variant and originally selected articles from Wikipedia as definitions.

Wen et al. [32] (henceforth **VTCG**) used the same definitions as in WS-BERT, so the same concerns are valid for their method. We fixed `new_topic` column and run experiments with the original code and parameters. We also test the method without target definitions (VTCG-NW) and with BART-base without modifications (VTCG-BO).

BS-RGCN [26] was not tested on Sem2016T6 and needed modifications to work for its shorter texts – we let more words be used for knowledge graph embeddings. We run BS-RGCN with the original code and configuration proposed for VAST, and our modified version on Sem2016T6.

4.4   Prompt Based Methods

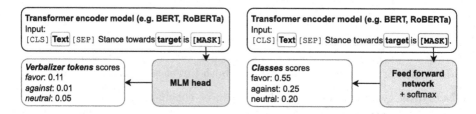

Fig. 1. Comparison of two approaches for using prompts with transformer encoder language models. Left: common approach proposed in [28]. Right: our modification.

For transformer encoder models, a masked language pretraining objective can be used to train the model for a downstream task, as shown by [10, 28, 29]. Compared to standard fine-tuning, this requires a task-specific prompt and verbalizer tokens (verbalizers), i.e. tokens from model's dictionary that would replace the [MASK] token following the prompt to indicate given class. Such an approach was not yet examined specifically for zero-shot stance detection, but was shown to be promising [24]. Several methods for selecting verbalizers were proposed in the aforementioned publications.

We use two prompt-based approaches, see Fig. 1. In our modification of [28] we replace the head used in masked language modelling (MLM) pretraining (returning scores for every token), by a feed forward network returning scores corresponding to 3 stance classes which eliminates the need to specify a verbalizer and doesn't restrict classification to predefined class tokens.

As a feed forward head we use 2 layers dim $h \times$ dim h and dim $h \times |C|$ (where C – a set of classes, dim h – a model hidden state size) with dropout and GELU activation layer between.

We propose slight modifications to the best prompt of [24] (*The stance towards [target] is [MASK].*, **P2**) making it written in the first person and using a word that is an indicator that it concludes the previous text: *Therefore my stance towards [target] is [MASK].* (**P1**).

We run prompt based method with prompts described above. We name the models as follows: `model-type-prompt_name`, where model is underlying encoder model, type is PV (with verbalizer) or PFF (our model with feed forward classification head) e.g. `BERT-PV-P1`.

We use gpt-3.5-turbo (Apr. 30, 2023) from OpenAI, based on GPT-3 architecture[2] and the FLAN-UL2 model – based on encoder-decoder transformer architecture, with 19.5B parameters, fine-tuned to various tasks, with zero-shot ability [25,31] – as LLM baselines for in zero-shot setting. For each dataset, we propose one prompt based on its annotation task (**P3**, **P4**) and a designed prompt (**P5**) aimed at improving results (we provide them in our source code repository). We use the online inference API for both models[3],[4]

Table 1. The numbers of samples in VAST train, in *+Discard amb. – list* type and ambiguous samples are excluded, see Sec. 4.2

	VAST train	**Discard** *list* type	**+Discard** amb.
Samples	13447	6922	6870
Unique texts	1845	1845	1844
favor samples	5327	2104	2082
against samples	5595	2385	2363
neutral samples	2555	2104	2425
Stance targets	5014	1797	1794

5 Results

Tables 2 & 4 show results for methods with models of size comparable to BERT-base, while Table 3 & 4 focus on larger models. In Table 4 we consider variations of VAST usually neglected in literature.

5.1 Baseline Results

We failed to reproduce the BERT baseline of [4] (Table 2) for Sem2016T6 zero-shot dataset. The average score is slightly lower, even if some targets scored higher. The most substantial difference is for *Atheism* (*A*) target. RoBERTa model scored even lower. It is worth to notice that both results are not much higher than random guessing. The *large* models are marginally better when fine-tuned in a standard way (Table 3). In Table 4, we found the differences between dataset configurations to be substantial, and the best are achieved with all text unprocessed and with discarding a large amount of training samples. The *base* size models achieved comparable results with both the full test set and the zero-shot part.

---

[2] https://platform.openai.com/docs/models/gpt-3-5.
[3] https://platform.openai.com/docs/api-reference.
[4] https://huggingface.co/inference-api.

Table 2. Sem2016T6 zero-shot setting with smaller models: [*] evaluation on the test set in training, [†] – originally not tested with Sem2016T6, [‡] – modified by us for Sem2016T6.

	Method	Per-target results (F_{1fa})						Average
		DT	HC	FM	LA	A	CC	(F_{1mfa})
Base (ours)	Random guessing	0.315	0.321	0.360	0.328	0.330	0.254	0.318
	BERT-base	**0.403**	0.549	**0.441**	**0.447**	**0.368**	**0.299**	**0.418**
	RoBERTa-base	0.279	**0.565**	0.327	0.401	0.268	0.248	0.348
Reported results	BERT [4]	0.401	0.496	0.419	0.448	0.552	0.373	0.448
	TOAD [4]	0.495	0.512	0.541	0.462	0.461	0.309	0.463
	JointCL[*] [20]	0.505	0.548	0.538	0.495	0.545	0.397	0.505
	TarTK [36]	0.508	0.551	0.538	0.487	0.562	0.395	0.507
	PT-HCL [19]	0.501	0.545	0.546	0.509	0.565	0.389	0.509
	FECL [35]	**0.516**	0.556	0.553	0.533	**0.573**	0.418	0.525
	MPCL [14]	0.512	**0.595**	**0.556**	**0.534**	0.567	**0.454**	**0.536**
Reproduced results	JointCL*	**0.453**	**0.551**	**0.451**	**0.470**	**0.464**	**0.280**	**0.445**
	WS-BERT[†] [13]	0.212	0.356	0.279	0.262	0.420	0.076	0.268
	BS-RGCN[†‡] [26]	0.214	0.325	0.257	0.253	0.396	0.100	0.258
	VTCG [32][†]	0.400	0.294	0.285	0.362	0.415	0.234	0.332
	VTCG-NW[†]	0.516	0.501	0.262	0.317	0.325	0.270	0.365
	VTCG-BO[†]	0.322	0.464	0.324	0.166	0.373	0.120	0.295
Prompt models	BERT-base-PV-P1	0.371	0.587	**0.468**	**0.488**	0.342	0.258	0.419
	BERT-base-PFF-P1	0.369	0.573	0.429	0.478	0.326	0.286	0.410
	RoBERTa-base-PV-P1	**0.599**	**0.710**	0.455	0.457	0.417	**0.375**	**0.502**
	RoBERTa-base-PFF-P1	0.537	0.654	0.426	0.474	**0.423**	0.303	0.470

Table 3. Evaluation on Sem2016T6 zero-shot setting with BERT-large and larger.

Method	Per-target results (F_{1mfa})						Average	Average
	DT	HC	FM	LA	A	CC	(F_{1mfa})	(F_{1m})
BERT-large	**0.422**	0.514	0.404	0.416	**0.382**	**0.399**	**0.423**	**0.463**
RoBERTa-large	0.233	**0.573**	**0.482**	**0.431**	0.252	0.112	0.347	0.404
OpenStance [33]	-	-	-	-	-	-	-	0.637
ChatGPT [34]	-	0.780	0.690	0.593	-	-	-	-
gpt-3.5-turbo-P3	0.684	0.821	0.715	0.547	0.126	0.732	0.588	0.589
gpt-3.5-turbo-P5	0.661	0.821	0.724	0.692	0.539	**0.707**	0.697	**0.670**
FLAN-UL2-P3	**0.700**	**0.824**	**0.729**	0.682	0.687	0.543	0.694	**0.670**
FLAN-UL2-P5	0.630	0.748	0.706	**0.742**	**0.763·**	0.692	**0.713**	0.634
RoBERTa-large-PV-P1	0.631	**0.788**	**0.679**	0.612	0.622	0.268	0.600	0.624
RoBERTa-large-PFF-P1	**0.641**	0.777	0.655	**0.623**	**0.716**	**0.365**	**0.634**	**0.653**
RoBERTa-large-PV-P2	0.624	0.779	0.646	0.612	0.687	0.196	0.591	0.622
RoBERTa-large-PFF-P2	0.634	0.761	0.656	0.598	0.606	0.308	0.594	0.629

5.2 Reproduced Results

Our reproduced results are generally lower for Sem2016T6 for all methods, with only two of the approaches better than random guessing. From the validation results we see that models actually learn stance detection, ($F_{1fa-val}$ was always higher than 0.65), but fail to generalise for a zero-shot target. On VAST (Table 4) our reproduced results are mostly lower than reported by the authors. Methods that use unprocessed texts tend to score higher, which is consistent with our analysis of the possible VAST configurations. Methods based on Wikipedia def-

Table 4. Smaller models on VAST: *UP* – unprocessed texts, *UT* – unprocessed stance targets, *DA* – without ambiguous samples, *DA* – no *list* samples, see Sect. 4.2. * "?" no information or lacking source code. † only if explicitly stated in publication. ‡ corrected Wikipedia target definitions (Sect. 4.3), § [2].

	Method	Dataset variant				Result	ZS Result
		UP*	UT†	DA†	DL†	(F_{1m})	(F_{1mZS})
Base (ours)	BERT-base	-	-	-	-	0.695	0.694
	BERT-base	-	-	+	+	0.716	0.717
	BERT-base	+	-	-	-	0.716	0.719
	BERT-base	+	+	-	-	0.707	0.707
	BERT-base	+	-	+	-	0.717	0.720
	BERT-base	+	+	+	-	0.719	0.722
	BERT-base	+	-	+	+	0.733	0.736
	BERT-base	+	+	+	+	0.735	0.739
	RoBERTa-base	+	+	+	+	**0.757**	**0.768**
Reported results	TGA-NET§	-	-	-	-	0.665	0.666
	CKE-Net [22]	?	?	-	-	0.701	0.702
	JointCL [20]	-	-	-	-	-	0.723
	TarTK [36]	?	?	-	-	-	0.736
	WS-BERT [13]	-	-	-	-	0.745	0.753
	DTCL [21]	?	?	-	-	0.712	0.708
	BS-RGCN [26]	+	-	-	-	0.713	0.726
	PT-HCL [19]	?	?	-	-	-	0.716
	FECL [35]	?	?	-	-	-	0.725
	MPCL [14]	?	?	-	-	-	0.724
	POLITICS [24]	?	?	-	-	0.763	-
	VTCG [32]	+	-	-	-	**0.773**	**0.764**
Reproduced results	JointCL	-	-	-	-	0.701	0.700
	VTCG‡	+	-	-	-	0.730	0.739
	VTCG-NW	+	-	-	-	**0.731**	**0.747**
	VTCG-BO	+	-	-	-	0.723	0.742
	WS-BERT‡	-	-	-	-	0.677	0.685
	BS-RGCN	+	-	-	-	0.694	0.716
Prompt models	BERT-base-PV-P1	+	+	+	+	0.730	0.732
	BERT-base-PFF-P1	+	+	+	+	0.720	0.728
	RoBERTa-base-PV-P1	+	+	+	+	**0.764**	**0.776**
	RoBERTa-base-PFF-P1	+	+	+	+	0.762	0.770

initions fail to achieve results better than baselines when corrected definitions are provided. Just one tested method, VTCG, but without target definitions (VTCG-NW) achieved better results than the language model itself (BART-base, VTCG-BO).

5.3   Prompt Models Results

Prompting RoBERTa models shows significant improvement in comparison to fine-tuning on Sem2016T6. Our prompting RoBERTa-base achieved the highest average score for all targets. RoBERTa-large variants (Table 3) achieved markedly better results than the BERT and RoBERTa baselines. Our approach of prompting with a feed-forward classification head shows the best average

Table 5. Results for VAST with models larger or equal to BERT-large.

Method	Result (F_{1m})	ZS result (F_{1mZS})
BERT-large	0.750	0.759
RoBERTa-large	**0.811**	**0.833**
RoBERTa-large [33]	**0.780**	–
TTS [18]	–	**0.801**
gpt-3.5-turbo-P4	0.643	–
gpt-3.5-turbo-P5	**0.772**	–
FLAN-UL2-P4	0.652	–
FLAN-UL2-P5	0.707	–
RoBERTa-large-PV-P1	0.813	0.827
RoBERTa-large-PFF-P1	0.812	**0.832**
RoBERTa-large-PV-P2	**0.815**	0.827
RoBERTa-large-PFF-P2	0.811	0.830

Table 6. Results for VAST for *true neutral* samples with RoBERTa-base-PV-P1 trained with *synthetic neutrals* included or excluded.

	Precision	Recall	F_1
full training set	0.050	0.222	0.081
synthetic neutrals excluded	0.400	0.267	**0.320**

results between all tested prompting variants and the best results for most targets. We observe only a small but statistically insignificant ($p > 0.05$) improvement with prompt models and VAST with *base* models, and practically no difference with *large* ones. There is a visible difference in favor of an approach with verbalizers with *base* models, but feed-forward classification gave better with *large* models.

We show in Table 6 that the model learned from the full VAST train part score much lower on *true neutral* samples compared to a model trained on a set with *synthetic neutrals* excluded, despite that there are only 114 training samples left and the training set is highly unbalanced. This points to a limitation of training sets with lacking *true neutrals*.

5.4   Large Language Models

There is significant variance in Sem2016T6 per target results of gpt-3.5-turbo (Table 3). Results for P3 are very low for *atheism* (A) and *legalization of abortion* (LA), but substantially rise with changing the prompt. We think that this may be due to the specific tunning of the model to not produce harmful content, which may interfere with the classification of controversial topics, but its influence depends on the prompt used. We observe a lower variation of results between targets and prompts used for FLAN-UL2. For VAST (Table 5), there is a visible advantage of prompt P5 for both gpt-3.5-turbo and FLAN-UL2. It leads to the conclusion that prompts based on a definition from an annotation task (P3, P4) may not be the best candidates for stance detection with LLMs. Comparing to our prompting encoders LLMs have a slight advantage on Sem2016T6, but their results for VAST are lower. It should be noted that comparing both approaches is problematic since LLMs have no opportunity to learn annotators bias for a given dataset, which could be significant for a relatively subjective task such as stance detection.

5.5   Comparison with the State of the Art

Our results shed new light on the current SOTA for VAST. As seen in Table 4 results for BERT-base with a subset of the training dataset are higher than other BERT-base methods. Also, our results with RoBERTa-large are higher than those of any other model with >140M parameters, including the current SOTA [18].

Regarding Sem2016T6 in zero-shot configuration (2), we show that prompting RoBERTa-base comes close to the highest average result in [14]. One target, (*climate change is a real concern*), seems to be problematic, but it is less compatible with the used prompts due to its character of a claim. Among >140M models, we could only compare to [33] (the current SOTA) and our poposed prompting approach with RoBERTa-large (RoBERTa-PFF-P1) scored higher (Table 3).

6   Conclusions

We touched on many aspects in target-phrase zero shot stance detection, focusing on the two most relevant datasets.

In literature, transformer-based models are the most effective ones due to their ability to jointly encode both target and text. In addition to BERT-base, the most popular one, other approaches like RoBERTa or BART with comparable sizes, were shown to be equally or more effective. BERT should perform better, due to the next sentence prediction task [8], but instead RoBERTa, trained on single sentences, achieves generally better results, especially when applied in a prompt-based approach.

There were many attempts to improve performance of transformers-only solutions, but from both literature and especially our careful reproductions we see that the improvements are small or marginal. Integrating additional knowledge into a transformer model naturally facilitates zero-shot stance detection, because good knowledge about the stance target is often needed. Current knowledge-enhanced methods do not present significant improvements, but this may signal that more work is still needed. We showed that simply introducing Wikipedia target definitions into the model's input actually worsens results for both datasets. We think that this is more easily explained for the Sem2016T6 dataset, when learning to utilize additional knowledge from just five stance target definitions may be too much to expect. On the other hand, this is not the case for VAST. We hypothesize that simply using just target definitions and not knowledge about classified text may not be enough, and often the text may be crucial to disambiguating the target (limiting misleading definitions). Still, utilizing additional knowledge requires understanding it and linking it to the sample's text, which makes it a complex language-understanding task, especially for considered models. Knowledge graph embedding utilizing whole text is used in BS-RGCN, but also introduces additional model parameters (GNN) that were not pretrained on a broad corpus. We think that this may be the main reason why it fails to generalise for Sem2016T6 considering that the training set focuses on just five themes. Including knowledge embedding by concatenation with transformer output is a limiting factor in aggregating it with pertaining knowledge.

Our analysis highlighted several aspects of VAST that are important for proper interpretation of the previous works, and that must be carefully considered in all future works to make the results reliable and comparable. Besides that, we also showed that discarding certain samples from the training set, up to half of it, leads to better results! This proofs the importance of training data quality. It let us achieve SOTA results using just RoBERTa-large, but also the best results with BERT-base.

We proposed a modification of an approach based on prompting a transformer encoder model. As a result designing a prompt is simpler (no need for verbalizer tokens), it works especially well with *large* variants of models and achieves the best average result for Sem2016T6 zero-shot (excluding LLMs). We showed that the gap between the results of transformers with prompts in comparison to standard fine-tuning is visible for Sem2016T6. However, we did not notice statistically significant improvements on VAST, that may signal that prompting is especially effective in the case of fewer stance targets. Anyway, transformers with prompts may be a new strong baseline. They improve results by better utilising MLM pretraining task.

LLMs appeared to be very effective in zero-shot stance detection, especially in Sem2016T6 zero-shot with small number of targets. However, our prompting approach presented only slightly worse results, but utilising an order of magnitude fewer parameters. Both tested LLMs could not beat RoBERTa-large in our experiments on VAST. However, we spent only limited time on tuning the prompts, while correct prompts may be crucial for a high performance. It is also interesting that prompts based on annotation tasks were not as good candidates in comparison to slightly more complex ones.

In all datasets, we observe shortage of ambiguous/neutral samples related to targets (*true neutrals*). During tests on VAST a model trained on *synthetic neutrals* practically does not recognise *true neutral* samples, that is a major problem. Considering VAST, higher results are usually obtained for the zero-shot setting. This may suggest that a more challenging dataset with zero-shot targets more distinct from the training ones is needed. A large set of diverse stance targets with high-quality annotation is crucial for the further development.

We experimented with performance of stance detection models for different types of *neutral* samples, but due to the small number of *true neutral* samples in VAST, our results are only estimation and starting point for future experiments.

Acknowledgements. The work was financed as part of the investment:

"CLARIN ERIC - European Research Infrastructure Consortium: Common Language Resources and Technology Infrastructure (period: 2024–2026) funded by the Polish Ministry of Science (Programme: "Support for the participation of Polish scientific teams in international research infrastructure projects"), agreement number 2024/WK/01."

References

1. Addawood, A., Schneider, J., Bashir, M.: Stance classification of twitter debates. In: Processing of the 8th International Conference on Social Media. ACM Press, pp .1–10 (2017)
2. Allaway, E., McKeown, K.: Zero-Shot Stance Detection: a dataset and model using generalized topic representations. In: Proc. of the 2020 EMNLP, pp. 8913–8931. ACL, Online (Nov 2020)
3. Allaway, E., McKeown, K.: Zero-shot stance detection: paradigms and challenges. Front. Artif. Intell. **5**, 1070429 (2023). https://doi.org/10.3389/frai.2022.1070429
4. Allaway, E., Srikanth, M., McKeown, K.: Adversarial learning for zero-shot stance detection on social media. In: Proc. of the 2021 NAACL: Human Language Technologies, pp. 4756–4767. ACL, Online (Jun 2021)
5. Barbieri, F., Camacho-Collados, J., Espinosa Anke, L., Neves, L.: TweetEval: Unified benchmark and comparative evaluation for tweet classification. In: Findings of the ACL: EMNLP 2020, pp. 1644–1650. ACL, Online (Nov 2020)
6. Brown, T., et al.: Language models are few-shot learners. In: Larochelle, H., Ranzato, M., Hadsell, R., Balcan, M., Lin, H. (eds.) Advances in Neural Information Processing Systems. vol. 33, pp. 1877–1901. Curran Associates, Inc. (2020)
7. Derczynski, L., Bontcheva, K., Liakata, M., Procter, R., Wong Sak Hoi, G., Zubiaga, A.: SemEval-2017 task 8: RumourEval: Determining rumour veracity and support for rumours. In: Proc. of the 11th (SemEval-2017), pp. 69–76. ACL, Vancouver, Canada (Aug 2017)
8. Devlin, J., Chang, M.W., Lee, K., Toutanova, K.: BERT: Pre-training of deep bidirectional transformers for language understanding. In: Proc. of the 2019 NAACL: Human Language Technologies, pp. 4171–4186. ACL, Minneapolis, Minnesota (Jun 2019)
9. Fan, L., White, M., Sharma, E., Su, R., Choubey, P.K., Huang, R., Wang, L.: In plain sight: Media bias through the lens of factual reporting. In: Proc. of the 2019 EMNLP-IJCNLP. pp. 6343–6349. ACL, Hong Kong, China (Nov 2019)
10. Gao, T., Fisch, A., Chen, D.: Making pre-trained language models better few-shot learners. In: Proceeding of the 59th ACL and the 11th IJCNLP, pp. 3816–3830. ACL, Online (Aug (2021)
11. Glandt, K., Khanal, S., Li, Y., Caragea, D., Caragea, C.: Stance detection in COVID-19 tweets. In: Proceeding of the 59th ACL and the 11th Inter. Joint Conference on Natural Language Processing, pp. 1596–1611. ACL, Online (Aug 2021)
12. Hardalov, M., Arora, A., Nakov, P., Augenstein, I.: Cross-domain label-adaptive stance detection. In: Proceeding of the 2021 EMNLP, pp. 9011–9028. ACL, Online and Punta Cana, Dominican Republic (Nov 2021)
13. He, Z., Mokhberian, N., Lerman, K.: Infusing knowledge from Wikipedia to enhance stance detection. In: Proceeding of the 12th Workshop on Computational Approaches to Subjectivity, Sentiment and Social Media Analysis, pp. 71–77. ACL, Dublin, Ireland (May 2022)

14. Jiang, Y., Gao, J., Shen, H., Cheng, X.: Zero-shot stance detection via multi-perspective contrastive learning with unlabeled data. Inf. Process. Manage. **60**(4), 103361 (2023). https://doi.org/10.1016/j.ipm.2023.103361

15. Kocoń, J., Cichecki, I., Kaszyca, O., Kochanek, M., Szydło, D., Baran, J., et al.: ChatGPT: Jack of all trades, master of none. Inf. Fusion **99**, 101861 (Nov 2023). 10.1016/j.inffus.2023.101861

16. Lewis, M., et al.: BART: Denoising sequence-to-sequence pre-training for natural language generation, translation, and comprehension. In: Proceeding of the 58th ACL, pp. 7871–7880. ACL, Online (Jul 2020)

17. Li, Y., Sosea, T., Sawant, A., Nair, A.J., Inkpen, D., Caragea, C.: P-stance: A large dataset for stance detection in political domain. In: Findings of the ACL: ACL-IJCNLP 2021, pp. 2355–2365. ACL, Online (Aug 2021)

18. Li, Y., Zhao, C., Caragea, C.: Tts: A target-based teacher-student framework for zero-shot stance detection. In: Proceeding of the ACM Web Conference. 2023, pp . 1500-1509. WWW '23, ACM, New York, NY, USA (2023)

19. Liang, B., Chen, Z., Gui, L., He, Y., Yang, M., Xu, R.: Zero-shot stance detection via contrastive learning. In: Proceeding of the ACM Web Conference. 2022, pp. 2738-2747. WWW '22, ACM, New York, NY, USA (2022)

20. Liang, B., et al.: JointCL: A joint contrastive learning framework for zero-shot stance detection. In: Proceeding of the 60th ACL, pp. 81–91. ACL, Dublin, Ireland (May 2022)

21. Liu, R., Lin, Z., Fu, P., Liu, Y., Wang, W.: Connecting targets via latent topics and contrastive learning: A unified framework for robust zero-shot and few-shot stance detection. In: ICASSP 2022 - 2022 IEEE International Conference on Acoustics, Speech and Signal Processing (ICASSP), pp. 7812–7816 (2022)

22. Liu, R., Lin, Z., Tan, Y., Wang, W.: Enhancing zero-shot and few-shot stance detection with commonsense knowledge graph. In: Findings of the ACL: ACL-IJCNLP 2021, pp. 3152–3157. ACL, Online (Aug 2021)

23. Liu, Y., et al.: Roberta: a robustly optimized BERT pretraining approach. CoRR **abs/1907.11692** (2019)

24. Liu, Y., Zhang, X.F., Wegsman, D., Beauchamp, N., Wang, L.: POLITICS: Pre-training with same-story article comparison for ideology prediction and stance detection. In: Findings of the ACL: NAACL 2022, pp. 1354–1374. ACL, Seattle, United States (Jul 2022)

25. Longpre, S., et al.: The flan collection: designing data and methods for effective instruction tuning, pp. 22631– 22648 (2023)

26. Luo, Y., Liu, Z., Shi, Y., Li, S.Z., Zhang, Y.: Exploiting sentiment and common sense for zero-shot stance detection. In: Proceeding of the 29th COLING, pp. 7112–7123. International Committee on Computational Linguistics, Gyeongju, Republic of Korea (Oct 2022)

27. Mohammad, S., Kiritchenko, S., Sobhani, P., Zhu, X., Cherry, C.: SemEval-2016 task 6: Detecting stance in tweets. In: Proceeding of the 10th (SemEval-2016), pp. 31–41. ACL, San Diego, California (Jun 2016)

28. Schick, T., Schütze, H.: Exploiting cloze-questions for few-shot text classification and natural language inference. In: Proceeding of the 16th Conference of EACL, pp. 255–269. ACL, Online (Apr 2021)

29. Schick, T., Schütze, H.: It's not just size that matters: Small language models are also few-shot learners. In: Proceeding of the 2021 NAACL: Human Language Technologies, pp. 2339–2352. ACL, Online (Jun 2021)

30. Sobhani, P., Inkpen, D., Zhu, X.: A dataset for multi-target stance detection. In: Proceeding of the 15th EACL, pp. 551–557. ACL, Valencia, Spain (Apr 2017)

31. Tay, Y., et al.: Unifying language learning paradigms. arXiv preprint arXiv:2205.05131 (2023)

32. Wen, H., Hauptmann, A.: Zero-shot and few-shot stance detection on varied topics via conditional generation. In: Proceeding of the 61st ACL, pp. 1491–1499. ACL, Toronto, Canada (Jul 2023)

33. Xu, H., Vucetic, S., Yin, W.: OpenStance: Real-world zero-shot stance detection. In: Proceeding of the 26th CoNLL, pp. 314–324. ACL, Abu Dhabi (Dec 2022)

34. Zhang, B., Ding, D., Jing, L.: How would stance detection techniques evolve after the launch of chatgpt? arXiv preprint arXiv:2212.14548 (2023)

35. Zhao, X., Zou, J., Zhang, Z., Xie, F., Zhou, B., Tian, L.:Feature enhanced zero-shot stance detection via contrastive learning, pp. 900–908 (2023)

36. Zhu, Q., Liang, B., Sun, J., Du, J., Zhou, L., Xu, R.: Enhancing zero-shot stance detection via targeted background knowledge. In: Proceeding of the 45th International ACM SIGIR Conference on Research and Development in IR, pp . 2070-2075. SIGIR '22, ACM, New York, NY, USA (2022)

Architectural Modifications to Enhance Steganalysis with Convolutional Neural Networks

Remigiusz Martyniak[ID] and Bartosz Czaplewski[✉][ID]

Faculty of Electronics, Telecommunications and Informatics, Gdańsk University of Technology, 11/12 Gabriela Narutowicza Street, 80-233 Gdańsk, Poland
bartosz.czaplewski@pg.edu.pl

Abstract. This paper investigates the impact of various modifications introduced to current state-of-the-art Convolutional Neural Network (CNN) architectures specifically designed for the steganalysis of digital images. Usage of deep learning methods has consistently demonstrated improved results in this field over the past few years, primarily due to the development of newer architectures with higher classification accuracy compared to their predecessors. Despite the advances made, further improvements are desired to achieve even better performance in this field. The conducted experiments provide insights into how each modification affects the classification accuracy of the architectures, which is a measure of their ability to distinguish between stego and cover images. Based on the obtained results, potential enhancements are identified that future CNN designs could adopt to achieve higher accuracy while minimizing their complexity compared to current architectures. The impact of modifications on each model's performance has been found to vary depending on the tested architecture and the steganography embedding method used.

Keywords: Convolutional neural network · Deep learning · Steganalysis · Steganography

1 Introduction

Every day, billions of images and graphics are transmitted over the Internet. However, this medium also provides the opportunity to conceal information within them in a way that is imperceptible to the human eye. This form of communication, where digital images act as carriers for hidden data, requiring specialized analysis for extraction, is considered a secure means of conveying information. In such scenarios, only the intended recipient, equipped with the necessary software to reveal the concealed information within the image, can decipher the true message being conveyed.

The field of techniques of concealing information is known as steganography. Steganography methods are used to conceal covert communication by embedding information within another medium known as the cover object, e.g. digital image.

L. Franco et al. (Eds.): ICCS 2024, LNCS 14833, pp. 50–67, 2024.
https://doi.org/10.1007/978-3-031-63751-3_4

Conversely, the development of techniques designed to detect hidden information falls under the field of steganalysis. It plays a crucial part in the information systems security field and is dedicated to detecting the presence of steganographic techniques. It is important to emphasize that steganalysis primarily focuses on confirming the presence of a hidden message within the cover object, rather than extracting it.

In research related to steganography and steganalysis, two embedding methods are often used: Wavelet Obtained Weights (WOW) [1] and Spatial Universal Wavelet Relative Distortion (S-UNIWARD) [2]. Both algorithms are content-adaptive, i.e. information is embedded primarily within the regions of an image where the complexity in terms of patterns and pixel value diversity is the highest. This approach makes the detection of hidden communication by steganalysis methods more difficult.

For many years, steganalysis has relied on algorithms grounded in advanced statistical models, referred to as SRM (Statistical Rich Models) [3]. Nevertheless, in recent years, the concept of using machine learning for this purpose has gained prominence. The primary goal of these algorithms is binary classification, determining whether a given digital image is containing a hidden information or not.

Currently, the most promising results in digital image steganalysis are achieved through the use of Deep Learning methods [4] employing Convolutional Neural Networks (CNNs). As noted in [5], steganalysis has traditionally followed a two-stage paradigm, involving manual feature extraction in the initial stage, followed by classification using Ensemble Classifiers or Support Vector Machines in the subsequent stage. However, the emergence of Deep Learning techniques has revolutionized steganalysis by unifying and automating these two distinct stages, leading to advancements that surpass the Rich Models with Ensemble Classifiers [5].

Therefore, the initial objective of this study was to identify the best-performing CNN architectures specifically designed for steganalysis. The second objective was to conduct experiments on these architectures by creating variants of them with structural modifications. The rationale behind conducting such experiments was to determine how specific modifications would impact the model's performance. Thus, structural changes that could increase the classification accuracy of state-of-the-art architectures would provide a contribution to the field of steganalysis.

The goal of this article is to present the research findings concerning the application of machine learning for image steganalysis. The research aims to investigate the impact of modifying CNNs on steganalysis performance. To the best of the authors' knowledge, similar results have not been previously documented in the literature.

The contributions of this paper are as follows. The paper explores the effects of various architectural modifications to the state-of-the-art CNNs on model performance. 17 variants for the Yedroudj-Net and 20 variants for the GBRAS-Net architectures were proposed. The study evaluates how individual modifications could introduce enhancements to existing networks that either improve classification efficiency or reduce the number of trainable parameters. The obtained results allowed to determine specific directions for further research.

The related work section provides an overview of prior research. The methodology section presents the research approach, the dataset, types of experiments, as well as the experimental setup details. The conclusions are presented in the last section.

2 Related Work

Firstly, autoencoders were used for steganalysis [6]. The study demonstrated that initializing a CNN with filters from a pre-trained stack of convolutional autoencoders, combined with feature pooling layers, yielded promising results.

Subsequently, a CNN was introduced in [7], which was capable of autonomously learning feature representations via multiple convolutional layers. This model unified the feature extraction and classification steps within a single architecture.

Another application of CNNs in steganalysis was explored in [8]. In this scenario, the authors consistently used the same embedding key. Experimental results for a proposed CNN and a fully connected NN outperformed the RM+EC approach.

In [9] the authors employed CNNs as base learners for ensemble learning and tested various ensemble strategies. Their methodology included recovery of lost information caused by spatial subsampling in the pooling layers during feature vector formation.

JPEG steganalysis research was presented in [10]. The authors attempted to adapt CNNs to a comprehensive feature set through network pre-training. The primary challenge in surpassing rich-model-based frameworks were training convergence issues.

Another JPEG research was proposed in [11]. The authors incorporated JPEG awareness into CNNs. They introduced the catalyst kernel allowing the network to learn kernels more relevant for detecting stego signals introduced by JPEG steganography.

A method against J-UNIWARD method was presented in [12]. The author confirmed the role of both the pooling method and the depth of CNNs. It was demonstrated that a 20-layer CNN generally outperforms the most advanced feature model-based methods.

The Xu-Net against S-UNIWARD and HILL methods was presented in [13]. The results show the usefulness of using absolute values in early feature maps and the negative impact of larger filter sizes in convolutional layers on network efficiency.

Next, the SR-Net [14] utilized four original types of layers. In contrast to other nets, SR-Net doesn't initially employ a high-pass preprocessing filtering layer. However, the depth of this network is very high, leading to high model complexity.

Another significant architecture is Yedroudj-Net [15]. Yedroudj-Net employs the Trunc activation function [16], which limits data values to a chosen range. This eliminates the occurrence of large values that would be processed by subsequent layers.

The repeated use of the stego-key was explored in [17]. The study led to the CNN with the state-of-the-art efficiency while having 20 times fewer learnable parameters, which means easier convergence and reduced memory and computing power demands.

The GBRAS-Net [18] incorporated the depthwise layer and the separable layer. The goal is to achieve an effect similar to a regular convolutional layer while reducing the number of parameters. This architecture does not utilize fully connected layers.

The work presented in this paper is based on two different deep neural networks: Yedroudj-Net [15] and GBRAS-Net [18]. These models were selected due to their high performance in terms of classification accuracy. Both models employ preprocessing that includes 30 predefined high-pass filters. The weights of these filters are not learned during the training process. The output signal from the preprocessing convolutional layer is propagated through the deep neural network, where further processing occurs.

3   Methodology

The steganalysis begins with cover and stego images serving as the input (Fig. 1). These images undergo processing by a modified CNN. The model's output is a binary result: '0' for a cover image prediction or '1' for a stego image prediction. Based on the obtained model's output and true labels assigned to the input images, the overall classification accuracy of the architecture variant can be calculated.

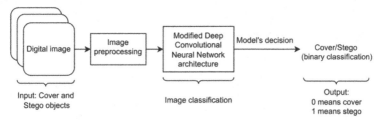

Fig. 1. Steganalysis workflow using a CNN.

3.1   Dataset

For this study, the BOSSBase 1.01 dataset [19] was used for training, validation, and testing. The rationale behind choosing BOSSBase stems from the fact that it is the most commonly used dataset in the steganalysis field. This facilitates the comparison of results with those described in other studies. The mentioned dataset consists of 10,000 grayscale images in PGM format. Following the approach presented in the aforementioned papers, each image was resized from 512×512 px to 256×256 px. It significantly reduces the computational resources required for image processing, thereby reducing the training time for each variant.

The dataset was divided into three subsets: 8,000 images for the training set, 1,000 images for the validation set, and the remaining 1,000 images for the testing set. This 8:1:1 split ratio ensures that the neural network receives ample training examples while maintaining a sufficient number of images for accurate testing. This particular split also proved to yield the best results in the test scenarios described in [20].

3.2   Steganographic Techniques

For the embedding algorithms, WOW and S-UNIWARD with embedding capacity set to 0.4 bpp were used. These algorithms are widely employed in steganalysis research work, allowing for meaningful comparison of obtained results with other studies.

All of the original 10,000 images (cover objects) from BOSSBase were embedded with a message using a different stego-key for each image. This process resulted in a dataset comprising a total of 20,000 images: 10,000 cover and 10,000 stego images. MATLAB implementations from [21] were used for the embedding process, with minor code modifications to load and embed the payload in large batches of images.

3.3 Architectures

The selection of models for the experiments was based on three primary criteria. Firstly, their selection was based on their classification accuracy stated in the papers. Secondly, it was based on the number of trainable parameters (architecture's complexity). Lastly, the chosen models were required to exhibit distinct architectural approaches. For these reasons, Yedroudj-Net and GBRAS-Net were selected. They exhibited high classification accuracy, relatively low complexity, and notable structural differences. A detailed comparison of both CNNs, including their architecture and training hyperparameters, is provided in Table 1.

Table 1. Comparison of Yedroudj-Net and GBRAS-Net.

Information	Yedroudj-Net	GBRAS-Net
No. of trainable parameters	445 k	166 k
Preprocessing stage	Convolutional layer with 30 5 × 5 SRM filters	Convolutional layer with 30 5 × 5 SRM filters
No. of depthwise-separable layers	0	4
No. of convolutional layers	5	8
No. of parameters added by depthwise-sep. and conv. layers	298 k	166 k
Filter (kernel) sizes	5 × 5 and 3 × 3	3 × 3 and 1 × 1
No. of filters in conv. layers	30, 30, 32, 64, 128	30, 30, 60, 60, 60, 60, 30, 2
Activation after conv. layers	Trunc, ReLU	ELU
Pooling type after conv. layers	Average Pooling with 5 × 5 kernel and (2,2) stride	Average Pooling with 2 × 2 kernel and (2,2) stride
Weight initialization method	Glorot Uniform	Glorot Uniform
No. of dense layers	3	0
No. of neurons in subsequent dense layers (input-output)	128–256, 256–1024, 1024–2	–
No. of trainable parameters in dense layers	263 k	0
Activation after dense layers	ReLU	–
Optimizer	SGD	Adam
Learning rate	0.01	0.001
Weight decay	0.0001	0
Momentum	0.95	0.2

(continued)

Table 1. (*continued*)

Information	Yedroudj-Net	GBRAS-Net
Pooling at the output	Global Average Pooling	Global Average Pooling
Activation function at the output	Softmax	Softmax

3.4 Types of Experiments

The changes introduced to the architectures can be categorized into three main types:

− Addition and removal of dense or convolutional layers. These alterations of this kind typically have the most significant impact on both the model's accuracy and complexity.
− Modifications in the structure of individual layers within the original architecture. This category primarily involves changes such as altering filter sizes in successive convolutional layers.
− Simple-to-implement changes that do not increase the model's complexity in terms of the number of trainable parameters but do affect the training process and may, for example, prolong it. Examples include using a different weight optimization algorithm, weight initialization method, and activation functions after each convolutional layer.

A detailed description of the variants is provided in the Results Section.

3.5 Metrics

For each experiment, the training and validation accuracy values for every epoch were calculated. It allowed to closely monitor the model's learning process. Moreover, the final classification accuracy for the testing dataset was calculated, which is the main performance evaluation criterion in this paper. It provides valuable information about the overall model's performance and its ability to generalize to unseen data.

Classification accuracy alone does not provide a comprehensive assessment of the model's performance. This is a very general measure that does not provide a complete picture of the behavior of the analyzed model. Therefore, two additional metrics were utilized: Receiver Operating Characteristic (ROC) curve and the confusion matrix.

ROC curve provides a comprehensive visualization of a model's True Positive Rate (TPR) against its False Positive Rate (FPR) across all classification thresholds. In steganalysis, this curve offers insights into the trade-off between correctly identifying stego images and incorrectly classifying cover images, which helps to visually compare the performance of different models. This is the best method for comparing models. One can compare TPRs at a given FPR or compare Area Under Curve (AUC) to indicate a better detector.

Confusion matrix complements ROC curve by providing a detailed breakdown of a model's predictions, including true positives, true negatives, false positives, and false negatives. This breakdown helps to identify specific areas for improvement in a model's performance. Moreover, all other measures, such as precision, recall, F1 score, or specificity, can be derived from the confusion matrix.

The ROC curve, confusion matrix, and overall classification accuracy results are the most important evaluation metrics to determine the performance of a deep neural network in binary classification problems like steganalysis.

3.6 Experimental Setup

The experiments were performed locally using NVIDIA GeForce RTX 2080 Ti GPU with the support of CUDNN v11.6. To implement the variants of Yedroudj-Net and GBRAS-Net architectures, Python v3.9.16 programming language, PyTorch v1.13.1+cu116, and PyTorch Lightning v2.0.0 libraries were used. The results of the experiments were recorded using Neptune.ai API and stored on its cloud platform.

4 Results

4.1 Modifications for Yedroudj-Net

The Yedroudj-Net architecture is more complex than GBRAS-Net, i.e. 445 thousand trainable parameters, which is 279 thousand more. This difference primarily stems from the inclusion of three fully connected layers in the classification stage. Therefore, the initial experiments focused on simplifying the original architecture.

Simplification Strategies. In the first two Yedroudj-Net variants, simplification strategies by removal of dense layers were explored:

– Variant 1: Removal of the last dense layer, resulted in a minor decrease in accuracy for WOW, but surprisingly led to an increase of 0.55 pp for S-UNIWARD. Notably, by removing a dense layer with 1024 neurons, a significant reduction of 264 thousand trainable parameters was achieved.
– Variant 2: Removal of the last two dense layers caused a decrease in accuracy by 1.45 pp for WOW and 1.6 pp for S-UNIWARD.

The obtained results raise a question regarding the necessity of utilizing as many as three dense layers from a complexity-to-performance trade-off point of view. A similar simplification approach was adopted for the convolutional layers:

– Variant 3: Removal of the last convolutional layer, which contains 64 filters at the input and 128 at the output, resulted in an increase in classification accuracy by 0.7 pp for WOW and a decrease of 0.15 pp for S-UNIWARD.
– Variant 4: Removal of the last two convolutional layers led to a loss of more than 4 pp in accuracy for both embedding algorithms, making this type of change not justified from a performance perspective.

Expanding the Architecture. In this set of experiments, the objective was to extend the original architecture by incorporating additional and dense layers. Note that after each added convolutional layer, batch normalization and ReLU function were applied.

Initially, one dense layer with 512 neurons was added (variant no. 5). This modification resulted in a minor decrease in accuracy while increasing the number of parameters. It was decided not to further experiment with the addition of more dense layers, as there was no indication that it would enhance the model's performance.

Similarly, one and two convolutional layers were integrated into the Yedroudj-Net architecture after the preprocessing layer:

− Variant 6: Added one convolutional layer with 30 filters, 1 × 1 kernel size, a stride of 1, and no padding, ensuring that the output feature map has the same dimensions as the input feature map. Classification accuracy increased for both embedding algorithms: by 1.75 pp for WOW and 1.35 pp for S-UNIWARD.

− Variant 7: Added two convolutional layers with 30 filters, but this time employing a 3 × 3 kernel size, with a stride and padding of 1. What is more, it was decided to maintain uniform filter sizes across all layers, thereby changing kernel sizes of the original first and second Yedrouj-Net layers from 5 × 5 to 3 × 3. This adjustment also helped to mitigate the increase in the number of trainable parameters caused by the addition of two convolutional layers. Accuracy increased by 4.5 pp for WOW, and 6.8 pp for S-UNIWARD. The training/validation accuracy plot for this variant is in Fig. 2, and the ROC curve is shown in Fig. 3. According to the results provided in Table 2, the model exhibited a high degree of accuracy in classifying both cover and stego images.

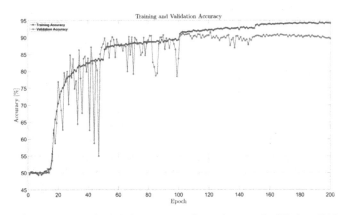

Fig. 2. Training and validation accuracy for variant no. 7 of Yedroudj-Net.

Convolutional Layer Parameter Tuning. The following set of experiments involved altering key parameters within the convolutional layers, such as kernel sizes and average pooling operation. The results of these experiments are as follows:

− Variant 8: Changing filter sizes from 3 × 3 to 5 × 5 in the last three convolutional layers resulted in a decrease in accuracy for both steganography algorithms. Furthermore, it significantly increased the model's complexity as the last three layers contain a higher number of filters compared to the initial layers.

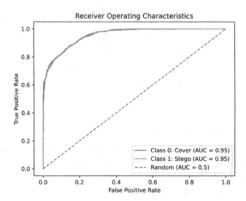

Fig. 3. ROC curve for variant no. 7 of Yedroudj-Net architecture.

Table 2. Normalized confusion matrix for variant no. 7 of Yedroudj-Net.

	Predicted Cover	Predicted Stego
Actual Cover	0.88 (True Negative)	0.12 (False Positive)
Actual Stego	0.15 (False Negative)	0.85 (True Positive)

– Variant 9: Altering filter sizes from 5×5 to 3×3 in the first two convolutional layers, resulting in a kernel size of 3×3 for each layer, leading to an increase in accuracy by 0.55 pp for WOW and a decrease by 0.55 pp for S-UNIWARD.
– Variant 10: Changing filter sizes from 5×5 to 3×3 in the first three convolutional layers and from 3×3 to 5×5 in the last two convolutional layers, led to an increase in classification accuracy by 0.55 pp for WOW and by 0.85 pp for S-UNIWARD. However, this change also significantly increased the number trainable parameters, because the last two convolutional layers of Yedroudj-Net feature a greater number of filters (64 and 128 at the output), compared to these at the beginning of the architecture.
– Variant 11: Altering filter sizes from 5×5 to 1×1 in the first convolutional layer resulted in increased accuracy for both WOW and S-UNIWARD by 0.85 and 1.4 pp, respectively. It also reduced the number of parameters by 36 thousand.

Based on the results obtained from this series of experiments, the most significant conclusion is that employing a smaller kernel size in the initial convolutional layer or layers can lead to improved classification accuracy.

In this series of experiments related to convolutional layers, the impact of average pooling was investigated with the following variants:

– Variant 12: Removal of average pooling led to a 1.7 pp increase in classification accuracy for WOW, and a decrease by 1.85 for S-UNIWARD.
– Variant 13: Adjustment of the average pooling settings from (5,2) to (3,1), where the first number represents the kernel size, and the second number denotes the stride,

resulted in a 1.8 pp increase in classification accuracy for WOW, and a 1.65 pp decrease for S-UNIWARD.

Optimization and Activation Strategies. The final four experiments focused on modifications to the optimization algorithm, weight initialization method, and activation functions. Changing the optimizer from SGD to Adam (variant no. 14), as well as the weight initialization method from Glorot Uniform to Kaiming Normal (variant no. 15) resulted in decreased classification accuracy.

However, more intriguing results emerged when the ReLU activation functions were replaced with ELU activation (variant no. 16). The classification accuracy decreased by 3.8 pp for the WOW algorithm but increased by 5 pp for S-UNIWARD, highlighting the significant role of the activation function in the context of using CNNs for steganalysis, in which case the steganography embedding algorithm used for cover communication is most often unknown.

Contrarily, changing the Truncated Linear Unit (TLU) activation function, which is used twice in Yedroudj-Net, to ReLU (variant no. 17 in Table 3) resulted in an increase of 1.3 pp for WOW and a decrease of 1.1 pp for S-UNIWARD.

Table 3. Classification accuracy results for Yedroudj-Net architecture variants.

#	Description	Change in classification accuracy (in pp)		Change in the number of parameters (in thousands)
		WOW	S-UNIWARD	
1	Removed the last dense layer	−0.25	+0.55	−264
2	Removed the last two dense layers	−1.45	−1.6	−297
3	Removed the last two conv. layers	−4.2	−6.65	−117
4	Added one dense layer	−0.35	−0.25	+524
5	Removed the last conv. layer	+0.7	−0.15	−90
6	Added one conv. layer	+1.75	+ 1.35	+1
7	Added two conv. layers	+4.5	+ 6.8	+138
8	5 × 5 kernel size in the last 3 conv. layers	−0.5	−0.8	+180
9	3 × 3 kernel size in the first 2 conv. layers	+0.55	-0.55	−28

(continued)

Table 3. (*continued*)

#	Description	Change in classification accuracy (in pp)		Change in the number of parameters (in thousands)
		WOW	S-UNIWARD	
10	3 × 3 kernel size in the first 3 conv 5 × 5 kernel size in the last 2 conv. layers	+0.55	+0.85	+136
11	1 × 1 kernel size in the first layer	+0.85	+1.4	−36
12	No average pooling	+1.7	−1.85	0
13	(3,1) average pooling instead of (5,2)	+1.8	−1.65	0
14	Adam optimizer instead of SGD	−5	−1.45	0
15	Kaiming Normal initialization instead of Glorot Uniform	−0.2	−0.3	0
16	ELU activation instead of ReLU	−3.8	+5	0
17	ReLU activation instead of Trunc	+1.3	−1.1	0

In Table 3, the results of the conducted experiments are shown, comprising seventeen different variants of the Yedroudj-Net architecture. Changes in the classification accuracy compared to the original architecture are expressed in percentage points.

4.2 Modifications for GBRAS-Net

Similar to the modifications for Yedroudj-Net, four types of experiments were conducted to assess their impact on the GBRAS-Net architecture performance.

Simplification Strategies. Convolutional and depthwise layers were removed, as the original GBRAS-Net architecture does not incorporate dense layers.

− Variant 1: Removal of five convolutional layers (3rd–7th) led to a decrease in classification accuracy by 1.95 pp for WOW and 1 pp for S-UNIWARD. This change involved the removal of a substantial portion of convolutional layers, resulting in a reduction of trainable parameters by 121.3 thousand.
− Variant 2: Removal of the five aforementioned convolutional layers, along with the 1st depthwise layer. The test of this variant was carried out to assess the significance of depthwise layers. This modification resulted in a major decrease in accuracy, exceeding

5 pp for both embedding algorithms. This underscores the critical role of depthwise layers within the GBRAS-Net.

Additionally, three more variants (no. 3, 4, 5) were explored, involving the removal of three, four and seven convolutional layers, all of which yielded similar results.

Expanding the Architecture. In the subsequent series of experiments, the impact of the following additions was investigated:

− Variant 6: Two convolutional layers at the beginning of the network. These layers consisted of 30 filters, kernel size of 3 × 3, stride and padding set to 1.
− Variant 7: Two convolutional layers following the 4th depthwise layer. Given that in the original architecture, the convolutional layer before the 3rd and 4th depthwise layer has 60 filters, the added convolutional layers also had 60 filters. They featured a 3 × 3 kernel size, stride, and padding of 1.

The addition of convolutional layers in both cases resulted in a decrease in accuracy for WOW and S-UNIWARD algorithms. Batch normalization and the ReLU activation function was applied after each additional convolutional layer.

Furthermore, the impact of incorporating dense layers at the end of the network was examined. Initially, two layers were added (variant no. 8) with 512 and 2 neurons, respectively. In the subsequent variant (no. 9) three dense layers were employed with 512, 1024 and 2 neurons. In both variants, a minimal increase in accuracy for WOW was observed, but conversely, a decrease for S-UNIWARD can be noticed: by 3.05 pp for variant no. 17 and by 0.3 pp in the case of variant no. 18.

Convolutional Layer Parameter Tuning. The impact of kernel size was explored:

− Variant 10: Decrease in filter sizes across subsequent layers, from 5 × 5 to 3 × 3, was implemented. The kernel size of the first three convolutional layers was increased from 3 × 3 to 5 × 5, while the following convolutional layers maintained a 3 × 3 kernel size, including the very last layer, which originally features 30 input and 2 output neurons with 1 × 1 kernel size.
− Variant 11: Further reduction in filter size, from 7 × 7 to 5 × 5 and then to 3 × 3, was applied. The first three convolutional layers had 7 × 7 kernels, the next two: 5 × 5, and the subsequent two: 3 × 3.

Both variants 10 and 11 resulted in reduced performance for both algorithms.
The next three experiments involved increasing filter sizes of the last conv. layers:

− Variant 12: Increase from 1 × 1 to 3 × 3 in the last two convolutional layers increased accuracy by 0.15 pp for WOW and by 1.05 pp for S-UNIWARD.
− Variant 13: Changing the filter size of the 5th-8th convolutional layer from 3 × 3 to 5 × 5. This alteration led to an increase in accuracy by 0.25 pp for WOW and a decrease of 1.4 pp for S-UNIWARD.
− Variant 14: Changing the filter size of the 5th and 6th convolutional layers from 3 × 3 to 5 × 5 and of the 7th and 8th from 3 × 3 to 7 × 7. As a result, the accuracy increased

by 0.7 pp for WOW and decreased by 0.5 pp for S-UNIWARD. The training/validation accuracy plot for this variant is in Fig. 4, and the ROC curve is shown in Fig. 5. Based on the results presented in Table 4, it can be stated that this variant demonstrates slightly better performance in the correct classification of images that are steganographic.

Fig. 4. Training and validation accuracy for variant no. 14 of GBRAS-Net architecture.

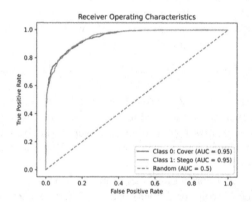

Fig. 5. ROC curve for variant no. 14 of GBRAS-Net architecture.

Table 4. Normalized confusion matrix for variant no. 14 of GBRAS-Net.

	Predicted Cover	Predicted Stego
Actual Cover	0.83 (True Negative)	0.17 (False Positive)
Actual Stego	0.11 (False Negative)	0.89 (True Positive)

The impact of altering the settings of the average pooling operation was explored:

− Variant 15: Removal of the average pooling operation, which was used three times in the architecture, resulted in a decrease by more than 2 pp for both WOW and S-UNIWARD. It is important to note that global average pooling, employed just before the softmax activation function, remained unchanged.
− Variant 16: Adjustment of the average pooling settings from (2,2) to (5,2) led to a decrease in accuracy by 1.6 pp for WOW and an increase by 0.1 pp for S-UNIWARD. Similar results were obtained in the subsequent variant (no. 17), where average pooling (3,1) was applied: a decrease of 1.55 for WOW and by 0.5 for S-UNIWARD.

Optimization and Activation Strategies. Switching from Adam to SGD optimizer (variant no. 18) resulted in a noticeable accuracy decrease for both algorithms.

Using the Kaiming Normal weight initialization method instead of Glorot Uniform (variant no. 19) resulted in an increase by 0.1 pp for WOW and decrease of 0.8 pp for S-UNIWARD.

Replacing ELU activation with ReLU (variant no. 20) resulted in improved accuracy by 0.5 pp for WOW, but also a 1.55 pp decrease for S-UNIWARD.

Table 5. Classification accuracy results for GBRAS-Net architecture variants.

#	Description	Change in classification accuracy (in pp)		Change in the number of parameters (in thousands)
		WOW	S-UNIWARD	
1	Removed five (3rd–7th) conv. layers	−1.95	−1	−121.3
2	Removed five (3rd–7th) conv. and the 1st depthwise layer	−5.35	−7.65	−121.1
3	Removed three (5th–7th) conv. layers	−0.7	−1.2	−97
4	Removed four (4th–7th) conv. layers	−0.65	+1.45	−113
5	Removed seven (2nd–8th) conv. layers	−2.6	−0.35	−129
6	Added two conv. layers	−0.7	−1.45	17
7	Added two conv. layers after the 4th depthwise layer	−0.35	−0.85	66
8	Added two dense layers	+0.25	v3.05	18
9	Added three dense layers	+0.3	−0.3	566

(*continued*)

Table 5. (*continued*)

#	Description	Change in classification accuracy (in pp)		Change in the number of parameters (in thousands)
		WOW	S-UNIWARD	
10	Decreasing filter size in subsequent layers: 5 × 5 → 3 × 3	−2.8	−4.8	81
11	Decreasing filter size in subsequent layers: 7 × 7 → 5 × 5 → 3 × 3	−2.65	−1.15	287
12	3x3 kernel size in the last two layers	+0.15	+1.05	15
13	Increasing filter size in subsequent layers: 3 × 3 → 5 × 5	+0.25	−1.4	216
14	Increasing filter size in subsequent layers: 3 × 3 → 5 × 5 → 7 × 7	+0.7	−0.5	346
15	No average pooling	−2.2	−2.75	0
16	(5,2) average pooling instead of (2,2)	−1.6	0.1	0
17	(3,1) average pooling instead of (2,2)	−1.55	−0.5	0
18	SGD optimizer instead of Adam	−0.7	−2.25	0
19	Kaiming Normal initialization instead of Glorot Unifom	+0.1	−0.8	0
20	ReLU activation instead of ELU	+0.5	−1.55	0

In Table 5, the results of the conducted twenty experiments for GBRAS-Net are presented. Changes in the classification accuracy compared to the original architecture are expressed in percentage points.

5 Conclusions

Firstly, it is evident that different modifications introduced to CNN architectures yield varying classification accuracies depending on the steganographic algorithms employed to embed the payload within images. Notable examples are as follows.

Yedroudj-Net: The replacement of the ReLU activation with ELU led to a significant decrease in accuracy (by 3.8 pp) for WOW and a substantial increase in classification accuracy for S-UNIWARD (by 5 pp). Notably, this was the only case in which the model exhibited better performance for the S-UNIWARD compared to WOW.

GBRAS-Net: Increasing filter sizes in subsequent layers resulted in improved classification accuracy for WOW but decreased it for S-UNIWARD, like in the case of changing the activation function from ReLU to ELU.

The impact of various architectural changes on the effectiveness of CNNs depends significantly on their fundamental structure and the embedding algorithms used. Consequently, identifying a universal modification that can consistently lead to improved classification performance for any given architecture and embedding algorithms is challenging. However, the following set of modifications could be advantageous.

For Yedroudj-Net: The addition of one or two extra convolutional layers after the preprocessing layer significantly improved the classification accuracy for both algorithms without a substantial increase in spatial complexity. This change can also be combined with adjustments to the filter sizes, employing smaller kernels in the initial convolutional layers and larger ones (5×5) in the final two layers. To reduce the spatial complexity of the model with these implemented changes, it is possible to remove the last dense layer to achieve a similar number of trainable parameters as in the original architecture with minimal impact on the neural network performance.

For GBRAS-Net: Increasing the kernel sizes in the last two convolutional layers from 1×1 to 3×3 resulted in an increase in classification accuracy for both algorithms, with only minimal additional spatial complexity. If there is a need to further reduce the complexity (to less than 166 thousand trainable parameters) with minimal impact on classification accuracy, up to five convolutional layers could be removed.

The change of activation function is a simple and cost-effective change that may significantly impact classification accuracy. It is worth noting that the popular ReLU activation function decreases the classification accuracy in the case of S-UNIWARD, but improves it in the case of WOW (variants 16 and 17 in Table 3, and variant 20 in Table 5), e.g. using ELU instead of ReLU decreases the accuracy by 3.8 pp for the WOW but increased by 5 pp for S-UNIWARD in Yedroudj-Net. In some cases, ReLU function is worse at dealing with the problems of vanishing gradient and dead neurons. Similarly, the addition of convolutional layers, especially at the beginning of the architecture with filters of small dimensions (e.g. 1×1 or 3×3), can have a minor impact on network complexity while substantially improving steganalysis results.

Further research could involve the creation of additional variants for the architectures and embedding algorithms studied in this work, as well as different steganographic techniques, such as J-UNIWARD. Future work could also involve experimentation with different variants of preprocessing methods, as the preprocessing stage is a crucial component of current state-of-the-art models. Image datasets other than BOSSBase are also worth investigating as a future work.

Acknowledgments. This study was funded by the Faculty of Electronics, Telecommunications, and Informatics of Gdańsk University of Technology, and by the research subsidy from the Polish Ministry of Science and Higher Education.

Disclosure of Interests. The authors have no competing interests to declare that are relevant to the content of this article.

References

1. Holub, V., Fridrich, J.: Designing steganographic distortion using directional filters. In: 2012 IEEE international workshop on information forensics and security (WIFS), pp. 234–239 (2012)
2. Holub, V., Fridrich, J., Denemark, T.: Universal distortion function for steganography in an arbitrary domain. EURASIP J. Inform. Secur. (2014)
3. Fridrich, J.: Steganography in digital media: principles, algorithms, and applications. Cambridge University Press, United Kingdom (2009)
4. Czaplewski, B.: Current trends in the field of steganalysis and guidelines for constructions of new steganalysis schemes. Telecommun. Rev. + Telecommun. News **10**(3), 1121–1125 (2017)
5. Reinel, T.-S., Raúl, R.-P., Gustavo, I.: Deep learning applied to steganalysis of digital images: a systematic review. IEEE Access **7**, 68970–68990 (2019)
6. Tan, S., Li, B.: Stacked convolutional auto-encoders for steganalysis of digital images. In: Proceedings of the signal and information processing association annual summit and conference, pp. 1–4. APSIPA, Siem Reap, Cambodia (2014)
7. Qian, Y., Dong, J., Wang, W., Tan, T.: Deep learning for steganalysis via convolutional neural networks. In: Proceeding of the media watermarking, security, and forensics 2015, MWSF'2015, Part of IS&T/SPIE Annual Symposium on Electronic Imaging, SPIE'2015, pp. 94090J–94090J–10. San Francisco, California, USA (2015)
8. Pibre, L., Pasquet, J., Ienco, D., Chaumont, M.: Deep learning is a good steganalysis tool when embedding key is reused for different images, even if there is a cover source-mismatch. In: Proceedings of the media watermarking, security, and forensics, MWSF, Part of I&ST international symposium on electronic imaging, pp. 1–11. EI, San Francisco, California, USA (2016)
9. Xu, G., Wu, H.Z., Shi, Y.Q.: Ensemble of CNNs for steganalysis: an empirical study. In: Proceeding of the 4th ACM workshop on information hiding and multimedia security, pp. 103–107. IH&MMSec 2016, Vigo, Galicia, Spain (2016)
10. Zeng, J., Tan, S., Li, B., Huang, J.: Pre-training via fitting deep neural network to rich-model features extraction procedure and its effect on deep learning for steganalysis. In: Proceedings of the media watermarking, security, and forensics 2017, MWSF'2017, Part of IS&T symposium on electronic imaging, p. 6. EI, Burlingame, California, USA (2017)
11. Chen, M., Sedighi, V., Boroumand, M., Fridrich, S.: JPEG-phase-aware convolutional neural network for steganalysis of JPEG images. In: Proceedings of the 5th ACM workshop on information hiding and multimedia security, pp. 75–84. IH&MMSec 2017, Drexel University in Philadelphia, PA (2017)
12. Xu, G.: Deep convolutional neural network to detect JUNIWARD. In: Proceedings of the 5th ACM workshop on information hiding and multimedia security, pp. 67–73. IH&MMSec 2017, Drexel University in Philadelphia, PA (2017)
13. Xu, G., Wu, H.Z., Shi, Y.Q.: Structural design of convolutional neural networks for steganalysis. IEEE Signal Process. Lett. **23**(5), 708–712 (2016)
14. Boroumand, M., Chen, M., Fridrich, J.: Deep residual network for steganalysis of digital images. IEEE Trans. Inf. Forensics Secur. **14**(5), 1181–1193 (2019)
15. Yedrouj, M., Comby, F., Chaumont, M.: Yedrouj-Net: an efficient CNN for spatial steganalysis. In: Proceedings of the IEEE international conference on acoustics, speech and signal processing, ICASSP 2018, Calgary, Alberta, Canada (2018)

16. Ye, J., Ni, J., Yi, Y.: Deep learning hierarchical representations for image steganalysis. IEEE Trans. Inf. Forensics Secur. **12**(11), 2545–2557 (2017)
17. Czaplewski, B.: An improved convolutional neural network for steganalysis in the scenario of reuse of the stego-key. In: 28th international conference on artificial neural networks and machine learning – ICANN 2019, Munich (2019)
18. Reinel, T.-S., et al.: GBRAS-Net: a convolutional neural network architecture for spatial image steganalysis. IEEE Access **9**, 14340–14350 (2021)
19. Binghamton's University DDE download section. https://dde.binghamton.edu/download. Accessed 22 Nov 2023
20. Tabares-Soto, R., et al.: Sensitivity of deep learning applied to spatial image steganalysis. PeerJ Comput. Sci. **7**, e616 (2021)
21. Binghamton's University DDE Steganographic Algorithms section. https://dde.binghamton. edu/download/stego_algorithms. Accessed 22 Nov 2023

Rotationally Invariant Object Detection on Video Using Zernike Moments Backed with Integral Images and Frame Skipping Technique

Aneta Bera$^{(\boxtimes)}$ ⓘ, Dariusz Sychel ⓘ, and Przemysław Klęsk ⓘ

Faculty of Computer Science and Information Technology, West Pomeranian University of Technology, ul. Żołnierska 49, 71-210 Szczecin, Poland
{abera,dsychel,pklesk}@zut.edu.pl

Abstract. This is a follow-up study on Zernike moments applicable in detection tasks owing to a construction of complex-valued integral images that we have proposed in [3]. The main goal of the proposition was to calculate the mentioned features fast (in constant-time). The proposed solution can be applied with success when dealing with single images, however it is still too slow to be used in real-time applications, for example in video processing. In this work we attempted to solve mentioned problem.

In this paper we propose a technique in order to reduce the detection time in real-time applications. The degree of reduction is controlled by two parameters: fs (related to the gap between frames that undergo a full scan) and nb (related to the size of neighborhood to be searched on non-fully scanned frames). We present a series of experiments to show how our solution performs in terms of both detection time and accuracy.

Keywords: Zernike moments · Complex-Valued Integral Images · Detection Time Reduction · Object Detection

1 Introduction

The classical approach to object detection is based on sliding window scans. It is computationally expensive, involves a large number of image fragments (windows) to be analyzed, and in practice precludes the applicability of advanced methods for feature extraction. In particular, many *moment* functions [13], commonly applied in image recognition tasks, are often precluded from detection, as they involve inner products, i.e., linear-time computations with respect to the number of pixels. This problem becomes more evident when detecting objects on video. Also, the deep learning approaches cannot be applied directly in dense detection procedures (sliding window-based), and require preliminary stages of prescreening or region-proposal.

There exist a few feature spaces (or descriptors) that have managed to bypass the mentioned difficulties owing to constant-time techniques discovered

L. Franco et al. (Eds.): ICCS 2024, LNCS 14833, pp. 68–83, 2024.
https://doi.org/10.1007/978-3-031-63751-3_5

for them within the last two decades. Haar-like features (HFs), local binary patterns (LBPs) and histogram of oriented gradients (HOG) descriptor are state-of-the-art examples from this category [1,5,17]. The crucial algorithmic trick that underlies these methods and allows for constant-time — $O(1)$ — feature extraction are *integral images*. They are auxiliary arrays storing cumulative pixel intensities or other pixel-related expressions. Having prepared them before the actual scan, one is able to compute fast the wanted sums the so-called 'growth' operations. Each growth involves two additions and one subtraction using four entries of an integral image.

In our previous work [3] we have introduced mentioned integral images in order to compute Zernike moments (ZMs). We prepared a set of special complex-valued integral images and an algorithm that allows to calculate Zernike moments fast, namely in constant time. Thanks to the proposed solution, Zernike moments become suitable for dense detection procedures, where the image is scanned by a sliding window at multiple scales, and where rotational invariance is required at the level of each window. In [8] we indicated numerically fragile places in our algorithm and identified their causes. Then, in order to reduce numerical errors, we propose piecewise integral images and derive a numerically safer formula for Zernike moments. Moreover, in [9] we enrich derived initial idea by proposing an extended space of Zernike invariants backed with integral images. This feature space includes not only the moduli of Zernike moments but also real and imaginary parts of suitable moment products. All mentioned solutions were supported by a series of experiments.

Recent literature confirms that ZMs are still popular and used in many applications. In [7] authors used ZMs with K-nearest Neighbors for leaf recognition, in [10] authors used selected ZMs to determine the rotation angle of the objects. Authors of [21] proposed to use ZMs and support vector machine for brain tumor diagnosis. ZMs are applied in many other image recognition tasks e.g.: human age estimation [12] or traffic signs recognition [19]. The authors of [11] proposed algorithm using image normalization and Zernike moments which allows to recognize stars based on telescope images. This solution allows to assign stars to their position in catalog. In 2020 the authors of [18] presented a way to match terrain using Zernike moments and HOG descriptors based on data from Synthetic Aperture Radar (SAR) and REM Radar. Yet, it is quite difficult to find examples of detection tasks applying ZMs directly.

Zernike moments are also used for detection task of objects in motion. E.g [16] shows a way for detection of doubtful or uncommon actions in video sequence based on Zernike moments and Canny edge detector. In [2] authors used motion energy image (MEI) with Zernike moments in order to detect humans actions. In [22], the authors proposed to use two particular Zernike moments (selected by them) in order to detect moving objects. It is worth noting that also in this task Zernike moments were not applied directly on video frames but after some kind of preprocessing or feature selection.

In this paper we address the problem of using Zernike moments for object detection on video, which requires to calculate them faster than according to the original algorithm proposed in [3].

2   Zernike Moments Theory and Calculation Using Integral Images

2.1   Zernike Moments

Zernike moments (ZMs) can be defined in both polar and Cartesian coordinates as: ïż£

$$M_{p,q} = \frac{p+1}{\pi} \int_0^{2\pi} \int_0^1 f(r,\theta) \sum_{s=0}^{(p-|q|)/2} \beta_{p,q,s} r^{p-2s} e^{-iq\theta} \, r \, dr \, d\theta, \tag{1}$$

$$= \frac{p+1}{\pi} \iint_{x^2+y^2 \leqslant 1} f(x,y) \sum_{s=0}^{(p-|q|)/2} \beta_{p,q,s} (x+iy)^{\frac{1}{2}(p-q)-s} (x-iy)^{\frac{1}{2}(p+q)-s} \, dx \, dy, \tag{2}$$

where:

$$\beta_{p,q,s} = \frac{(-1)^s (p-s)!}{s!((p+q)/2-s)!((p-q)/2-s)!}, \tag{3}$$

i is the imaginary unit ($i^2 = -1$), and f is a mathematical or an image function defined over unit disk [3,20]. p and q indexes, represent moment order, hence they must be simultaneously even or odd, moreover $p \geqslant |q|$.

ZMs are in fact the *coefficients* of an *expansion* of function f, given in terms of Zernike polynomials $V_{p,q}$ as the orthogonal base:[1]

$$f(r,\theta) = \sum_{\substack{0 \leqslant p \leqslant \infty \\ p-|q| \text{ even}}} \sum_{-p \leqslant q \leqslant p} M_{p,q} V_{p,q}(r,\theta), \tag{4}$$

where $V_{p,q}(r,\theta) = \sum_{s=0}^{(p-|q|)/2} \beta_{p,q,s} r^{p-2s} e^{iq\theta}$. Note that, $V_{p,q}$ combines a standard polynomial defined over radius r and a harmonic part defined over angle θ. In practical applications, finite partial sums of expansion (4) are used. Suppose ρ denotes imposed maximum polynomial order and ϱ denotes imposed maximum harmonic order, additionally $\rho \geqslant \varrho$. Then, the partial sum that approximates f can be written down as:

$$f(r,\theta) \approx \sum_{\substack{0 \leqslant p \leqslant \rho \\ p-|q| \text{ even}}} \sum_{-\min\{p,\varrho\} \leqslant q \leqslant \min\{p,\varrho\}} M_{p,q} V_{p,q}(r,\theta). \tag{5}$$

ZMs are invariant to scale transformations, but only their absolute value is invariant to rotation. Proof of these properties were presented in [3].

---

[1] ZMs expressed by (1) arise as inner products of the approximated function and Zernike polynomials: $M_{p,q} = \langle f, V_{p,q} \rangle / \|V_{p,q}\|^2$.

2.2   Zernike Moments in Detection Task

In practical tasks it is more convenient to work with rectangular, rather than circular, image fragments. Singh and Upneja [15] proposed a workaround to this problem: a square of size $w \times w$ pixels (w is even) becomes *inscribed* in the unit disc, as shown in Fig. 1, and zeros are "laid" over the square-disc complement. This reduces integration over the disc to integration over the square.

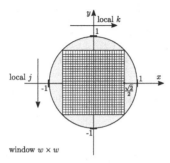

Fig. 1. The trick from [15] to calculate OFMMs: square image window inscribed in the unit circle, zero values laid in the complement of square.

2.3   Constant-Time Calculation of Zernike Moments

In [3] we proposed a way to calculate Zernike moments using set of multiple integral images in order to use those features in reasonable time, during detection procedure based on sliding window technique.

Let's concentrate on a scenario of a computer detection procedure. Suppose a digital image of size $n_x \times n_y$ is traversed by a sliding window of size $w \times w$, where w is even (for clarity we discuss only a single-scale scan within the detection procedure). The situation is presented in Fig. 2. Let (j, k) denote global coordinates of a pixel in the image. For each window under analysis, its offset — the top left corner of the window — will be denoted by (j_0, k_0). Therefore, the indexes of pixels that belong to the window are: $j_0 \leqslant j \leqslant j_0 + w - 1$, $k_0 \leqslant k \leqslant k_0 + w - 1$. Additionally, it will be convenient to introduce a notation (j_c, k_c) for the *central index* of the window:

$$j_c = \frac{1}{2}(2j_0 + w - 1), \quad k_c = \frac{1}{2}(2k_0 + w - 1). \tag{6}$$

Let $\{ii_{t,u}\}$ denote a set of **complex-valued integral images**[2]:

$$ii_{t,u}(l, m) = \sum_{\substack{0 \leqslant j \leqslant l \\ 0 \leqslant k \leqslant m}} f(j, k)(k - ij)^t (k + ij)^u, \quad \begin{smallmatrix} 0 \leqslant l \leqslant n_y - 1 \\ 0 \leqslant m \leqslant n_x - 1 \end{smallmatrix}; \tag{7}$$

---

[2] In [3] we have proved that integral images $ii_{t,u}$ and $ii_{u,t}$ are complex conjugates at all points, which allows for computational savings.

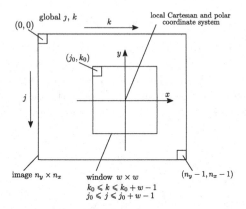

Fig. 2. Illustration of detection procedure using sliding window.

Proposition 1. *Suppose a set of integral images* $\{ii_{t,u}\}$, *defined as in* (7), *has been prepared prior to the detection procedure. Then, for any square window in the image, each of its Zernike moments can be calculated in constant time —* $O(1)$, *regardless of the number of pixels in the window, as follows:*

$$
\widehat{M}_{2p+o,2q+o} = \frac{4p+2o+2}{\pi w^2} \sum_{2q+o \leqslant 2\,s+o \leqslant 2p+o} \beta_{2p+o,2q+o,p-s} \left(\frac{\sqrt{2}}{w}\right)^{2\,s+o}
$$

$$
\cdot \sum_{t=0}^{s-q} \binom{s-q}{t} (-k_c+ij_c)^{s-q-t} \sum_{u=0}^{s+q+o} \binom{s+q+o}{u} (-k_c-ij_c)^{s+q+o-u} \mathop{\Delta}_{\substack{j_0,j_0+w-1 \\ k_0,k_0+w-1}} (ii_{t,u}). \quad (8)
$$

The proof of this is presented in detail in [3].

2.4 Numerical Errors and Their Reduction

Formula (8) contains two numerically fragile places. The first one are integral images themselves, defined by (7) where global pixel indexes j, k present in power terms $(k-ij)^t(k+ij)^u$ vary within: $0 \leqslant j \leqslant n_y - 1$ and $0 \leqslant k \leqslant n_x - 1$. Hence, for an image of size, e.g., 640×480, the summands vary in magnitude roughly from $10^{0(t+u)}$ to $10^{3(t+u)}$. Obviously, the rounding-off errors amplify as the $ii_{t,u}$ sum progresses towards the bottom-right image corner.

The second fragile place are: $(-k_c+ij_c)^{s-q-t}$ and $(-k_c-ij_c)^{s+q+o-u}$, involving the central index, see (8). Their products can too become very large in magnitude as computations move towards the bottom-right image corner.

The solution to this numerical problem, presented in [8], is based on integral images that are defined *piecewise*. We partitioned every integral image into a number of adjacent pieces, say of size $W \times W$ (border pieces may be smaller due to remainders), where W was chosen to exceed the maximum allowed width for the sliding window. Each piece obtains its own "private" coordinate system. Informally speaking, the (j, k) indexes that are present in formula (7) become reset to $(0,0)$ at top-left corners of successive pieces. Similarly, the values accumulated so far in each integral image $ii_{t,u}$ become zeroed at those points. For more details see [8].

2.5 Extended Feature Space of Zernike Invariants

During our previous research [9] we proposed a technique to extend the feature space. The central role is played by expression for generating the invariants:

$$M_{p,q}{}^n \, M_{v,s}, \quad nq + s = 0.$$

In that context we considered which tuples (n, p, q, v, s) should be allowed into the final collection of feature indexes and what information they carry. We distinguished and presented several groups among them.

Remembering that $M_{p,q}{}^n \, M_{v,s}$ is a complex number, we used both its *real* and *imaginary* parts as separate features, if it was possible. For that purpose we extended the tuples to consist of six members: (n, p, q, v, s, i), where the last index $i \in \{0,1\}$ indicates, whether we used real or imaginary part of $M_{p,q}{}^n \, M_{v,s}$.

Knowing the indexation scheme, we presented the actual extraction procedure to be invoked for each analyzed window in (see [9] for details).

Table 1 shows counts of features in both extended and non-extended spaces. Both spaces may constitute a useful input information for machine learning and rotationally-invariant detection.

Table 1. Number of features in extended (black) and non-extended (gray) feature spaces.

ρ \ ϱ	0	1	2	3	4	5	6
0	1						
	1						
1	1	2					
	1	2					
2	2	3	6				
	2	3	4				
3	2	6	11	16			
	2	4	5	6			
4	4	8	20	25	34		
	3	5	7	8	9		
5	4	13	29	45	56	63	
	3	6	8	10	11	12	
6	7	16	43	59	87	94	111
	4	7	10	12	14	15	16

Proposed solution demonstrates how to generate a large number of constant-time Zernike invariants using computations supported by integral images (complex-valued). Thanks to that, we can provide many useful features which is a beneficial for machine learning.

3 Frame Skipping Technique

Zernike invariants, even with the computationally fast form (with use of integral images), are still not fast enough to be applied for object detection on video. In this section we present an approach that allows to reduce the detection time for that purpose.

The idea is based on performing a full scan on the image from the camera, but not on each frame. For better explanation, let's assume that every 5th frame is scanned fully in order to find (detect) some objects. Once this is done we need to memorize their positions in the image. With this information at disposal, in the next frame the sliding window shall only be placed in close neighborhood of detected objects, and once they are redetected their positions can be updated. For yet another frame, the sliding window shall take advantage of neighborhoods from two previous frames, etc. After 4 frames, the image would be re-scanned, which would enable, e.g., finding objects that just appeared on it.

We will now proceed to a more detailed description of the proposed solution.

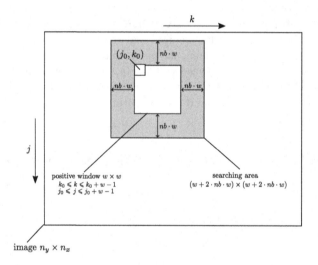

Fig. 3. Illustration of nb parameter for detection procedure using sliding window and frame skipping technique.

For the purpose of presented solution we introduce two parameters fs and nb. The first of them fs determines how many frames the full search of the entire image takes place, e.g., if the value of this parameter is 4, there are 3 frames among the full scans for which the full scan will not be performed. When $fs = 1$, a full scan will occur every frame. The second parameter, neighborhood (radius), determines size of area around the found windows with objects that will be searched — in case of images from camera without a full search. For Example, if window with found object has size 100×100 of pixels, nb is set to

0.5, it means the searched area will be of size 200×200 — we are adding 50 pixels (50% from 100) to each side of window marked as positive. Note that, the size of searched area depends on the window size marked as positive. The bigger the window, the bigger the area that will be searched.

Figure 3 presents a situation where one positive window was found in a full image scan. It is easy to see the advantages of using the presented solution. Instead of searching the entire $n_y \times n_x$ image with a sliding window (which comes in various sizes and scales), we will only search the area around the positive window, i.e., $w \cdot (1 + 2 \cdot nb) \times w \cdot (1 + 2 \cdot nb)$, which will save us a lot of time. This partial scan will be performed for $fs - 1$ frames.

Algorithm 1. Detection procedure with frame skipping technique

---

procedure DETECTOBJECT($\mathcal{I}$, $\mathcal{W}_d$, fr_n, nb, fs, clf)
    ▷ $\mathcal{W}_d$ contain coordinates of windows classified as positive on previous step.
    Create list $\mathcal{W}_r$ for storing coordinates of windows classified as positive.
    if fr_n mod $fs = 0$ **then**
        for $w \in$ GENDETCOORDS($\mathcal{I}_w$, $\mathcal{I}_h$, detection_parameters) **do**
            Use cls to classify fragment of $\mathcal{I}$ at coordinates w.
            if cls return 1 **then**
                Append w to $\mathcal{W}_r$.
    else
        if Areas contained in $\mathcal{W}_d$ intersect with each other **then**
            Merge areas to avoid redundancy.
        for $w \in \mathcal{W}_d$ **do**
            Expand area represented by w by adding margin equal to nb
            for coords $\in$ GENDETCOORDS(w_{expand}, detection_parameters) **do**
                Use cls to classify fragment of $\mathcal{I}$ at coordinates coords.
                if cls return 1 **then**
                    Append coords to $\mathcal{W}_r$.
    return $\mathcal{W}_r$

Algorithm 1 presents a detection procedure using frame skipping technique. clf represents a classifier, fr_n is frame number, nb neighborhood, fs skip, I video frame. Wd is a coordinate vector in the form of $[x, y, w, h]$, which describes the area that needs to be processed by standard detection procedure. In code below you can see two *GenDetCoords* procedure calls. The first one is a standard detection procedure, where whole the video frame is processed. The second one generates coordinates only in indicated areas — specified by positive frames and neighborhood.

4 Experiments

4.1 Learning Algorithms and General Settings

In experiments we apply *RealBoost+bins* as the main learning algorithm producing ensembles of weak classifiers. Each weak classifier is based on a single selected feature. Bins are equally wide and set up regularly, once 1% of outliers has been removed. The responses of classifiers are real-valued and calculated using the logit transform. For more details we address the reader, e.g., to [6,14].

It is worth remarking that we use Jaccard index (ratio of intersection and union areas) in two places in experiments: (1) to postprocess detected windows and (2) to check positive indications against the ground truth in tests. Typically, a detector produces a cluster of many positive windows around each target. At the postprocessing stage, we group such clusters into single indications. This means that at each step within the postprocessing loop, two windows with the highest Jaccard index become averaged. Later, when comparing positive indications against the ground truth, we expect each indication to have an index of at least 0.5 with respect to some target position in order to be counted as a true positive. Otherwise, it becomes a false alarm.

In experiment we have arranged a data set containing capital letters from the modern English alphabet. Pictures containing the characters from computer fonts were retrieved from the set prepared by T.E. de Campos et al. [4]. We have limited the subset representing the letter 'A' to several fonts with similar characteristics and treated it as our base for creating positive examples. Subsets with other letters were combined to prepare the negative examples.

Figure 4 depicts the source graphical material used in the experiment. Images, for both training and testing, were generated by randomly placing objects over random backgrounds.

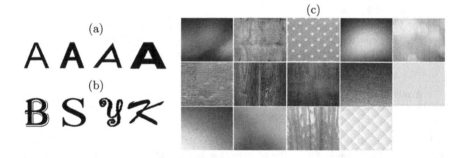

Fig. 4. Sample images and all backgrounds used to generate the data. Positives: letters A (a), negatives: other letters (b), backgrounds (c).

In training images, letters were allowed to rotate randomly within a limited range of ±45°. In the testing material letters were allowed to rotate randomly within the full range of 360°.

For detection procedure we created a video sequence from an image. We set a certain frame size, and move it around the given image and with the offset set according to pattern presented in Fig. 5.

Table 2 presents the experimental setup.

When reporting results of experiments, we use the following names: M, M-NER, E and E-NER — they define the type of features that was used. M stands

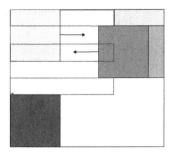

Fig. 5. Illustration of how video from image was generated.

Table 2. Experimental setup.

quantity / parameter	value	additional information
train data		
no. of positive examples	20 384	windows with letter 'A'
no. of negative examples	323 564	windows with letters other than 'A' plus random samples of backgrounds
train set size	343 948	positive and negative examples in total
test data		
no. of images	100	
no. of positive examples	213	windows with letter 'A'
no. of negative examples	1 858 587	implied by detection procedure
test set size	1 858 800	positive and negative examples in total
no. of negative examples	100 419	sampled on random from negatives
detection procedure (scanning with a sliding window)		
video frame resolution	640 × 480	imposed resolution
no. of detection scales	5	
window growing coefficient	1.2	
smallest and largest window size	100 × 100, 208 × 208	
window jumping coefficient	0.05	
$frame\_skip$ (fs)	$\{1, 5, 10\}$	
$radius$ (nb)	$\{15, 30, 50, 100\}$	

for the moduli of Zernike moments, E extended product invariants. The '-NER' suffix stands for numerical errors reduction. r parameter suggests that ring-based variation to increase number of features (by a factor of $2R - 1$) was used. Experiments were conducted parallel on Intel Xeon E5-2699 v4 CPU, 22/44 core/thread, 55MB cache.

4.2   Detection Performance

Tables 3, 4, 5, 6 present results of detection performance. FPS stands for 'frames per second', *total* is the total time of detection on whole video and *avg* is the average time of detection for single video frame.

As can be seen, both nb and fs have influence on time performance. The lower the value of nb is, the more significantly the time of the detection procedure decreases. The higher the value of fs, the lower the time of detection. It is worth paying a particular attention to FPS parameter. For the traditional detection

Table 3. Detection performance for moduli of Zernike moments and $p = 8$, $q = 8$, $r = 8$.

fs	$nb = 0.1$			$nb = 0.2$			$nb = 0.3$			$nb = 0.4$		
M	FPS	total	avg	FPS	total	avg	FPS	total	avg	FPS	total	avg
1	1.5	32352	652	1.5	34081	686	1.5	33462	676	1.6	31942	643
5	7.4	8756	135	6.4	9142	156	5.8	9444	171	5.3	9953	188
10	13.7	6183	73	12	6155	83	10.2	5970	98	8.6	6498	116
M-NER	FPS	total	avg	FPS	total	avg	FPS	total	avg	FPS	total	avg
1	0.4	128685	2658	0.3	149028	3085	0.3	151250	3127	0.4	129931	2683
5	1.6	31627	611	1.6	31079	623	1.6	31771	638	1.4	35805	726
10	3.4	16683	293	3.3	15841	303	2.7	188893	374	2.5	20481	406

Table 4. Detection performance for extended product invariants of Zernike moments and $p = 8$, $q = 8$, $r = 8$.

fs	$nb = 0.1$			$nb = 0.2$			$nb = 0.3$			$nb = 0.4$		
E	FPS	total	avg	FPS	total	avg	FPS	total	avg	FPS	total	avg
1	0.5	97185	2000	0.5	100212	2067	0.5	103255	2127	0.5	97824	2012
5	2.2	23936	457	2.2	23521	463	2	24984	498	1.8	27708	560
10	4.9	12249	204	4.1	13457	245	3.5	14595	286	3.2	16042	316
E-NER	FPS	total	avg	FPS	total	avg	FPS	total	avg	FPS	total	avg
1	0.3	157316	3252	0.3	166613	3446	0.3	178216	3690	0.3	157205	3251
5	1.5	33360	652	1.4	36835	740	1.3	37729	768	1.2	41399	841
10	2.7	20094	366	2.5	20865	403	2.5	20437	408	2.1	23746	477

procedure with sliding window and Zernike moments, its value is never bigger than 2. When using the proposed solution with frame skipping technique, the FPS was even 13.7 for moduli of Zernike moments.

Table 3 presents the results obtained for moduli of Zernike moments (with and without numerical error reduction) and $p = 8$, $q = 8$, $r = 8$. For features of type M we can observe the greatest gain in terms of time performance. The algorithm was able to process 13.7, 12, 10.2 and 8.6 frames per second, respectively for nb being set to 0.1, 0.2, 0.3 or 0.4 and fs equals to 10. For fs set to 5, we can also see a noticeable decrease in window processing time.

The time needed to generate features in $M - NER$ version is greater, and thus the profit from introducing modification is smaller, but still visible.

Table 5. Detection performance for moduli of Zernike moments and $p = 10$, $q = 10$, $r = 8$.

fs	$nb = 0.1$			$nb = 0.2$			$nb = 0.3$			$nb = 0.4$		
M	FPS	total	avg	FPS	total	avg	FPS	total	avg	FPS	total	avg
1	1	49575	1013	1	51485	1052	0.9	51718	1054	1	48780	996
5	4.6	12632	218	4.1	12878	243	3.9	13482	257	3.6	14223	276
10	9.3	7790	107	7.8	7782	129	6.8	7992	146	5.9	8976	169
M-NER	FPS	total	avg	FPS	total	avg	FPS	total	avg	FPS	total	avg
1	0.3	187889	3892	0.2	198526	4110	0.2	196523	4070	0.3	179710	3725
5	1.3	33334	790	1.2	42119	860	1.1	43122	878	1	48386	984
10	1.9	27711	523	1.9	25793	520	2	24524	493	1.8	27313	551

Table 4 presents the results obtained for extended product invariants of Zernike moments (with and without numerical error reduction) and $p = 8$, $q = 8$, $r = 8$. In this case, the base time was lower, and therefore the time after the modification also. However, you can see that for the parameter set to 10, the number of windows processed per second increased in every case.

Tables 5 and 6 present detection performance for moduli and extended product of Zernike moments, for $p = 10$, $q = 10$, $r = 8$. The results in the tables confirm the previously described observations.

Table 6. Detection performance for extended product invariants of Zernike moments and $p = 10$, $q = 10$, $r = 8$.

fs	$nb = 0.1$			$nb = 0.2$			$nb = 0.3$			$nb = 0.4$		
E	FPS	total	avg	FPS	total	avg	FPS	total	avg	FPS	total	avg
1	0.3	150883	3119	0.3	147192	3045	0.3	152779	3131	0.3	1568289	3240
5	1.4	35619	701	1.4	35215	711	1.3	37181	755	1.2	39936	812
10	2.8	19649	353	2.6	19637	379	2.3	21447	426	2	25020	497
E-NER	FPS	total	avg	FPS	total	avg	FPS	total	avg	FPS	total	avg
1	0.2	249028	5161	0.2	266057	5517	0.2	265249	5500	0.2	256278	5313
5	1	51050	1024	0.9	54793	1121	0.9	56915	1164	0.8	63597	1305
10	1.6	28700	540	1.7	28998	576	1.5	32045	650	1.4	35738	720

4.3 Detection Accuracy

Tables 7, 8, 9, 10 show accuracy of our detection experiments.

Table 7. Detection results for moduli of Zernike moments and $p = 8$, $q = 8$, $r = 8$.

fs	$nb = 0.1$			$nb = 0.2$		
M	accuracy	sensitivity	FAR	accuracy	sensitivity	FAR
1	0.999993831	1	$6.169 \cdot 10^{-6}$	0.999993831	1	$6.169 \cdot 10^{-6}$
5	0.999948604	0.25	$1.028 \cdot 10^{-6}$	0.99999486	0.95833333	$1.028 \cdot 10^{-6}$
10	0.999937297	0.11458333	$1.028 \cdot 10^{-6}$	0.999977384	0.72916667	$1.028 \cdot 10^{-6}$
M-NER	accuracy	sensitivity	FAR	accuracy	sensitivity	FAR
1	0.999997944	1	$2.056 \cdot 10^{-6}$	0.999997944	1	$2.056 \cdot 10^{-6}$
5	0.99995066	0.27	$1.028 \cdot 10^{-6}$	0.999996916	0.97916667	$1.028 \cdot 10^{-6}$
10	0.99993825	0.125	$1.028 \cdot 10^{-6}$	0.999986635	0.875	$1.028 \cdot 10^{-6}$
fs	$nb = 0.3$			$nb = 0.4$		
M	accuracy	sensitivity	FAR	accuracy	sensitivity	FAR
1	0.999993831	1	$6.169 \cdot 10^{-6}$	0.999993831	1	$6.169 \cdot 10^{-6}$
5	0.999996916	0.97916667	$1.028 \cdot 10^{-6}$	0.999996916	0.979	$1.028 \cdot 10^{-6}$
10	0.999986635	0.875	$1.028 \cdot 10^{-6}$	0.999986635	0.875	$1.028 \cdot 10^{-6}$
M-NER	accuracy	sensitivity	FAR	accuracy	sensitivity	FAR
1	0.999997944	1	$2.056 \cdot 10^{-6}$	0.999997944	1	$2.056 \cdot 10^{-6}$
5	0.999996916	0.97916667	$1.028 \cdot 10^{-6}$	0.999996916	0.97916667	$1.028 \cdot 10^{-6}$
10	0.999986635	0.875	$1.028 \cdot 10^{-6}$	0.999986635	0.875	$1.028 \cdot 10^{-6}$

Table 7 presents the results obtained for moduli of Zernike moments (with and without numerical error reduction) and $p = 8$, $q = 8$, $r = 8$. For the version of features without numerical error reduction (M) it can be seen that the sensitivity stabilizes for $nb = 0.3$. For lower values, especially for 0.1, there is a significant decrease in sensitivity compared to the version without using the proposed solution. For version with numerical error reduction (M-NER) sensitivity stabilizes for $nb = 0.2$, and further increasing the value of this parameter is no longer beneficial.

Table 8. Detection results for extended product invariants of Zernike moments and $p = 8$, $q = 8$, $r = 8$.

fs	$nb = 0.1$			$nb = 0.2$		
E	accuracy	sensitivity	FAR	accuracy	sensitivity	FAR
1	0.999990744	1	$9.256 \cdot 10^{-6}$	0.999990744	1	$9.256 \cdot 10^{-6}$
5	0.999956827	0.39583333	$2.057 \cdot 10^{-6}$	0.999994859	0.97916667	$3.085 \cdot 10^{-6}$
10	0.999941408	0.20833333	$2.057 \cdot 10^{-6}$	0.999998458	0.875	$3.085 \cdot 10^{-6}$
E-NER	accuracy	sensitivity	FAR	accuracy	sensitivity	FAR
1	0.999993831	1	$6.169 \cdot 10^{-6}$	0.999993831	1	$6.169 \cdot 10^{-6}$
5	0.999958832	0.41666667	$1.028 \cdot 10^{-6}$	0.999996916	0.97916667	$1.028 \cdot 10^{-6}$
10	0.999942437	0.20833333	$1.028 \cdot 10^{-6}$	0.999986636	0.875	$1.028 \cdot 10^{-6}$
fs	$nb = 0.3$			$nb = 0.4$		
E	accuracy	sensitivity	FAR	accuracy	sensitivity	FAR
1	0.999990744	1	$9.256 \cdot 10^{-6}$	0.999990744	1	$9.256 \cdot 10^{-6}$
5	0.999995887	0.97916667	$2.056 \cdot 10^{-6}$	0.999995887	0.97916667	$2.056 \cdot 10^{-6}$
10	0.999985604	0.875	$2.057 \cdot 10^{-6}$	0.999985604	0.875	$2.057 \cdot 10^{-6}$
E-NER	accuracy	sensitivity	FAR	accuracy	sensitivity	FAR
1	0.999993831	1	$6.169 \cdot 10^{-6}$	0.999993831	1	$6.169 \cdot 10^{-6}$
5	0.999996916	0.97916667	$1.028 \cdot 10^{-6}$	0.999996916	0.97916667	$1.028 \cdot 10^{-6}$
10	0.999986636	0.875	$1.028 \cdot 10^{-6}$	0.999986636	0.875	$1.028 \cdot 10^{-6}$

Table 8 presents the results obtained for extended product invariants of Zernike moments (with and without numerical error reduction) and $p = 8$, $q = 8$, $r = 8$. For both versions of features, with and without numerical error reduction it can be seen that the sensitivity stabilizes for $nb = 0.2$.

Table 9. Detection results for moduli of Zernike moments and $p = 10$, $q = 10$, $r = 8$.

fs	$nb = 0.1$			$nb = 0.2$		
M	accuracy	sensitivity	FAR	accuracy	sensitivity	FAR
1	0.999995887	0.96180556	0	0.999995887	0.96180556	0
5	0.999947575	0.23263889	0	0.999995888	0.96180556	0
10	0.999936269	0.09722222	0	0.999985607	0.85763889	0
M-NER	accuracy	sensitivity	FAR	accuracy	sensitivity	FAR
1	0.999995888	1	$4.113 \cdot 10^{-6}$	0.999995888	1	$4.113 \cdot 10^{-6}$
5	0.999947576	0.23958333	$1.028 \cdot 10^{-6}$	0.999996916	0.97916667	$1.028 \cdot 10^{-6}$
10	0.999937297	0.11458333	$1.028 \cdot 10^{-6}$	0.999986635	0.875	$1.028 \cdot 10^{-6}$
fs	$nb = 0.3$			$nb = 0.4$		

continued

Table 9. continued

fs	$nb = 0.1$			$nb = 0.2$		
M	accuracy	sensitivity	FAR	accuracy	sensitivity	FAR
1	0.999995887	0.96180556	0	0.999995887	0.96180556	0
5	0.999995888	0.96180556	0	0.999995888	0.96180556	0
10	0.999985607	0.85763889	0	0.999985607	0.85763889	0
M-NER	accuracy	sensitivity	FAR	accuracy	sensitivity	FAR
1	0.999995888	1	$4.113 \cdot 10^{-6}$	0.999995888	1	$4.113 \cdot 10^{-6}$
5	0.999996916	0.97916667	$1.02 \cdot 10^{-6}8$	0.999996916	0.97916667	$1.028 \cdot 10^{-6}$
10	0.999986635	0.875	$1.028 \cdot 10^{-6}$	0.999986635	0.875	$1.028 \cdot 10^{-6}$

Table 9 presents the results obtained for moduli of Zernike moments (with and without numerical error reduction) and $p = 10$, $q = 10$, $r = 8$. For both versions of features, with and without numerical error reduction, it can be seen that the sensitivity stabilizes for $nb = 0.2$.

Table 10 presents the results obtained for extended product invariants of Zernike moments (with and without numerical error reduction) and $p = 10$, $q = 10$, $r = 8$. As in previous cases, for both versions of features, with and without numerical error reduction, it can be seen that the sensitivity stabilizes for $nb = 0.2$.

Table 10. Detection results for extended product invariants of Zernike moments and $p = 10$, $q = 10$, $r = 8$.

fs	$nb = 0.1$			$nb = 0.2$		
E	accuracy	sensitivity	FAR	accuracy	sensitivity	FAR
1	0.999995887	0.96180556	0	0.999995887	0.96180556	0
5	0.999956827	0.38888889	0	0.999995887	0.96180556	0
10	0.999941408	0.19097222	0	0.999985608	0.87563889	0
E-NER	accuracy	sensitivity	FAR	accuracy	sensitivity	FAR
1	0.99999383	1	$6.17 \cdot 10^{-6}$	0.99999383	1	$6.17 \cdot 10^{-6}$
5	0.999956827	0.38541667	$1.028 \cdot 10^{-6}$	0.999996916	0.97916667	$1.028 \cdot 10^{-6}$
10	0.999942437	0.20933333	$1.028 \cdot 10^{-6}$	0.999986636	0.875	$1.028 \cdot 10^{-6}$
fs	$nb = 0.3$			$nb = 0.4$		
E	accuracy	sensitivity	FAR	accuracy	sensitivity	FAR
1	0.999995887	0.96180556	0	0.999995887	0.96180556	0
5	0.999995887	0.96180556	0	0.999995887	0.96180556	0
10	0.999985605	0.87563889	0	0.999985605	0.87563889	0
E-NER	accuracy	sensitivity	FAR	accuracy	sensitivity	FAR
1	0.99999383	1	$6.17 \cdot 10^{-6}$	0.99999383	1	$6.17 \cdot 10^{-6}$
5	0.999996916	0.97916667	$1.028 \cdot 10^{-6}$	0.999996916	0.97916667	$1.028 \cdot 10^{-6}$
10	0.999986636	0.875	$1.028 \cdot 10^{-6}$	0.999986636	0.875	$1.028 \cdot 10^{-6}$

5 Conclusion

We have proposed an algorithm that allows for faster rotationally-invariant object detection based on Zernike moments in comparison to standard detection procedure involving full scans by the sliding window. The presented technique takes advantage of frame skipping and is applicable only for video sequences. Although the detection time can be significantly reduced, it is fair to remark that the sensitivity measure can drop when number of skipped frames is set to a too large value. Of course, everything depends on the speed of movement, whether it's the camera or the object in the scene. Selecting a larger nb parameter provides more flexibility, but of course it takes more computation time.

References

1. Acasandrei, L., Barriga, A.: Embedded face detection application based on local binary patterns. In: 2014 IEEE International Conference on High Performance Computing and Communications (HPCC,CSS,ICESS), pp. 641–644 (2014)
2. Aly, S., sayed, A.: An effective human action recognition system based on zernike moment features. In: 2019 International Conference on Innovative Trends in Computer Engineering (ITCE), pp. 52–57 (2019)
3. Bera, A., Klęsk, P., Sychel, D.: Constant-Time Calculation of Zernike Moments for Detection with Rotational Invariance. IEEE Trans. Pattern Anal. Mach. Intell. **41**(3), 537–551 (2019)
4. de Campos, T.E., et al.: Character recognition in natural images. In: Proceedings of the International Conference on Computer Vision Theory and Applications, Lisbon, Portugal, pp. 273–280 (2009)
5. Dalal, N., Triggs, B.: Histograms of Oriented Gradients for Human Detection. In: Conference on Computer Vision and Pattern Recognition (CVPR'05) – Vol 1, pp. 886–893. IEEE Computer Society (2005)
6. Friedman, J., Hastie, T., Tibshirani, R.: Additive logistic regression: a statistical view of boosting. Ann. Stat. **28**(2), 337–407 (2000)
7. Jia, Z., Liao, S.: Leaf recognition using k-nearest neighbors algorithm with zernike moments. In: 2023 8th International Conference on Image, Vision and Computing (ICIVC), pp. 665–669 (2023)
8. Klęsk, P., Bera, A., Sychel, D.: Reduction of Numerical Errors in Zernike Invariants Computed via Complex-Valued Integral Images. In: Krzhizhanovskaya, V.V., et al. (eds.) ICCS 2020. LNCS, vol. 12139, pp. 327–341. Springer, Cham (2020). https://doi.org/10.1007/978-3-030-50420-5_24
9. Klęsk, P., Bera, A., Sychel, D.: Extended zernike invariants backed with complex-valued integral images for detection tasks. Proc. Comput. Sci. **192**, 357–368 (2021)
10. Lai, W., et al.: Single-pixel detecting of rotating object using zernike illumination. Opt. Lasers Eng. **172**, 107867 (2024)
11. Liang, X., Ma, W., Zhou, J., Kong, S.: Star identification algorithm based on image normalization and zernike moments. IEEE Access **8**, 29228–29237 (2020)
12. Malek, M.E., Azimifar, Z., Boostani, R.: Facial age estimation using Zernike moments and multi-layer perceptron. In: 22nd International Conference on Digital Signal Processing (DSP), pp. 1–5 (2017)
13. Mukundan, R., Ramakrishnan, K.: Moment Functions in Image Analysis — Theory and Applications. World Scientific (1998)
14. Rasolzadeh, B., et al.: Response binning: improved weak classifiers for boosting. In: IEEE Intelligent Vehicles Symposium, pp. 344–349 (2006)
15. Singh, C., Upneja, R.: Accurate computation of orthogonal Fourier-Mellin moments. J. Math. Imaging Vis. **44**(3), 411–431 (2012)

16. Vengurlekar, S.G., Jadhav, D., Shinde, S.: Object detection and tracking using zernike moment. In: 2019 International Conference on Communication and Electronics Systems (ICCES), pp. 12–17 (2019)
17. Viola, P., Jones, M.: Rapid Object Detection using a Boosted Cascade of Simple Features. In: Conference on Computer Vision and Pattern Recognition (CVPR'2001), pp. 511–518. IEEE (2001)
18. Wang, K., Wang, H., Wang, J.: Terrain matching by fusing hog with zernike moments. IEEE Trans. Aerosp. Electron. Syst. **56**(2), 1290–1300 (2020)
19. Xing, M., et al.: Traffic sign detection and recognition using color standardization and Zernike moments. In: 2016 Chinese Control and Decision Conference (CCDC), pp. 5195–5198 (2016)
20. Zernike, F.: Beugungstheorie des Schneidenverfahrens und seiner verbesserten Form, der Phasenkontrastmethode. Phys. **1**(8), 668–704 (1934)
21. Zheng, N., Zhang, G., Zhang, Y., Sheykhahmad, F.R.: Brain tumor diagnosis based on zernike moments and support vector machine optimized by chaotic arithmetic optimization algorithm. Biomed. Signal Process. Control **82** (2023)
22. Zhou, Z., Liu, P., Chen, G., Liu, Y.: Moving object detection based on zernike moments. In: 2016 5th International Conference on Computer Science and Network Technology (ICCSNT), pp. 696–699 (2016)

Reduced Simulations for High-Energy Physics, a Middle Ground for Data-Driven Physics Research

Uraz Odyurt[1,2]([✉]) [ID], Stephen Nicholas Swatman[3,4] [ID],
Ana-Lucia Varbanescu[3,5] [ID], and Sascha Caron[1,2] [ID]

[1] High-Energy Physics, Radboud University, Nijmegen, The Netherlands
[2] National Institute for Subatomic Physics (Nikhef), Amsterdam, The Netherlands
{uodyurt,scaron}@nikhef.nl
[3] Informatics Institute, University of Amsterdam, Amsterdam, The Netherlands
s.n.swatman@uva.nl
[4] European Organisation for Nuclear Research (CERN), Geneva, Switzerland
[5] CAES, University of Twente, Enschede, The Netherlands
a.l.varbanescu@utwente.nl

Abstract. Subatomic particle track reconstruction (tracking) is a vital task in High-Energy Physics experiments. Tracking is exceptionally computationally challenging and fielded solutions, relying on traditional algorithms, do not scale linearly. Machine Learning (ML) assisted solutions are a promising answer. We argue that a complexity-reduced problem description and the data representing it, will facilitate the solution exploration workflow. We provide the REDuced VIrtual Detector (REDVID) as a complexity-reduced detector model and particle collision event simulator combo. REDVID is intended as a simulation-in-the-loop, to both generate synthetic data efficiently and to simplify the challenge of ML model design. The fully parametric nature of our tool, with regards to system-level configuration, while in contrast to physics-accurate simulations, allows for the generation of simplified data for research and education, at different levels. Resulting from the reduced complexity, we showcase the computational efficiency of REDVID by providing the computational cost figures for a multitude of simulation benchmarks. As a simulation and a generative tool for ML-assisted solution design, RED-VID is highly flexible, reusable and open-source. Reference data sets generated with REDVID are publicly available. Data generated using REDVID has enabled rapid development of multiple novel ML model designs, which is currently ongoing.

Keywords: Reduced-order modelling · Simulation · High-energy physics · Synthetic data

1 Introduction

In many computational sciences, the adoption of ML-assisted solutions can lead to serious gains in computational efficiency and data processing capacity,

L. Franco et al. (Eds.): ICCS 2024, LNCS 14833, pp. 84–99, 2024.
https://doi.org/10.1007/978-3-031-63751-3_6

resulting from algorithmic advantages intrinsic to ML. Computational efficiency can also be achieved by paving the way for the utilisation of dedicated hardware, i.e., GPUs, FPGAs and purpose-built accelerators. ML algorithms are highly compatible with the use of such specialised hardware. In this work, we explore the use of ML-assisted techniques in high-energy physics.

ML-assisted solution design is an explorative and data-demanding endeavour. One of the effective approaches to achieve a suitable design is Design-Space Exploration (DSE). Complex problems involve many parameters, contributing to a space with many dimensions, which in turn deems the exploration expensive. There is often a need for simplification of the problem domain, i.e., search-space reduction, to facilitate the initial steps within this explorative process.

Generative elements are often needed as part of the explorative process, to enable synthetic data generation in large quantities, at will. Furthermore, designing and training models with better rigour requires total control over all aspects of data generation. Providing sufficient control and on demand ability to synthesise data that is representative of corner cases contributes to achieving effective models. Such corner cases seldom/disproportionally appear in real-world data or highly accurate, i.e., *physics-accurate*, simulation data.

Use-case. Our focus is a major use-case from the field of High-Energy Physics (HEP), *the critical task of subatomic particle track reconstruction (tracking)*, present in data processing for experiments performed at the Large Hadron Collider (LHC). Detectors such as ATLAS, record interaction data of subatomic particles with detector sensors, allowing physicists to reconstruct particle trajectories through tracking algorithms and to gain knowledge on how subatomic particles behave. Current solutions rely largely on traditional, computationally expensive statistical algorithms, with Kalman filtering as their most demanding block. Even with constant efforts channelled into better parallelisation schemes for these algorithms, the data consumption capability is rather limited. The challenge will be even greater with the upcoming High-Luminosity LHC upgrade [3], given its increased data volume generation and experiment frequency.

Although physics-accurate simulators, such as Geant4 [1], are readily available, applying such levels of accuracy to generative elements comes at a hefty computational cost. Accordingly, these simulators are not suitable for frequent *timely* executions and constant data generation, as required for DSE iterations. As such, we propose an exploration methodology that can be much faster, through the informed simplification of the design-space for the ML-assisted solution. Our methodology is specifically being considered for the tracking use-case. To this end, we have *designed and implemented* the *REDuced VIrtual Detector (REDVID)*, to both simplify the problem at hand and act as an efficient tool for frequent simulations and synthetic data generation. While our tool is not a fully physics-accurate one, it does respect the high-level relations present in subatomic particle collision events and detector interactions. REDVID is fully (re)configurable, allowing definition of experiments through varying detector models, while preserving the *cascading effects* of every change.

Considering possible complexity reduction strategies, the spectrum varies from physics-accurate data manipulations, e.g., dimensionality/granularity reduction, to omitting the scenario interactions beforehand. A strategy solely based on data reduction will fail to preserve the behavioural integrity of the system, as it will fail to propagate cascading effects resulting from reductions. Even simplified examples such as the TrackML data [2] are too complex.

Contribution. REDVID, as an experiment-independent, fully (re)configurable, and complexity-reduced simulation framework for HEP [20], is provided. Simulations consist of complexity-reduced detector models, alongside a particle collision event simulator with reduced behavioural-space. REDVID is intended as a simulation-in-the-loop for ML model design workflows, providing:

- ML model design - Problem simplification facilitates ML solution design, as opposed to real-world use-case definitions, which are often too complex to negotiate directly.
- Parametric flexibility - The model generator is capable of spawning detectors based on reconfigurable geometries.
- Computational efficiency - Behavioural-space reductions directly improve event simulation and processing times.

Our other contributions include:

- Supporting pedagogical tasks in higher education by presenting complex interactions from HEP experiments through understandable data.
- Publishing open reference data sets, which are of independent interest for physicists and data scientists alike [18,21].

Outline. Section 2 provides the background on HEP experiments and similar simulators. In Sect. 3, we provide the design details considered for REDVID. Notable implementation techniques are elaborated in Sect. 4. Data set related results are given in Sect. 5, followed by Sects. 6 and 7, covering the relevant literature and our conclusions, respectively.

2   Background and Motivation

In this section we elaborate the premise of HEP experiments, as well as the role of simulation in these, to get familiar with the context of our use-case.

2.1   HEP Experiments

When talking about *HEP experiments*, we refer to high-energy particle collision events. Two types of collision experiments are performed at LHC: proton-proton and ion-ion collisions. Protons are extracted from hydrogen atoms, while ions are actually heavy lead ions. Beams of particles are sent down the beam pipe in opposing directions and made to collide at four specific spots. These four spots are the residing points of the four major detectors installed at LHC, namely, ALICE [9], ATLAS [10], CMS [11] and LHCb [12].

Take the ATLAS detector for instance. The role played by ATLAS in the study of fundamental particles and their interactions, rely on two main tasks, *tracking* and *calorimetry*. Through tracking, i.e., particle track reconstruction, the momentum, p, of a particle can be calculated, while the energy, E, is calculated through calorimetry. Having the momentum and the energy for a given particle, its mass, m, can be determined, following the *energy-momentum relation* expressed as,

$$E^2 = (mc^2)^2 + (pc)^2.$$

In the above equation, c represents the speed of light and is a constant. The mass measurement allows the study of the properties for known particles, as well as potentially discovering new unknown ones. As such, it is fair to state that *particle track reconstruction is one of the major tasks in high-energy physics.*

2.2 Role of Simulation in HEP

Simulation allows for, amongst others, the validation and training of particle track reconstruction algorithms. Two distinguished stages are considered for HEP event simulations, i.e., *physics event generation* and *detector response simulation* [15]. Event generation as the first stage, involves the simulation of particle collision events, encompassing the processes involved in the initial proton-proton or ion-ion interactions. Event generation is governed by intricate sets of physical rules and is performed by software packages such as Herwig [13] and Pythia [23], i.e., physics-accurate simulations.

Detector response simulation, the second stage, integrates the movement of the particles generated by the first stage through a detector geometry, simulating the decay of unstable particles, the interactions between particles and matter, electromagnetic effects, and further physical processes such as hadronisation. Common event simulators providing such functionality include Geant4 [1], FLUKA [5] and MCNP [16]. In accelerator physics applications, event simulators are used to simulate the interactions between particles and sensitive surfaces in an experiment, as well as with so-called passive material, such as support beams. Interactions with sensitive surfaces may undergo an additional *digitisation* step, simulating the digital signals that can be read out of the experiment. Considering the example of ATLAS, three data generating simulators are notable, namely, Geant4, FATRAS [15] and ATLFAST [22].

Following the Monte Carlo simulation approach, FATRAS has been designed to be a fast simulator. It is capable of trajectory building based on a simplified reconstruction geometry and does provide support for material effects, as well as particle decay. FATRAS also generates hit data.

ATLFAST follows a different approach towards trajectory simulation and doesn't generate hit data, making it unsuitable for tracking studies. ATLFAST relies on hard-coded smearing functions based on statistics from full simulations. These functions are dependent on particle types, momentum ranges and vertex radii. Such details are specific to the design elements of the virtual detector geometry. A change in the design will require finding new functions.

REDVID fills the gap for a reconfigurable framework, suitable for first-phase solution exploration and design. This is due to the deliberate reduction in complexity, for both the generated data and the problem description, while keeping the high-level causal relations in place. REDVID is end-to-end parametric, i.e., all the generated data is built upon the detector geometry and randomised particle trajectories, both reconfigurable. REDVID has been developed in Python, making its integration with Python-based ML design workflows seamless. Figure 1 positions REDVID versus other well-known tools, as we consider it.

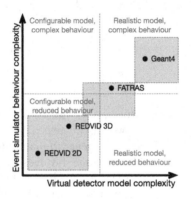

Fig. 1. Different simulators are capable of providing different levels of complexity, depicted as grey areas. ATLFAST is not included for lack of hit data generation.

3   Simulation Application and Design

The underlying question here is what is a good strategy for designing and training a capable and rigorous ML model to predict the behaviour of a (complex) real system? For our HEP use-case, the system is already complex; and when considering the upcoming High-Luminosity LHC upgrade [3], this complexity will increase even further. As such, when looking for an ML-assisted solution for tracking, we need to efficiently explore a large set of options.

Addressing complex real-world tasks directly will require synthesising close to real-world data, which can be performed by high-accuracy simulations. High-accuracy simulations in general, and physics-accurate simulations in particular, are extremely expensive computational tasks. Having such tools as part of an exploration workflow, e.g., ML model design, triggering frequent executions of the simulation with altered configuration, will inevitably turn into a serious challenge. Even if there are accommodating hardware resources available, algorithmic limitations will turn software tools into workflow bottlenecks. Yet another notable drawback is the high cost of energy when running frequent computationally expensive tasks. To alleviate this massive challenge, it is highly beneficial, and perhaps necessary, to not only design reduced models and simulators[1], but *to provide parametric (re)configurability to support automated exploration.*

However, the initial testing of new ML-assisted solutions, i.e., ML model designs, does not require the ground truth, which physics-accurate simulations are designed to produce. Instead, we argue that a cost-effective and reduced simulation, preserving the behavioural relations of the complex system (proton-proton/ion-ion collision event experiments), can be better and more effectively integrated in ML model design workflows, as shown in Fig. 2.

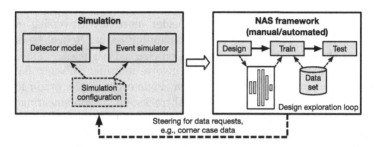

Fig. 2. Reduced simulations in ML model design workflows, e.g., a Neural Architecture Search (NAS), with this paper's focus on the area with the yellow fill. (Colour figure online)

3.1 Reduction Approach

Having a validly approximate representation is achieved through the reduction of the behavioural-space to a minimal subset, best encapsulating the complex system. Both model complexity and simulator complexity can be targets of such a reduction. The first and foremost effect of an approximate simulation is better computational efficiency. Note that there can be many such approximations, depending on the intended balance between computational efficiency and behavioural approximation level. The other advantage, especially when it comes to ML model design processes, is facilitation of an effective model design by providing a middle ground that has a lower complexity and can be used for better understanding of the challenge and testing of the early designs.

Both actual experiments and physics-accurate simulations for our use-case, i.e., proton-proton/ion-ion collision events inside a detector such as ATLAS, are immensely complex. Removing (some of) the physics-accurate constraints results in major behavioural-space reductions. This applies to both the detector model and the behaviour affecting the event simulator. While moving away from physics-accuracy, our aim has been to conserve logical, mathematical and geometrical relations, which would provide the basis for a flexible parameterisation. Preserving relations between interacting elements of a system preserves occurrence of *cascading effects* when the system is being steered through reconfiguration. For instance, a change in the structural definition of the detector

---

[1] A model and a simulator go hand in hand to form a simulation.

model will affect the recorded hit points during the event simulation. It must be noted that we have intentionally avoided the time dimension complexities. Accordingly, a list of major reductions that we have considered follows.

- Simplified detector geometry: Compared to the real detector, we have considered much simpler elements for the geometry of our virtual detector model, consisting of elements with disk or cylinder shapes, ultimately arriving at a Reduced-Order Model (ROM).
- Particle types: Currently, we do not consider explicit particle types in our event simulator. The track type variation however, could be seen as a consequence of differing particle types.
- Simplified tracks: Currently, we consider particles traversing a linear (straight), helical uniform, or helical expanding paths. Helical tracks could be seen as the effect of a magnetic field on charged particles.
- Collision points: The real experiments involve multiple collisions happening almost at the same time. We consider a single event at the origin for linear tracks and a non-aligning one for helical tracks, i.e., origin smearing.
- Hit coordinates smearing: We introduce noise in our hit calculations and hit coordinate parameters by drawing random samples from a Gaussian distribution. We also consider the noise standard deviation as proportional to the variable range. The noise ratio can be adjusted by the user.

3.2   Detector Model

At its core, a detector model is comprised of the geometric definitions of the included elements, shapes, sizes, and placements in space. Although we can support a variety of detector geometries, the overall structure, especially for our experimental results, is based on the ATLAS detector. Accordingly, there are four sub-detector types, *Pixel*, *Short-strip*, *Long-strip* and *Barrel*. The pixel and the barrel types have cylindrical shapes with the pixel being a filled cylinder, while the barrel being a cylinder shell with open caps. These are not hard requirements, as the geometry is fully parametric, and differing definitions can be opted for, e.g., a pixel as a cylinder shell. The long-strip and the short-strip types are primarily intended as flat disks, but can be defined as having a thickness, rendering them as cylinders. Sub-detector types can be selectively present or absent. Figure 3 depicts a representative variation of the detector geometry involving the aforementioned elements.

Structurally speaking, in a real-world detector, like ATLAS, the internals of short-strip and long-strip sub-detector types are different. We on the other hand, reduce such complexities to placement location and size, i.e., distance from the origin and sub-detector disk radius. Note that our geometric model does support disk thickness, which basically would turn disks into shallow cylinders. However, we have considered flat disks for our experiments.

3.3   Particle Collision Event Simulation

One of the simplifications for our complexity reduction approach is to consider a single collision per event. However, the list of complexities, even without the

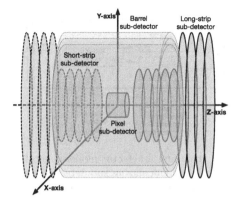

Fig. 3. Parametric detector geometry, allowing for inclusion/exclusion of different sub-detector types, with full control over sub-layer counts, sizes and placements.

polluting effects of multiple collisions, is extensive. Particles travelling through the detector matter could lead to secondary collisions, resulting in drastic changes in their trajectory. Such secondary collisions could also lead to the release of particles not originating from the collision event itself. These will show up as tracks with unusual starting points within the detector space. Some particles could also come to a halt, which would be seen as abruptly terminating tracks.

Such physics-accurate behaviour of particles interacting with the present matter in detectors is not considered for our simulator. It must be noted that the generation of tracks originating far away from the origin and prematurely terminating tracks, can be emulated in our simulator in a randomised fashion.

4 Implementation

Though both two-dimensional (2D) and three-dimensional (3D) spaces are supported, we will focus on the implementation details relevant to the 3D case. REDVID is open source [19] and has been developed in Python.

4.1 Modules

Considering the tasks at hand, detector spawning and event simulation, our software can be divided into three main logical modules: *Detector generator*, *Event simulator* and *Reporting*, depicted in Fig. 4. The current implementation considers the sequential execution of modules in the order given. However, one can easily generate detectors without simulating events, or simulate events with previously generated detectors, or even calculate hits based on previously generated tracks. Such input/output capability will allow our software to interact with other commonly utilised tools. The main configuration parameter defining the execution path is the `detector_type`, which can be 2D or 3D.

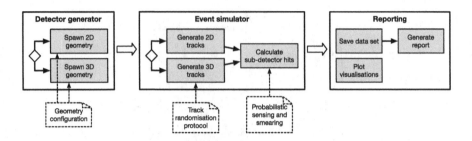

Fig. 4. REDVID modules, including a detector model generator, an event simulator, generating randomised tracks and calculating sub-detector hit points based on tracks and geometric data, as well as different reporting elements.

4.2   Coordinate Systems

We have opted for the cylindrical coordinate system to represent sub-detectors, tracks and hits. This is convenient, as we are considering the Z-axis as the beam pipe in LHC experiments and all geometric shapes defined within a detector, whether disks or cylinders, are actually of the type cylinder. The three parameters to define any point in the cylindrical coordinate system are the radial distance from the Z-axis, the azimuthal angle between the X-axis and the radius, and the height of the point from the XY-plane, i.e., r, θ and z, respectively.

In this coordinate system, hit points can be precisely defined given the tuple $(r_{hit}, \theta_{hit}, z_{hit})$. Geometric shapes can also be defined with boundaries for r_{sd} and z_{sd}, e.g., a disk will have fixed z_{sd}, unbounded θ and bounded r_{sd}. Here sd stands for sub-detector. Our software does support partial disks, i.e., a disk with a hole in the middle, which can be considered when the beam pipe is expected to be part of the geometry. Disks with thickness (cylinders) will have a small boundary for the parameter z_{sd}. As previously explained, short-strip and long-strip sub-detector types are defined as disks. For the pixel type, as it is a filled cylinder, both r_{sd} and z_{sd} will be bounded. When it comes to the barrel type, as it is a cylinder shell, there will be a fixed r_{sd} with bounded z_{sd}.

To implement linear tracks and to define them in the cylindrical coordinate system, both a direction vector and a point, P_0, that the track (line) goes through are needed. The direction vector, V_d, is considered as a vector from the origin, landing on a point in space, represented with a tuple (r_d, θ_d, z_d). The direction vector is randomised and then normalised for the z parameter, meaning that the direction vector will either have $z_d = 1$ or $z_d = -1$. The boundaries of this randomisation depend on the track randomisation protocol, explained in the next section. If we consider all linear tracks as starting from the detector origin, the point $(0,0,0)$ is considered on the track. However, this is rarely the case. The resulting parametric form of a track (line) is,

$$r = r_0 + t \cdot r_d,$$
$$\theta = \theta_0 + \theta_d,$$
$$z = z_0 + t \cdot z_d,$$

with (r, θ, z) representing a point on the track, (r_0, θ_0, z_0) as the origin point, $\langle r_d, \theta_d, z_d \rangle$ as the direction vector, and t as free variable. Similarly, the parametric form for helical track definitions is,

$$r = r_0 + a \cdot t \,,$$
$$\theta = \theta_0 + d \cdot t \,,$$
$$z = z_0 + b \cdot t \,,$$

with (r, θ, z) representing a point on the track and (r_0, θ_0, z_0) as the origin point, while a, d and b represent radial, azimuthal and pitch coefficients, respectively.

Regarding both linear and helical tracks, our software supports origin smearing, i.e., the starting point of helical tracks is in a randomised vicinity of the point $(0, 0, 0)$.

4.3   Track Randomisation Protocols

As seen in Fig. 4, the track randomisation step directly affects sub-detector hit calculation and is totally dependent on the randomisation protocol indicated in the configuration. Focusing on the implementation for the 3D space, different track randomisation protocols can be considered. We list four base protocols and five combination protocols, mixing the characteristics of base protocols:

Protocol 1 - Last layer hit guarantee Hits are guaranteed to occur on the farthest layer of every sub-detector type, which means the farthest layer of every sub-detector type is the randomisation domain for the landing points of tracks. A hit guarantee on the last layer will also guarantee hits on the previous layers for that sub-detector type. This protocol is designed to maximise the number of hits per sub-detector type within the data set.

In principle, our implementation applies *Protocol 1* per each available sub-detector type and randomly selects from the total generated track pool. Since for instance, if a track lands on the last layer of strip sub-detector types, it might not necessarily result in hit points on barrel layers.

Protocol 2 - Spherically uniform distribution To have a more uniform distribution of randomised tracks, without imposing any geometric conditions, is to have the track end points land on a sphere. Note that tracks do not have actual end points as these are unbounded lines.

Protocol 3 - Conical jet simulation Tracks are randomised in distinct subsets, bundled in a close vicinity within a narrow cone, representing a jet(s). This protocol on its own may not be a sensible choice and it would work best in combination with other protocols.

Protocol 4 - Beam pipe concentration Tracks have a higher concentration around the beam pipe, i.e., higher track generation probability as the radius gets smaller.

Without giving exhaustive and repetitive descriptions, feasible combination protocols are: *Protocols 1 and 3, Protocols 1 and 4, Protocols 2 and 3, Protocols 3 and 4* and *Protocols 1, 3 and 4.*

For data generation we have only considered protocol 1 to increase recorded hit points for all tracks and to have hit points for all sub-detector types. Additional track randomisation protocols focusing on specific corner cases, can be easily defined and added to the tool. To implement protocol 1, i.e., to guarantee that tracks land on the last layer of a sub-detector type, we consider the coordinate domain of the last layer as the randomisation domain for track direction vectors. Thus, before normalisation, all randomised V_d will land on the last layer.

Not every combination is allowed. For instance, protocols 1 and 2 cannot be applied at the same time, as it is self-evident that a spherical uniform distribution and a last layer hit guarantee cannot be true at the same time. Accordingly, we can consider the base protocols within two main categories, *distribution protocols*, affecting how tracks are distributed in space, and *feature protocols*, defining special forms of localised distribution. Currently, protocol 3 is the only feature protocol defined. While feature protocols can be combined with any distribution protocol, most distribution protocols are mutually exclusive. A combination of two or more base distribution protocols will also lead to another, more specific, distribution protocol, e.g., protocols 1 and 4. Figure 5 provides a visual overview.

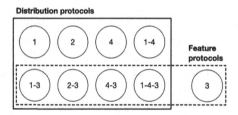

Fig. 5. Visualising how different base distribution and feature protocols can be combined to achieve more complex track randomisation behaviour.

4.4   Hit Point Calculation

Regarding hit point coordinates, i.e., $(r_{hit}, \theta_{hit}, z_{hit})$, depending on the sub-detector shape, we are dealing with either a fixed z_{sd} or a fixed r_{sd}, for disks and barrels, respectively. Here, we consider the disks as being flat and to have no thickness, while the barrels consist only of cylinder shells. Shapes with thickness are supported, for which the techniques involved will be similar.

Considering the set of track equations, we are to calculate the free variable t at the sub-detector layer of interest, i.e., t_{sd}. For hit coordinates at disks,

$$z_{hit} = z_{sd},$$

$$\theta_{hit} = \theta_d,$$

$$t_{sd} = \frac{z_{sd}}{z_d} = \frac{z_{sd}}{1},$$

$$\Rightarrow t_{sd} = z_{sd},$$

$$r_{hit} = t_{sd} \cdot r_d = z_{sd} \cdot r_d.$$

Note that in the above calculation z_d and z_{sd} must have matching signs, rendering $t_{sd} > 0$. In other words, tracks extruding towards the positive or the negative side of the Z-axis can hit sub-detector layers present at the positive or the negative side of the Z-axis, respectively. We also know that $z_d \neq 0$.

A similar calculation considering the r_{sd} as fixed will result in the hit coordinates for a barrel sub-detector layer, which we will not repeat here. General approach towards calculation of hits resulting from helical tracks follows the same principles, which we will not repeat here. Figure 6 depicts a simple event with five tracks, including separate views of the full event (Fig. 6a) and calculated hits (Fig. 6b), for demonstration purposes. Note that the detector orientation is vertical.

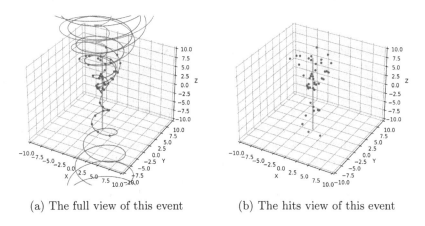

(a) The full view of this event   (b) The hits view of this event

Fig. 6. An example event with five tracks

5   Data Set Generation

We have followed simulation recipes with 10 000 events and varying track counts of $[1, 10\,000]$ per event for each experiment, for both linear and helical tracks. Hit recording is performed with smearing enabled and the detector geometry is the same for all recipes. These generated data sets for linear and helical tracks are intended as reference for physicists and data scientists alike and are publicly accessible over Zenodo open repository [18,21]. Data set schema alongside data header descriptions are included in the accompanying README files.

In order to evaluate the performance of REDVID, we have benchmarked the execution of simulations with a lower event count, 1 000 events per simulation and similar variations of track concentrations per event as before, i.e., [1, 10 000]. For our metric collections, including CPU-time and execution duration, high-precision counters from the time library available in Python have been used. The collected CPU-time results are provided in Table 1.

Table 1. REDVID execution CPU-time cost for simulations of 1 000 events with various track concentrations. All values are in milliseconds. Full simulation times are provided in minutes as well.

Track(s) per event	3D detector spawning	Track randomisation Mean	Hit discovery Mean	Full simulation 1 000 events (minutes)
1	0.025	0.043	1.463	2 731.17 (0.05)
10	0.025	0.083	13.429	15 418.589 (0.26)
100	0.025	0.465	129.864	137 623.954 (2.29)
1 000	0.025	4.582	1 285.989	1 353 396.641 (22.56)
10 000	0.024	43.765	12 496.208	13 591 628.526 (226.53)

Simulations have been performed on the DAS-6 compute cluster [4]. The machines used are each equipped with a single 24-core AMD EPYC 7402P processor and 128 GB of main memory. Note that the mean CPU-time calculations do not include the first event of each recipe batch. This is due to the presence of cold-start effect for the first event and delays resulting from it.

Though we have enforced single-threaded operation for our benchmarks, workload parallelisation is rather trivial. The number of events to be generated can be divided into any desired number of batches and distributed amongst multiple threads. Considering the timing results, we observe that the CPU-time values scale linearly, i.e., a tenfold increase in the track concentration per event results in roughly a tenfold increase in the full simulation CPU-time.

6   Related Work

While the overall available data is abundant, corner case data is rather scarce. Real-world data, or data synthesised with accurate (in our case physics-accurate) simulations is complex in terms of data dimensionality and granularity. This complexity is directly resulting from the complexity of the real system, or the accurate model of the system in case of simulations. Within the HEP landscape, we touched upon the complexity of simulators such as Geant4 in Sect. 2, as well as the dependence on these simulators by tools like ATLFAST.

The first challenge, lack of annotated data for one or more specific scenarios, has been recognised in the literature [14]. The second challenge though, the issue of complexity, is not as well known. A closely related acknowledgement has been made regarding the complexity level of models for simulations [8].

The two main shortcomings of the previous efforts towards the use of ML in physics problems have been use-case specificity [24] and the lack of user-friendly tools [6]. As noted by Willard et al. [24], the efforts surrounding the use of ML for physics-specific problems are focused on sub-topics, or even use-cases. Although our methodology and synthetic data focuses on the domain of tracking for detector data, we could claim that it is independent of the chosen detector.

The point from [24] regarding the computational efficiency of ROMs matches our motivation. Where our work differs is in the placement of our ROM within our methodology. Our reduced model of a detector is considered as the model for simulations aimed at synthetic data, which is different than ML-based surrogate models as ROMs [7,17], or ML-based surrogate models built from ROMs [25].

7 Conclusion and Future Work

With many computational science applications exploring the use of ML-assisted solutions, there is a need for reduced complexity simulations to facilitate the design process. We show how such a reduction through ROMs and a smaller behavioural-space for the simulator, can result in a lower complexity for synthesised data. This is particularly relevant for our HEP use-case.

We have presented the design and implementation details of REDVID (REDuced VIrtual Detector), our simulation framework fulfilling such a reduction. To demonstrate REDVID's feasibility, we executed it with relevant workload recipes, and have made available the resulting data sets over Zenodo open repository. We further analysed the computational cost figures for these experiments, and we conclude that, even though our tool is developed in Python, computational cost figures (case in point, 15 s, 138 s and 22 min of CPU-time for 1 000 events with 10, 100 and 1 000 tracks per event, respectively) indicate efficiency for frequent executions. Accordingly, the lightweight nature of REDVID simulations makes our tool a suitable choice as a simulation-in-the-loop with data-driven workflows for HEP, e.g., searching for a ML-assisted solution to address the challenge of tracking.

We have explained that, to opt for such an approximation, is a deliberate act, positioning REDVID as a suitable middle ground amongst other available tools, not as exact as physics-accurate simulations, and not as synthetic as dummy data generators. The reduced complexity especially allows for early problem formulation and testing at early stages, when dealing with ML-assisted solution design workflows. Yet another advantage of reduced complexity data that still respects the high-level relations, is in its pedagogical merit, enabling problem solving practices in higher education.

Acknowledgement. This project is supported by the Nederlandse Organisatie voor Wetenschappelijk Onderzoek (NWO), a.k.a., the Dutch Research Council (grant no. 62004546).

References

1. Agostinelli, S., et al.: Geant4-a simulation toolkit. Nucl. Instrum. Methods Phys. Res., Sect. A **506**, 250–303 (2003)
2. Amrouche, S., et al.: The tracking machine learning challenge: accuracy phase. In: The NeurIPS '18 Competition (2020)
3. Apollinari, G., Brüning, O., Rossi, L.: High Luminosity LHC Project Description. Tech. rep, CERN (2014)
4. Bal, H., et al.: A medium-scale distributed system for computer science research: infrastructure for the long term. Computer (2016)
5. Böhlen, T., et al.: The FLUKA Code: developments and challenges for high energy and medical applications. Nuclear Data Sheets (2014)
6. Carleo, G., et al.: Machine learning and the physical sciences. Rev. Mod. Phys. **91**, 045002 (2019)
7. Chen, G., Zuo, Y., Sun, J., Li, Y.: Support-vector-machine-based reduced- order model for limit cycle oscillation prediction of nonlinear aeroelastic system. Math. Probl. Eng. (2012)
8. Chwif, L., Barretto, M., Paul, R.: On simulation model complexity. In: 2000 Winter Simulation Conference Proceedings (Cat. No.00CH37165) (2000)
9. Collaboration, T.A., et al.: The ALICE experiment at the CERN LHC. J. Instrum. (2008)
10. Collaboration, T.A., et al.: The ATLAS Experiment at the CERN Large Hadron Collider. J. Instrum. (2008)
11. Collaboration, T.C., et al.: The CMS experiment at the CERN LHC. J. Instrum. (2008)
12. Collaboration, T.L., et al.: The LHCb detector at the LHC. J. Instrum. (2008)
13. Corcella, G., et al.: HERWIG 6: an event generator for hadron emission reactions with interfering gluons (including supersymmetric processes). J. High Energy Phys. (2001)
14. de Melo, C.M., et al.: Next-generation deep learning based on simulators and synthetic data. Trends Cogn. Sci. (2022)
15. Edmonds, K., et al.: The Fast ATLAS Track Simulation (FATRAS). Tech. rep, CERN (2008)
16. Forster, R., et al.: MCNP^{TM} Version 5. Nucl. Instrum. Methods Phys. Res. Sect. B (2004)
17. Kasim, M.F., et al.: Building high accuracy emulators for scientific simulations with deep neural architecture search. Sci. Technol. Mach. Learn. (2021)
18. Odyurt, U.: REDVID collision event data - helical tracks and hits. Zenodo (2024). https://doi.org/10.5281/zenodo.10514246
19. Odyurt, U.: REDVID repository (2023). https://VirtualDetector.com/redvid/repository.html
20. Odyurt, U.: REDVID simulation framework (2023). https://VirtualDetector.com/redvid/
21. Odyurt, U., Swatman, S.N.: REDVID collision event data - linear tracks and hits. Zenodo (2023). https://doi.org/10.5281/zenodo.8322302
22. Richter-Was, E., Froidevaux, D., Poggioli, L.: ATLFAST 2.0 a fast simulation package for ATLAS. Tech. rep., CERN (1998)
23. Sjöstrand, T., Mrenna, S., Skands, P.: PYTHIA 6.4 physics and manual. J. High Energy Phys. (2006)

24. Willard, J., Jia, X., Xu, S., Steinbach, M., Kumar, V.: Integrating scientific knowledge with machine learning for engineering and environmental systems. ACM Comput. Surv. (2022)
25. Xiao, D., et al.: A reduced order model for turbulent flows in the urban environment using machine learning. Build. Environ. (2019)

Modeling 3D Surfaces with a Locally Conditioned Atlas

Przemysław Spurek[1,4], Sebastian Winczowski[1], Maciej Zięba[2],
Tomasz Trzciński[3,4], Kacper Kania[3], and Marcin Mazur[1(✉)]

[1] Faculty of Mathematics and Computer Science, Jagiellonian University,
Krakow, Poland
{przemyslaw.spurek,marcin.mazur}@uj.edu.pl
[2] Wroclaw University of Science and Technology, Wroclaw, Poland
[3] Warsaw University of Technology, Warsaw, Poland
[4] IDEAS NCBR, Warsaw, Poland

Abstract. Recently proposed methods for reconstructing 3D objects use a mesh with an atlas consisting of planar patches that approximate the object's surface. However, in real-world scenarios, the surfaces of reconstructed objects exhibit discontinuities that degrade the mesh's quality. Therefore, conducting additional research on methods to overcome discontinuities and improve mesh quality is always advantageous. This paper proposes to address the limitation by maintaining local consistency around patch vertices. We present LoCondA, a Locally Conditioned Atlas that represents a 3D object hierarchically as a generative model. The model initially maps the point cloud of an object onto a sphere and subsequently enforces the mapping to be locally consistent on both the sphere and the target object through the use of a spherical prior. Using this method, the mesh can be sampled on the sphere and then projected back onto the manifold of the object, yielding diverse topologies that can be seamlessly connected. The experimental results demonstrate that this approach produces structurally coherent reconstructions with meshes of comparable quality to those of competitors.

Keywords: Mesh · Point cloud · Atlas · Hypernetwork

1 Introduction

Efficient 3D object representations are crucial for various computer vision and machine learning applications, from robotic manipulation [17] to autonomous driving [32]. Modern 3D registration devices, such as LIDARs and depth cameras, produce representations in the form of sparsely sampled, unordered sets of 3D points on objects' surfaces, which are known as *point clouds* [24,25]. While a single point cloud may offer surface details for object reconstruction [12], it does not provide enough information about the neighborhood structure of 3D points. This lack of information makes it difficult to reconstruct a smooth, high-fidelity manifold for the entire object's surface.

L. Franco et al. (Eds.): ICCS 2024, LNCS 14833, pp. 100–115, 2024.
https://doi.org/10.1007/978-3-031-63751-3_7

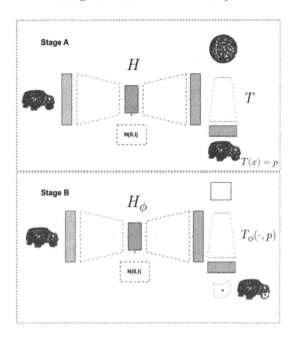

Fig. 1. Our approach extends the current hypermodels [27,28] (Stage A) by adding a mesh generation module (Stage B). First, we generate a seamless representation of the object by training a hypernetwork H that produces weights for a target network T that transforms a prior (usually uniform) distribution on the 2D unit sphere to the object surface S. This results in the ability to reach any point of S as a value of T. Next, we train another hypernetwork H_ϕ designed to produce weights for a target network T_ϕ conditioned on a given point $p = T(x) \in S$, which maps the 2D unit square to the patch on S, aligned with the neighborhood of p.

Recently, researchers have proposed using polygonal meshes to model object surfaces for improved accuracy [14,27,36]. A mesh is a collection of vertices connected by edges to form triangles, which create the surface of an object. This representation is efficient and easily rendered, while also providing additional advantages such as the ability to sample the surface at a desired resolution and apply texturing in any 3D computer graphics software. To obtain such a representation, it is necessary to use advanced methodologies that utilize deep learning models based on an autoencoder architecture [27,28,31], or utilize an ensemble of parametric mappings from 2D rectangular patches to 3D primitives, commonly known as an *atlas* [14,36]. Previous methods were limited by the topology of the autoencoder's latent space distribution, which made them unable to model intricate structures with a non-spherical topology [27,28,31]. In contrast, atlas-based approaches offer a greater flexibility and can model virtually any surface. However, these methods often result in discontinuities in the reconstructed shapes and their deformation due to the inconsistency of individual mappings.

Although [6,11] propose adjustments that improve the quality of the outcome, their aim is to correct distortions resulting from the combination of different mappings. It is postulated that the continuous nature of neural architectures may be leveraged to prevent the formation of distortions by enforcing local consistency among patch vertices within a model's objective function. Therefore, we present a new method called *Locally Conditioned Atlas* (LoCondA) that can generate and reconstruct high-quality 3D meshes. This method enhances the existing base hypermodels [27,28] by introducing a new module for mesh generation that relies on local surface parametrization, as shown in Fig. 1. It is based on the concept of the *continuous atlas*, which provides a new approach that extends existing atlas methods and allows for adaptive sampling of any number of patches to cover any part of the reconstructed object. Specifically, we transform the point cloud embedding obtained from the base model to parameterize the objective function represented by the multilayer perceptron (MLP) network[1]. This function aims to map a canonical 2D patch to the 3D patch on the surface of the target object. The positioning and shape of a 3D patch are conditioned using a single point from a point cloud generated by the base model. The process is repeated for each generated point while maintaining the local neighborhoods between the point cloud and the points located in the generated mesh. This procedure allows us to include the patch's stitching and reshaping in our framework's training objective, reducing the likelihood of shape discontinuities.

We summarize our contribution as follows:

- We propose a comprehensive framework for patch-based reconstruction methods that generate high-fidelity meshes from raw point clouds.
- We present the continuous atlas, a new paradigm that expands on existing atlas methods and enables adaptive sampling of any number of patches to cover any part of the reconstructed object.
- We introduce the Locally Conditioned Atlas (LoCondA), a new method for conditioning atlas-based approaches that effectively shares information between patches and resolves issues related to self-intersections and holes in reconstructed meshes.

2   Related Work

The literature presents a wide range of 3D shape reconstruction models, such as dense pixel-wise depth maps [4,5], normal maps [30], point clouds [12,25,36], meshes [4,31], implicit functions [8,21], voxels [9,16], shape primitives [7,10], parametric mappings [6,14], or combinations thereof [22,23]. All of the aforementioned representations have advantages and disadvantages depending on their memory requirements and accuracy in fitting the surface.

Our paper focuses on one of the most popular representations, which is based on polygonal meshes. A mesh is a set of vertices connected by edges that allows

---

[1] This is a neural network in which every node is connected to each node of the subsequent layer.

for a piecewise planar approximation of a surface. The object's mesh can be then accessed by transforming it onto a unit sphere [27,28,31]. However, this method is limited to reconstructing objects that are topologically similar to spheres.

Patch-based approaches, proposed by [6,11], offer a greater flexibility and can model virtually any surface, even those with non-disk topology. To achieve this, parametric mappings are used to transform 2D patches into 3D shapes, also known as 2D manifolds. An example of this is FoldingNet [36], which utilizes a single patch to model an object's surface.

AtlasNet [14] is a method that simultaneously trains a number of functions, which together form an atlas, to obtain multiple patches that model a mesh. Each function transforms a unit square into a neighborhood of a point on the surface. The atlas elements are trained independently, resulting in maps that are not stitched together. This can cause discontinuities, such as holes or intersecting patches (see, e.g., Fig. 3 in [14]).

To address the aforementioned problem, many methods extend the Chamfer loss function of the basic AtlasNet by adding extra terms. In [6], the authors included terms to prevent patch collapse, reduce patch overlap, and analytically calculate the exact surface properties instead of approximating them. In [11], two new terms were introduced to improve the overall consistency of the local mappings. One of these terms uses surface normals to ensure local consistency in estimation within and across individual mappings. The other term minimizes stitching errors to enforce better spatial configuration of the mappings. Although these modifications improve the quality of the obtained results, their objective is to correct deformations after patch stitching. In this paper, we propose a different approach to solve this problem by reformulating the classical definition of an atlas to obtain maps that are correctly connected. Therefore, our method aims to prevent the issue from occurring in the first place.

Given a data set $\mathcal{X}$ containing point clouds $X_1, \ldots, X_n$, an autoencoder aims to transport the data through a latent space $\mathcal{Z} \subset \mathbb{R}^D$ while minimizing the reconstruction error. This is accomplished by identifying an encoder $\mathcal{E} \colon \mathcal{X} \to \mathcal{Z}$ and a decoder $\mathcal{D} \colon \mathcal{Z} \to \mathcal{X}$ that minimize the reconstruction error between each X_i and the reconstruction $\mathcal{D}(\mathcal{E}(X_i))$. In the generative framework, we also ensure that the data sent to the latent space is drawn from a predetermined prior distribution [18,19]. To effectively represent point clouds, it is essential to define an appropriate reconstruction loss. In the literature, two distance measures are commonly used: Earth Mover (Wasserstein) Distance [26] and Chamfer Distance [29]. Earth Mover Distance (EMD) is a metric between two distributions based on the minimum cost to transform one distribution into the other. Specifically, for two equally sized sets $X, Y \subset \mathbb{R}^3$, EMD is defined as $\mathrm{EMD}(X, Y) = \min_{\phi: X \to Y} \sum_{x \in X} c(x, \phi(x))$, where ϕ is a bijection and $c(x, \phi(x)) = \frac{1}{2}\|x - \phi(x)\|_2^2$. On the other hand, Chamfer Distance (CD) measures the squared distance between each point in one set and the nearest neighbor in the other set, i.e., $\mathrm{CD}(X, Y) = \sum_{x \in X} \min_{y \in Y} \|x - y\|_2^2 + \sum_{x \in Y} \min_{y \in X} \|x - y\|_2^2$.

Point clouds may contain a variable number of data points that correspond to a single object and are registered at various angles. Therefore, methods that process them must be permutation and rotation invariant. The PointNet [24] framework enables the processing of 3D point clouds of varying sizes as input for neural networks. However, the output size remains a challenge. Hypernetworks [15] are a potential solution as they are neural models that generate weights for a separate target network capable of solving a specific task. Instead of producing a fixed-size 3D point cloud, a hypernetwork creates a target network to parametrize the object's surface and generate any desired number of points. The parameters of the target network are not directly optimized; only the weights of the hypernetwork are optimized during the training procedure.

3   Local Parametrization of a Surface

This section introduces the continuous atlas, a new approach for creating meshes from patches. It also discusses the limitations of current methods based on discrete atlas representations and presents our model as a solution to these limitations.

A set $S \subset \mathbb{R}^3$ is defined as a 2-manifold[2] (also called a surface) if for every point $p \in S$ there exists an open set U in $\mathbb{R}^2$ and an open set V in $\mathbb{R}^3$ containing p such that $S \cap V$ is homeomorphic to U. A corresponding homeomorphism is referred to as a *chart*. An atlas is a collection of charts that cover the 2-manifold.

By a *discrete atlas* of a 3D object surface S supported by a set of points $P = \{p_1, \ldots, p_k\} \subset S$ we mean a collection of charts $\{\phi_1, \ldots, \phi_k\}$, such that each ϕ_i maps the open square $U = (0,1) \times (0,1)$ onto a neighborhood $V(p_i)$ of a point p_i and the following condition is satisfied: $\bigcup_{i=1}^{k} \phi_i(U) = \bigcup_{i=1}^{k} V(p_i) = S$. In practice, the charts are represented by MLP networks $\phi_1(\cdot; \theta_1), \ldots, \phi_k(\cdot; \theta_k)$ trained together to minimize the global reconstruction error $\mathcal{L}\left(\bigcup_{i=1}^{k} \phi_i(U; \theta_i), S\right)$, where $\mathcal{L}$ is either the Chamfer Distance or the Earth Mover Distance. Using such a formulation, the charts are trained to induce the *patches* $\phi_1(U; \theta_1), \ldots, \phi_k(U; \theta_k)$, which together create a mesh that approximates the target object surface $\mathcal{S}$ as closely as possible. In theory, the method should produce a single seamless mesh. In practice, however, the result obtained is not ideal and requires additional post-processing to be used in real-world applications. This is because it works with a given number of charts (each generating a single patch) and does not take into account the stitching process itself, as no information is shared between patches. If a patch does not correctly cover the underlying neighborhood of a point, no other patch fixes that part, resulting in empty spaces on the object's surface. To mitigate the issues of the discrete atlas, we propose an approach that leverages the local structure of 3D objects [3,34,35,37].

By a *continuous atlas* of a 3D object surface S we define a continuous mapping $\phi \colon U \times S \to S$ which transforms the open square $U = (0,1) \times (0,1)$ and a point $p \in S$ onto a neighborhood $V(p) \subset S$ of the point p. Note that, unlike a discrete atlas, instead of using a finite set of charts, we use only a transformation ϕ

---

[2] We adhere to the concept presented in [14] (see also [6,11]).

that locally models the surface of the object, leading to a potentially infinite number of charts $\phi(\cdot, p)$. In contrast to a traditional conditioning mechanism in the AtlasNet, p is not a global descriptor but a direct point of S. In consequence, we can produce an arbitrary number of patches $\phi(U, p)$, which may be located in any place on the object's surface.

The proposed approach overcomes the limitations of previous discrete methods. Specifically, we theoretically solve the problem of stitching partial meshes since every chart is informed about the local neighborhood. Moreover, using a continuous atlas approach we can easily fill the missing spaces in the final mesh by adding a new mapping for the region of interest. Since we can create an arbitrary number of patches, we can locate a point in the empty space neighborhood and create an additional patch using ϕ function conditioned on the selected point.

Fig. 2. Sample airplane patches created using our method, with structurally similar parts placed next to each other to create smooth surfaces.

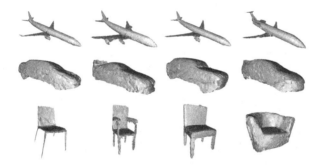

Fig. 3. Sample mesh representations generated by our method. Note that our model is capable of reconstructing complex shapes such as chair legs and backs.

We present in Fig. 2 that the resulting model can stitch patches on a sample airplane object. Note that the obtained surface is smooth due to the continuity of a continuous atlas. Additionally, Fig. 3 displays complete sample mesh representations produced by our method. In contrast, Fig. 4 shows mesh representations generated using varying numbers of patches.

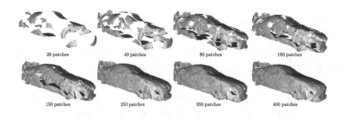

Fig. 4. Mesh representations with a different number of patches that were created by our method. Note that we need at least 200 patches to get high-quality reconstructions.

4   Local Approximation of a Surface

This section introduces the Locally Conditioned Atlas (LoCondA), a two-stage framework for generating and reconstructing meshes of object surfaces using the point cloud representation and the introduced notion of the continuous atlas (see Fig. 1). Both stages utilize a hypernetwork paradigm, which has an advantage over autoencoder architectures in processing inputs of varying sizes. In our approach, each point cloud is parametrized individually, unlike traditional approaches where conditioning parameters are shared among different point clouds. This can be beneficial for retrieval purposes.

Stage A: Generative autoencoder using hypernetworks As we operate directly on points located on the surfaces of 3D objects, we utilize two state-of-the-art generative autoencoder models based on hypernetworks, namely HyperCloud [27] and HyperFlow [28], instead of the traditional generative autoencoder architecture. These solutions use hypernetworks to generate weights of small neural networks that map a known prior distribution of points onto 3D objects. Thus, after training, we can implicitly sample any point $p \in S$ without accessing the actual target object surface S.

In our framework, we use a hypernetwork $H(\cdot; W_H)$ which, for a given point cloud $X \subset S$, returns the weights of the corresponding target network $T(\cdot; W_T)$. Thus X is represented by the function $T(\cdot; W_T) = T(\cdot; H(X; W_H))$. As T we use a classical MLP in the case of the HyperCloud implementation, and a Continuous Normalizing Flow (CNF) [13] in the case of the HyperFlow implementation. The hypernetwork H is pre-trained beforehand[3] using the following procedure: we take an input point cloud X and pass it to H to obtain the weights W_T of the target network T, which reconstructs X from a uniform noise $X_{S^2} \subset \mathbb{R}^3$ on the 2D unit sphere S^2, and then we minimize the reconstruction loss, expressed either by the Chamfer Distance or the Earth Mover Distance (as far as the HyperCloud architecture is considered), or by the negative log-likelihood (as far as the HyperFlow architecture is considered).

---

[3] So it does not contribute to the total training time.

Stage B: Locally Conditioned Atlas (LoCondA) Our LoCondA model imple-
ments the introduced approach of *continuous atlas*. Specifically, we represent
the underlying mapping ϕ with a neural architecture consisting of the following
parts: (1) a (small) target network $T_\phi(\cdot; W_{T_\phi})$ that transforms a sample X_U from
the uniform distribution on the square $U = (0,1) \times (0,1)$ and a point p from
the (reconstructed) point cloud $X \subset S$ into the object surface S, (2) a hypernetwork $H_\phi(\cdot; W_{H_\phi})$ that generates weights W_{T_ϕ} conditioned on a given point p
to ensure that the image of T_ϕ forms its neighborhood $V(p) \subset X$. Note that due
to a conditioning mechanism, we do not need to model each of the charts $\phi(\cdot, p)$
separately, which significantly reduces the number of parameters.

Training the LoCondA model can be done by minimizing the following loss
function: $\mathcal{L}\left(T_\phi(X_U, p; H_\phi(X, p; W_{H_\phi})), V(p)\right)$, where $\mathcal{L}$ is either the Chamfer
Distance or the Earth Mover Distance, $p = T(x; W_T)$ is a random point from
the reconstructed point cloud X (see Stage A), and $V(p)$ is taken as the set of
k elements closest to p in X (here k is treated as a hyperparameter). This leads
to the point cloud (the image of T_ϕ) that reconstructs $V(p)$, which can be easily
transformed into a mesh by connecting the points with edges, as in a natural
mesh built on X_U. The problem however is that such a formulation causes many
of the generated patches to have unnecessarily long edges, and the network folds
them so that the patch fits the surface of an object. To mitigate this problem,
we add an edge length regularization motivated by [31]. Specifically, we include
a regularization term of the form: $l_{loc} = \sum_e \|e\|_2^2$, where $\|e\|_2^2$ is the squared
norm of the length of an edge e. So we get the total loss $\mathcal{L} + \lambda l_{loc}$, where λ is a
hyperparameter of the model.

5   Experiments

This section presents the experimental results of the proposed method. Firstly,
we evaluate the model's generative capabilities. Secondly, we provide the reconstruction results in comparison to reference approaches. Finally, we compare the
quality of the generated meshes to baseline methods. The models were trained
using the Chamfer Distance as $\mathcal{L}$ and λ was set to 0.0001. LoCondA-HC and
LoCondA-HF are acronyms used to refer to different autoencoder architectures
(see Stage A). LoCondA-HC uses HyperCloud, while LoCondA-HF uses Hyper-
Flow. A grid search was employed to optimize the hyperparameters of all models.
The source code is available at https://github.com/gmum/LoCondA.

It is important to note that our method has a significant advantage in its
ability to generate an unlimited number of patches. To estimate the minimum
number of patches required, divide the number of points in a point cloud representation of the entire object by the number of points in one patch used in
training (i.e., hyperparameter k). From a practical standpoint, there is a trade-
off between reconstruction quality and time performance. If we prioritize quality,
we should produce as many patches as possible. However, if we prioritize time,
we should decrease the number of patches. In our experiments, we used 400
patches per object.

Generative capabilities. In this experiment, the generative capabilities[4] of the LoCondA model are compared to existing reference approaches. The evaluation protocol provided in [33] is followed, and standard measures such as the Jensen-Shannon Divergence (JSD), the Coverage (COV), and the Minimum Matching Distance (MMD) are used. The last two measures are calculated separately for the Chamfer Distance (CD) and the Earth Mover Distance (EMD). The study compares the results of various existing solutions for point cloud generation, including l-GAN [1], PC-GAN [20], PointFlow [33], HyperCloud(P) [27] and HyperFlow(P) [28]. Additionally, two baselines, HyperCloud(M) and Hyper-Flow(M), capable of generating meshes from the unit sphere, are considered in the experiment. Each model is trained using point clouds from one out of the three categories in the ShapeNet dataset: *airplane*, *chair*, and *car*.

Table 1. (Generative capabilities) Metric scores for different generative models performed on the ShapeNet dataset according to the evaluation protocol described in [33]. To simplify the notation, MMD-CD (lower is better) scores are multiplied by 10^3, MMD-EMD (lower is better) and JSD (lower is better) scores are multiplied by 10^2, and COV (greater is better) scores are expressed on a percentage scale. HC refers to the use of the HyperCloud autoencoder in LoCondA (our), and HF refers to the use of the HyperFlow autoencoder. For the HyperCloud and HyperFlow models, we use both variants that generate point clouds (P) or meshes (M). The most successful outcomes are indicated in bold.

Method	Airplane JSD	MMD CD	MMD EMD	COV CD	COV EMD	Chair JSD	MMD CD	MMD EMD	COV CD	COV EMD	Car JSD	MMD CD	MMD EMD	COV CD	COV EMD
Point cloud generation															
l-GAN	**3.61**	0.26	3.29	**47.90**	50.62	2.27	2.61	7.85	40.79	41.69	2.21	1.48	5.43	39.20	39.77
PC-GAN	4.63	0.28	3.57	36.46	40.94	3.90	2.75	8.20	36.50	38.98	5.85	1.12	5.83	23.56	30.29
PointFlow	4.92	**0.21**	3.24	46.91	48.40	1.74	2.42	7.87	**46.83**	**46.98**	**0.87**	**0.91**	5.22	44.03	46.59
HyperCloud(P)	4.84	0.26	3.28	39.75	43.70	2.73	2.56	**7.84**	41.54	46.67	3.09	1.07	5.38	40.05	40.05
HyperFlow(P)	5.39	0.22	**3.16**	46.66	**51.60**	**1.50**	2.30	8.01	44.71	46.37	1.07	1.14	5.30	45.74	47.44
Mesh generation															
HyperCloud(M)	9.51	0.45	5.29	30.60	28.88	4.32	2.81	9.32	40.33	40.63	5.20	1.11	6.54	37.21	28.40
HyperFlow(M)	6.55	0.38	3.65	40.49	48.64	4.26	3.33	8.27	41.99	45.32	5.77	1.39	5.91	28.40	37.21
LoCondA-HC	16.1	0.66	4.71	30.37	32.59	4.45	3.03	8.55	42.45	38.22	1.91	1.13	5.50	**53.69**	**50.56**
LoCondA-HF	4.80	0.22	3.20	44.69	47.91	2.54	**2.23**	7.94	43.35	42.60	1.16	0.92	**5.21**	44.88	47.72

Table 1 presents the results. LoCondA-HF achieves results comparable to those of the reference methods dedicated to point cloud generation. The evaluated measures for HyperFlow(P) and LoCondA-HF (which uses HyperFlow(P) as a base model in Stage A) are on the same level. Incorporating an additional module dedicated to mesh generation (Stage B) does not have a negative impact on our model's generative capabilities. However, when we use HyperFlow to directly generate meshes according to the procedure described in [28] (see the

---

[4] These refer to the ability of the model to produce fake samples that are indistinguishable from real data.

results for HyperFlow(M)), the generative capabilities are significantly inferior for the evaluated metrics.

Reconstruction capabilities The goal of this experiment is to evaluate the ability of our model to encode different shapes, measured by comparing the original objects with their reconstructions. The autoencoding task was conducted on 3D point clouds from three categories in the ShapeNet dataset, namely *airplane*, *chair*, and *car*. We compare LoCondA with AtlasNet [14] (the current state-of-the-art), where the prior shape is either a sphere or a set of patches. Additionally, we compare it with l-GAN [2] and PointFlow [33]. The experimental setup of PointFlow was followed.

Table 2. (Reconstruction capabilities) Metric scores for shape reconstruction by different models obtained on the ShapeNet dataset according to the evaluation protocol described in [14]. To simplify the notation, the scores are multiplied by 10, CD (lower is better) is multiplied by 10^4, and EMD (lower is better) is multiplied by 10^2. In LoCondA (our), HC refers to the use of the HyperCloud autoencoder, while HF refers to the use of the HyperFlow autoencoder. The most successful outcomes are indicated in bold.

Dataset	Metric	l-GAN		AtlasNet		PointFlow	LoCondA-HC	LoCondA-HF	Oracle
		CD	EMD	Sphere	Patches				
Airplane	CD	1.020	1.196	1.002	**0.969**	1.208	1.135	1.513	0.837
	EMD	4.089	**2.577**	2.672	2.612	2.757	2.881	2.990	2.062
Chair	CD	9.279	11.21	**6.564**	6.693	10.120	10.382	12.519	3.201
	EMD	8.235	6.053	5.790	**5.509**	6.434	6.738	6.973	3.297
Car	CD	5.802	6.486	**5.392**	5.441	6.531	6.575	7.247	3.904
	EMD	5.790	4.780	4.587	**4.570**	5.138	5.126	5.275	3.251
ShapeNet	CD	7.120	8.850	5.301	**5.121**	7.551	7.781	–	3.031
	EMD	7.950	5.260	5.553	5.493	**5.176**	5.826	–	3.103

The results are presented in Table 2, including the performance in both the Chamfer Distance (CD) and the Earth Mover Distance (EMD). It is worth noting that these two metrics depend on the point clouds' scale. The 'oracle' column shows the upper bound, which is the error between two different point clouds with the same number of points sampled from the same ground truth meshes. The results show that LoCondA-HC achieves competitive results compared to reference solutions. All competitors were trained using an autoencoding framework. However, LoCondA-HC and LoCondA-HF were also able to preserve generative capabilities during the experiment.

Figure 5 presents a comparison between AtlasNet and LoCondA-HC on the reconstruction task. The quality of the meshes is similar, but our method better describes objects that contain holes, such as empty spaces in the back of chairs. On the other hand, reconstruction results obtained by decoding the linearly

LoCondA-HC

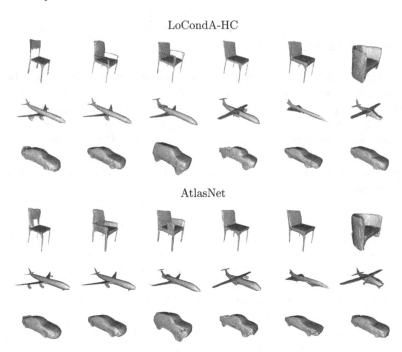

AtlasNet

Fig. 5. AtlasNet and LoCondA-HC (our) comparison for reconstruction task. Note that both meshes have similar quality, but our method provides a more accurate description of objects that have holes, such as empty spaces in the back of chairs.

interpolated latent vectors of two objects from each class are shown in Fig. 6. Note that LoCondA-HC generates coherent and semantically plausible objects for all interpolation steps. However, during the middle stage of the interpolation, we encountered patches that were not properly stitched together.

Mesh quality evaluation Empirical evidence demonstrates that the proposed framework generates high-quality and seamless meshes, solving the initial problem of disjoint patches occurring in the AtlasNet model. To evaluate the continuity of the output surfaces, we recommend using the measure described below.

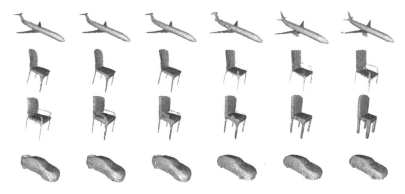

Fig. 6. Mesh interpolations generated by LoCondA-HC (our). Note that for all interpolation steps, our method produces coherent and semantically plausible objects. However, in the middle stage of the interpolation, we ran into patches that were not stitched together in the correct way.

A mesh is typically described either as watertight or not watertight, without a clear measure to define the degree of discontinuities in the object's surface. To address this issue, we propose the parity test, which provides an approximate check of whether a mesh is watertight. According to the test, for a ray cast from infinity towards the object, it must enter and leave the object. This is determined by checking the number of crossings the ray makes with all triangles in the mesh, which should be an even number for the ray to pass the parity test. The mesh is considered watertight if all rays pass the parity test.

To apply this knowledge, we express the measure of watertightness as the ratio of rays that passed the parity test to the total number of casted rays. We begin by sampling N points $p \in \hat{S}$ from all triangles of the reconstructed object $\hat{S}$. Since each point is associated with the triangle it was sampled from, we use the corresponding normal $\hat{n}$ and negate it to obtain the direction of a ray $R(\hat{S}) \ni r = -\hat{n}p$ towards the object[5]. The calculation of the number of crossings $c(r)$ with all triangles is performed. A value of 1 is assigned to each ray that passes a test, and 0 is assigned otherwise. The measure of watertightness is determined by summing the test results for all rays and dividing by the total number of rays: $\mathrm{WT}(\hat{S}) = \frac{\sum_{r \in R(\hat{S})} \mathbb{I}_{\{c(r) \bmod 2=0\}}}{|R(\hat{S})|}$, where $\mathbb{I}$ denotes a characteristic function.

Experiments were conducted to compare AtlasNet with LoCondA-HC and LoCondA-HF in terms of the watertightness of generated meshes. The results are presented in Table 3. Note that AtlasNet is unable to generate watertight meshes for any of the considered classes (*airplane*, *chair*, and *car*), which limits its applicability. On the other hand, LoCondA creates meshes in which all sampled rays pass the test.

---

[5] It is important to note that if a random direction of sampling were employed, confusion would arise due to the possible presence of tangent rays.

Table 3. Comparison of the LoCondA (our) and AtlasNet approaches in terms of watertightness (greater is better). Note that all meshes produced by our method are watertight.

Method	Airplane	Chair	Car	Average
AtlasNet (25 patches)	0.516	0.507	0.499	0.507
LoCondA-HC	**1.00**	**1.00**	**1.00**	**1.00**
LoCondA-HF	**1.00**	**1.00**	**1.00**	**1.00**

6   Conclusions

This paper introduces the Locally Conditioned Atlas (LoCondA), a new method for generating high-quality 3D meshes composed of 2D patches directly from raw point clouds. The proposed approach is based on the continuous atlas paradigm, which allows our model to produce an arbitrary number of patches to form a watertight mesh.

The framework presented here addresses the limitations of previous methods by resolving the challenge of stitching partial meshes. This is achieved by ensuring that each chart is aware of the local neighborhood. Furthermore, our approach can effectively fill in any missing regions in the final mesh by creating a new mapping for the relevant area. The empirical evaluation of LoCondA in several extensive experiments confirms its validity and competitive performance.

The main limitation of the proposed method is the use of a two-stage training procedure. In future work, we plan to extend this approach to an end-to-end training algorithm. Furthermore, we consider the possibility of working with colored meshes.

Acknowledgments. This work was supported by the National Centre of Science (Poland), grants no. 2021/43/B/ST6/01456 (P. Spurek), 2020/39/B/ST6/01511 (T. Trzciński), 2022/45/B/ST6/02817 (T. Trzciński), 2020/37/B/ST6/03463 (M. Zięba), and 2021/41/B/ST6/01370 (M. Mazur). Some experiments were performed on servers purchased with funds from the flagship project entitled "Artificial Intelligence Computing Center Core Facility" from the DigiWorld Priority Research Area within the Excellence Initiative - Research University program at Jagiellonian University in Krakow. This paper has been supported by the Horizon Europe Programme (HORIZON-CL4-2022-HUMAN-02) under the project "ELIAS: European Lighthouse of AI for Sustainability", GA no. 101120237.

References

1. Achlioptas, P., Diamanti, O., Mitliagkas, I., Guibas, L.: Learning representations and generative models for 3D point clouds. In: Proceedings of the 35th International Conference on Machine Learning. Proceedings of Machine Learning Research, vol. 80, pp. 40–49. PMLR (2018)
2. Achlioptas, P., Diamanti, O., Mitliagkas, I., Guibas, L.: Learning representations and generative models for 3d point clouds. In: International Conference on Machine Learning, pp. 40–49. PMLR (2018)
3. Ao, S., Guo, Y., Tian, J., Tian, Y., Li, D.: A repeatable and robust local reference frame for 3d surface matching. Pattern Recogn. **100**, 107186 (2020)
4. Bansal, A., Russell, B., Gupta, A.: Marr revisited: 2D-3D alignment via surface normal prediction. In: Proceedings of the IEEE Conference on Computer Vision and Pattern Recognition, pp. 5965–5974 (2016)
5. Bednarik, J., Fua, P., Salzmann, M.: Learning to reconstruct texture-less deformable surfaces from a single view. In: 2018 International Conference on 3D Vision (3DV), pp. 606–615. IEEE (2018)
6. Bednarik, J., Parashar, S., Gundogdu, E., Salzmann, M., Fua, P.: Shape reconstruction by learning differentiable surface representations. In: Proceedings of the IEEE/CVF Conference on Computer Vision and Pattern Recognition, pp. 4716–4725 (2020)
7. Chen, Z., Tagliasacchi, A., Zhang, H.: BSP-Net: generating compact meshes via binary space partitioning. In: Proceedings of the IEEE/CVF Conference on Computer Vision and Pattern Recognition, pp. 45–54 (2020)
8. Chen, Z., Zhang, H.: Learning implicit fields for generative shape modeling. In: Proceedings of the IEEE/CVF Conference on Computer Vision and Pattern Recognition, pp. 5939–5948 (2019)
9. Choy, C.B., Xu, D., Gwak, J.Y., Chen, K., Savarese, S.: 3D-R2N2: a unified approach for single and multi-view 3D object reconstruction. In: Leibe, B., Matas, J., Sebe, N., Welling, M. (eds.) ECCV 2016. LNCS, vol. 9912, pp. 628–644. Springer, Cham (2016). https://doi.org/10.1007/978-3-319-46484-8_38
10. Deng, B., Genova, K., Yazdani, S., Bouaziz, S., Hinton, G., Tagliasacchi, A.: CvxNet: learnable convex decomposition. In: Proceedings of the IEEE/CVF Conference on Computer Vision and Pattern Recognition, pp. 31–44 (2020)
11. Deng, Z., Bednařík, J., Salzmann, M., Fua, P.: Better patch stitching for parametric surface reconstruction. arXiv preprint arXiv:2010.07021 (2020)
12. Fan, H., Su, H., Guibas, L.J.: A point set generation network for 3D object reconstruction from a single image. In: Proceedings of the IEEE Conference on Computer Vision and Pattern Recognition, pp. 605–613 (2017)
13. Grathwohl, W., Chen, R.T.Q., Bettencourt, J., Sutskever, I., Duvenaud, D.: FFJORD: free-form continuous dynamics for scalable reversible generative models. In: International Conference on Learning Representations (2019)
14. Groueix, T., Fisher, M., Kim, V.G., Russell, B.C., Aubry, M.: A papier-mâché approach to learning 3D surface generation. In: Proceedings of the IEEE Conference on Computer Vision and Pattern Recognition, pp. 216–224 (2018)
15. Ha, D., Dai, A., Le, Q.V.: Hypernetworks. In: International Conference on Learning Representations (2017)
16. Häne, C., Tulsiani, S., Malik, J.: Hierarchical surface prediction for 3D object reconstruction. In: 2017 International Conference on 3D Vision (3DV), pp. 412–420. IEEE (2017)

17. Kehoe, B., Patil, S., Abbeel, P., Goldberg, K.: A survey of research on cloud robotics and automation. IEEE Trans. Autom. Sci. Eng. **12**(2), 398–409 (2015)
18. Kingma, D.P., Welling, M.: Auto-encoding variational Bayes. In: International Conference on Learning Representations (2014)
19. Knop, S., Spurek, P., Tabor, J., Podolak, I., Mazur, M., Jastrzebski, S.: Cramer-wold auto-encoder. J. Mach. Learn. Res. **21** (2020)
20. Li, C.L., Zaheer, M., Zhang, Y., Poczos, B., Salakhutdinov, R.: Point cloud GAN. In: International Conference on Learning Representations (Workshop Track) (2019)
21. Mescheder, L., Oechsle, M., Niemeyer, M., Nowozin, S., Geiger, A.: Occupancy networks: learning 3D reconstruction in function space. In: Proceedings of the IEEE/CVF Conference on Computer Vision and Pattern Recognition, pp. 4460–4470 (2019)
22. Muralikrishnan, S., Kim, V.G., Fisher, M., Chaudhuri, S.: Shape Unicode: a unified shape representation. In: Proceedings of the IEEE/CVF Conference on Computer Vision and Pattern Recognition, pp. 3790–3799 (2019)
23. Poursaeed, O., Fisher, M., Aigerman, N., Kim, V.G.: Coupling explicit and implicit surface representations for generative 3D modeling. In: Vedaldi, A., Bischof, H., Brox, T., Frahm, J.-M. (eds.) ECCV 2020. LNCS, vol. 12355, pp. 667–683. Springer, Cham (2020). https://doi.org/10.1007/978-3-030-58607-2_39
24. Qi, C.R., Su, H., Mo, K., Guibas, L.J.: PointNet: deep learning on point sets for 3D classification and segmentation. In: Proceedings of the IEEE Conference on Computer Vision and Pattern Recognition, pp. 652–660 (2017)
25. Qi, C.R., Yi, L., Su, H., Guibas, L.J.: PointNet++: deep hierarchical feature learning on point sets in a metric space. In: Advances in Neural Information Processing Systems, pp. 5099–5108 (2017)
26. Rubner, Y., Tomasi, C., Guibas, L.J.: The earth mover's distance as a metric for image retrieval. Int. J. Comput. Vis. **40**(2), 99–121 (2000)
27. Spurek, P., Winczowski, S., Tabor, J., Zamorski, M., Zieba, M., Trzciński, T.: Hypernetwork approach to generating point clouds. Proc. Mach. Learn. Res. **119** (2020)
28. Spurek, P., Zieba, M., Tabor, J., Trzcinski, T.: General hypernetwork framework for creating 3D point clouds. IEEE Trans. Pattern Anal. Mach. Intell. (2021)
29. Tran, M.P.: 3D contour closing: a local operator based on chamfer distance transformation (2013)
30. Tsoli, A., Argyros, A., et al.: Patch-based reconstruction of a textureless deformable 3D surface from a single RGB image. In: Proceedings of the IEEE/CVF International Conference on Computer Vision Workshops (2019)
31. Wang, N., Zhang, Y., Li, Z., Fu, Y., Liu, W., Jiang, Y.G.: Pixel2Mesh: generating 3D mesh models from single RGB images. In: Proceedings of the European Conference on Computer Vision (ECCV), pp. 52–67 (2018)
32. Yang, B., Luo, W., Urtasun, R.: PIXOR: real-time 3D object detection from point clouds. In: Proceedings of the IEEE Conference on Computer Vision and Pattern Recognition, pp. 7652–7660 (2018)
33. Yang, G., Huang, X., Hao, Z., Liu, M.Y., Belongie, S., Hariharan, B.: PointFlow: 3D point cloud generation with continuous normalizing flows. In: Proceedings of the IEEE International Conference on Computer Vision, pp. 4541–4550 (2019)
34. Yang, J., Zhang, Q., Cao, Z.: The effect of spatial information characterization on 3D local feature descriptors: a quantitative evaluation. Pattern Recogn. **66**, 375–391 (2017)
35. Yang, J., Zhang, Q., Xiao, Y., Cao, Z.: Toldi: an effective and robust approach for 3D local shape description. Pattern Recogn. **65**, 175–187 (2017)

36. Yang, Y., Feng, C., Shen, Y., Tian, D.: FoldingNet: point cloud auto-encoder via deep grid deformation. In: Proceedings of the IEEE Conference on Computer Vision and Pattern Recognition, pp. 206–215 (2018)
37. Zhao, H., Tang, M., Ding, H.: HoPPF: a novel local surface descriptor for 3D object recognition. Pattern Recogn. **103**, 107272 (2020)

DAI: How Pre-computation Speeds up Data Analysis

Kira Duwe[1]([⊠])[iD] and Michael Kuhn[2][iD]

[1] EPFL, Lausanne, Switzerland
kira.duwe@epfl.ch
[2] Otto von Guericke University, Magdeburg, Germany

Abstract. As data sizes continue to expand, the challenge of conducting meaningful analysis grows. Utilizing I/O (Input/Output) libraries, such as HDF5 (Hierarchical Data Format) and ADIOS2 (Adaptable IO System), facilitates the filtering of raw data, with prior research highlighting the advantages of dissecting these formats for enhanced metadata management. Our study introduces a novel data management technique aimed at boosting query performance for HPC analysis applications through the automatic precomputation of commonly used data characteristics, as identified by our user survey. The Data Analysis Interface (DAI), developed for the JULEA storage framework, not only enables querying this enriched metadata but also shows how domain-specific features can be integrated, demonstrating a potential improvement in query times by up to 22,000 times.

Keywords: Scientific data management · Self-describing data formats · HPC applications · Pre-computation

1 Introduction

Growing computing power and sophisticated data analysis techniques have advanced high-resolution climate research, despite storage and network limitations emerging as significant bottlenecks. While strategies like data compression and I/O optimizations can offer relief, the uneven pace of hardware advancements demands a rethinking of storage systems [5,8,23]. Emerging technologies like NVRAM and NVMe SSDs promise enhanced performance but face constraints in cost and capacity, underscoring the need for efficient data management in massive archives like the 300 PB at the German Climate Computing Centre. Addressing these challenges is vital for sustaining long-term access to critical data.

Besides the hierarchy of storage hardware, the I/O software stack also interferes with efficient data sifting and the optimal utilisation of the cluster. In climate research, the applications typically directly employ I/O libraries such as NetCDF (Network Common Data Form) [17], HDF5 (Hierarchical Data Format) [2] and ADIOS2 (Adaptable IO System) [13]. They offer rich metadata

© The Author(s), under exclusive license to Springer Nature Switzerland AG 2024
L. Franco et al. (Eds.): ICCS 2024, LNCS 14833, pp. 116–130, 2024.
https://doi.org/10.1007/978-3-031-63751-3_8

and optional hints for the I/O libraries to allow performance tuning. Due to the complex interplay of different components and optimisations in the I/O stack, performance issues are common.

No Application Changes. Developing efficient management techniques for HPC applications faces significant challenges, primarily due to the legacy code base. Scientific simulations are often developed by domain scientists, prioritizing domain knowledge over software engineering principles. Combined with its large size, this poses difficulties in adapting and implementing novel management techniques. Furthermore, asking for application changes is often met with resistance of varying degrees. This is not done out of bad faith but to preserve the considerable time and financial investments made into the code development often over decades. Solutions requiring application changes are unlikely to be tested, much less employed, on a large scale. Unfortunately, this heavily limits the design space for new approaches. In order to offer improvements that can be employed realistically, we focused on making our approach transparent for the application layer.

Contributions. Following the quote from Jim Gray, "Metadata enables data access", we examined the I/O libraries and their corresponding data formats for improvements in the data and metadata management [7]. We think there is great unused potential in these formats, especially ADIOS2. In this paper, we offer a new data management technique to tackle the challenge of finding new insights into PB of data. We build on previous work and dissect the file format inside the I/O libraries, making the changes transparent to the application layer. We extend the previous work to the automatic pre-computation of additional data characteristics that are common in post-processing across different scientific domains, as shown in our user survey. Because we focus on reductions, like sum and mean, the computation and storage overhead is minimal. To query the new metadata, we design the data analysis interface (DAI). Besides retrieval functionality, we show how offering domain-specific functionality like the climate indices can be incorporated, as well as tagging interesting data parts. We make the following contributions:

- Identifying opportunities to improve data management in HPC systems without requiring application changes
- Designing the DAI to allow for the pre-computation of data characteristics, enabling metadata tagging and access at varying levels of granularity, implemented in two ADIOS2 engines (KV-engine and DB-engine)[1].
- Implemented the DAI interface within the JULEA framework, significantly enhancing analysis application query times by up to 22,000 times[2].

2 Observations

The design of the proposed system was influenced by several key observations we made regarding HPC systems and their applications.

---

[1] https://github.com/parcio/adios2.

[2] https://github.com/kiraduwe/julea.

Data Access in HPC Systems. In HPC, checkpointing is a common practice where applications save their current state to resume computations later, especially since scientific simulations may run longer than the time allocated on a cluster. Besides, many HPC workloads follow a write-once-read-many (WORM) pattern, contrasting with database systems that frequently update data. A study of CERN workloads, analyzing over 2.49 billion events, highlighted this pattern, showing minimal updates among the events and a substantial difference between total written data (over 150 PB) and read data (more than 300 PB), emphasizing the predominance of reading over writing [16].

Observation 1: HPC data is predominantly written once, barely updated and read often [19]. *Therefore, we focus on improving the read performance, specifically emphasising the post-processing and analysis phase.*

HPC Applications. These applications fall into two main categories: large-scale simulations, like climate models, and analysis applications that process data from these simulations. Analysis applications, often crafted by small teams or individual developers, tend to be less complex, have fewer lines of code, and lack extensive optimization or parallelization. This makes them notably more adaptable to code modifications and the adoption of new APIs. *Observation 2: The stark reluctance to accept application changes does not apply to analysis applications, making them an ideal leverage point for trying new approaches and using novel interfaces.*

Self-describing Data Formats. I/O libraries like HDF5, NetCDF, and ADIOS2, which facilitate easy data management and exchange through self-describing data formats (SDDFs), enable researchers to annotate data with additional attributes, e.g. about experiment runs. SDDFs can be split into *file data* and *file metadata*, as shown in Fig. 1. The latter encompasses annotations, such as user-set attributes, as well as structural information about the data, such as variable dimensions and hierarchical ordering within groups. Separating file metadata from data is essential due to their differing access patterns; data servers optimized for large, contiguous I/O are less efficient with the small, random accesses required for metadata, which is currently stored within files, limiting performance.

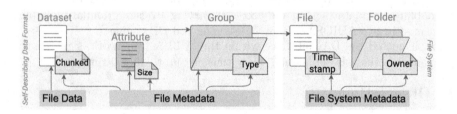

Fig. 1. Different types of metadata are indicated: File system metadata and file metadata. The former is further split into attributes set by the user directly (orange) and structural information encoded in the format (yellow). (Color figure online)

Previous studies have investigated various methods, including using relational databases, duplicating metadata in databases alongside files (e.g., EMPRESS 1 and EMPRESS 2), and eliminating the file concept to exclusively store data in databases and object stores, aiming to improve the accessibility of the file metadata [4,10,11]. Further details on these approaches will be discussed in Sects. 3 and 4.

Observation 3: Although the separate treatment of file metadata is a positive step, we believe there is untapped potential, particularly in ADIOS2. Expanding upon its capability and pre-computing additional statistics can bring significant benefits.

Typical Post-processing Operations. To determine appropriate statistics for pre-computation, we examined climate research simulations and their analysis applications. In particular, we looked at the widely used climate data operators (CDOs) [18]. Additionally, we conducted a survey among scientists from different domains to identify common mathematical operations used in data analysis.

Observation 4: Through our analysis, we observed that various domains share typical mathematical operations like reductions, e.g. sum and mean. Therefore, we pre-compute these statistics to improve data retrieval and analysis efficiency.

3   Format Dissection with JULEA

Designing and implementing a distributed storage solution for HPC clusters is complex, often involving custom client/server architectures like EMPRESS [11]. To streamline our efforts and capitalize on existing functionality, we chose to build upon the JULEA framework. JULEA not only reduced implementation work but also provided benchmarking tools, extensive testing, and continuous integration. Operating entirely in user space, JULEA simplifies development and debugging, offering various components like clients, servers, and backends to construct custom storage systems, as depicted in Fig. 2. JULEA provides versatile interfaces for interacting with clients directly or integrating into I/O libraries like HDF5 and ADIOS2 through VOL plugins or engines. Supporting object, key-value, and relational database servers, JULEA enables easy adaptation to various access patterns without altering the application code, allowing users to focus on functionality while seamlessly transitioning between backend technologies like LMDB and LevelDB with minimal configuration changes.

The dissection of self-describing data formats using the JULEA storage framework is illustrated in Fig. 2. The parallel application communicates across the different compute nodes through MPI. It uses an I/O library such as ADIOS2 or HDF5 for writing and reading its data. The data format is then dissected inside the library and its parts are directed forward to the storage solution by either the engine or the VOL plugin, depending on the used library.

4   Related Work

The Extensible Metadata PRovider for Extreme-scale Scientific Simulations (EMPRESS) stands out as a notable approach to enhanced metadata management, offering a metadata service tailored for self-describing data formats like

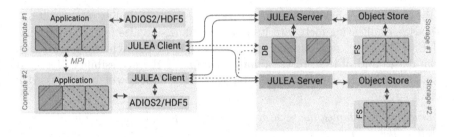

Fig. 2. JULEA setup with two compute nodes and two storage nodes. The key-value store manages the small file metadata pieces (green with fine-grained pattern), while the object store backend handles the large data chunks (grey with coarse-grained pattern). It also distributes the chunks over several storage nodes. (Color figure online)

HDF5 and ADIOS2 [11]. However, EMPRESS and its successor, EMPRESS 2, necessitate explicit specification of all desired metadata information, requiring application changes. In contrast, our solution builds upon the data model introduced by I/O libraries, concealing changes and reducing user involvement significantly.

Moreover, the drawbacks of maintaining a separate database for metadata, as discussed by Zhang et al., underscore the challenges inherent in such approaches [3, 25]. Projects like *BIMM* [9] and *SPOT* [22] also grapple with similar issues, while solutions like *JAMO* attempt to mitigate them using document databases like MongoDB. However, these approaches often encounter duplication of metadata and struggle to model the hierarchical nature of self-describing data formats effectively. By leveraging JULEA, our solution avoids these pitfalls, enabling seamless metadata management without the need for complex transformations or duplication of metadata. Additionally, innovations like *MIQS* [25] aim to address these challenges by introducing schema-free indexing solutions, but they still face limitations such as redundant storage of attribute-file path relations.

The Fast Forward Storage and IO (FFSIO) project aims to revolutionize exascale storage systems to meet the demands of HPC applications and large-scale data workloads. Despite offering a rich I/O interface, DAOS faces limitations due to extensive requirements for NVRAM and NVMe devices, making it unsuitable for cost-constrained environments [12].

The European Centre for Medium-Range Weather Forecasts (ECMWF) contributes to data archival through the ECMWF's File Storage system (ECFS) and Meteorological Archival and Retrieval System (MARS) systems, catering to the needs of weather modeling researcherss [6]. While ECFS manages files primarily with write calls, MARS operates as an object store, providing efficient storage management through a database-like interface.

In the realm of metadata management, projects like DAMASC [1], SoMeta [21] and HopsFS [15] present diverse approaches. DAMASC focuses on redefining the FS interface using database mechanisms, while SoMeta provides metadata infrastructure using a distributed hash table. HopsFS overcomes scalability limitations with distributed metadata services. Furthermore, object-based storage systems have shown efficiency and scalability advantages over traditional I/O stacks, with object-centric storage systems improving HDF5 performance when datasets are stored in object stores [14]. Additionally, leveraging Proactive Data Containers (PDC) as a tuning technique can significantly enhance performance compared to highly optimized HDF5 implementations [20].

5   Design and Implementation

As discussed in Sect. 2, the pre-computation of common mathematical operations in post-processing and analysis applications is a promising approach to improve the performance of HPC workflows without requiring any application changes. More specifically, we demonstrate how moving the computations of the analysis phase to the writing phase can offer multiple benefits.

1) The data is already present in RAM during the writing phase. So, it is the ideal point to compute derived characteristics. For the functions we find to be most suitable, the computational and storage overheads are negligible and therefore do not impact the writing process much. But even if there is an impact, given the read-mostly nature of most data in these systems, speeding up the reading significantly has a bigger impact than marginally slowing down the writing process. 2) By storing the derived characteristics that are much smaller in size they can be stored on the faster and smaller hardware tiers like SSDs or NVRAM. 3) By storing the data in a database we can use optimizations to improve the retrieval performance such as creating indexes on popular columns.

To select suitable computations, we decided to perform a user survey among researchers and study the functions offered by the library very commonly used to analyse climate data, namely the climate data operators (CDOs). By including a subset of the CDOs which can show how analysis libraries can be tied to the I/O engines.

5.1   Survey

The survey investigated the usage of self-describing data formats, e.g. the number of variables and time steps or the typical hierarchy depth, on the one hand, and interface preferences and used programming languages on the other hand. Furthermore, typical post-processing operations were evaluated as well to inform the pre-computation options. The survey targeted scientists across various institutes, including Sandia National Laboratories (SNL), University of Tennessee, Lawrence Berkeley National Laboratory (LBNL), German Climate Computing Center (DKRZ), Max-Planck-Institute of Meteorology (MPI-M), German Electron Synchrotron (DESY), Center for Bioinformatics (ZBH), University of Hamburg (UHH), Helmut-Schmidt University (HSU), German Aerospace Center (DLR), DDN, and McGill University. In addition to direct outreach, the survey was shared on the HDF5 user forum and the ADIOS discussions on GitHub, facilitating valuable input from library developers. The survey engaged a total of 38 participants, with 22 providing complete responses. Note that it is not intended to be a representative study of one or more scientific domains, it simply serves as a way to make a more informed decision about what operations to chose. Given the paper's focus on the merits of pre-computation, only the results for typical post-processing and analysis operations are included and discussed here for brevity.

The survey results illustrated in Fig. 3a show that reductions like computing averages, extrema, sums and variances are most common. Aside from that, researchers often compare simulation data to observational data, create histograms and identify outliers. As the reductions are used across various fields from climate research to astrophysics we decided to focus on them to show how this pre-computation can benefit different domains.

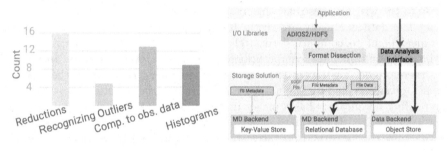

(a) Typical post-processing operations   (b) Data Analysis Interface (DAI)

Fig. 3. Survey results and the architecture of the DAI

5.2   Climate Data Operators

The climate data operators (CDO) are a command line tool offering various operations on climate data [18]. These range from file operations such as copying or splitting to arithmetic functions. Besides, modifications such as setting the time or the grid info are also numerous. The arithmetic functions contain operations to evaluate an expression or compare simulation data to constants. Additionally, CDOs offer the computation of correlation and covariance in the grid space or over time, as well as regressions to find trends in the data. Interpolation and remapping, as well as spectral and wind transformation, are common as well. We chose to focus on operations with a low overhead in terms of computation and storage space. Reductions like the sum and mean are ideal candidates as they reduce a potentially large dataset to a single result without requiring complex calculations. *Climate Indices* To highlight how the option for pre-computation can also help more specific scenarios and not just generic derived data characteristics, the DAI offers functions to trigger the automatic computation of climate indices. They are part of the European Climate Assessment (ECA) and are used to evaluate temperature and precipitation extremes and their variability. The indices are computed for a time period, typically years. An example of a climate index is the frost day index which counts the number of days per year that have a minimum temperature below 0°C.

5.3   Data Analysis Interface

Given the results of the user survey and the analysis of CDO functionality, we chose to support the following functions: various selection operations for data retrieval, statistics like the mean, the sum and the variance as well as the climate indices.

Retrieval. Our choice of selection operators for the DAI is motivated by the 6 typical access patterns for a variable in a self-describing data format [24]. A variable can be read as the whole domain, an xz-plane, xy-plane, yz-plane, a sub-area or a partial area of a plane. Thus, the DAI supports the selection of single database fields, time steps or a subset of a variable. The statistics offered by the CDOs contain various ranges, e.g. values over an entire ensemble, a field, and a time range. Thus, the DAI supports different granularity levels, namely the block level, step level or the variable level.

Pre-computation. Most functions of the DAI are related to specifying pre-computations and respective settings. Only statistics that are implemented manually in the engine can be precomputed. There is currently no option to define arbitrary mathematical operations as the support is more of a technical challenge in this setup than a research question. So, we focused on showing that offering these functions even in the simplest way can greatly benefit the analysis performance. For the same reasons, we assumed that the data uses a structured and uniform grid as this makes the calculation straightforward. Otherwise, a suitable interpolation must be used which could also involve remapping the data.

Tagging. We also offer to tag specific features for a variable, that fulfill a condition specified by the user. For example, all blocks for the temperature variable could be tagged that have a maximum temperature above 25°C. This allows to mark and later retrieve interesting data quickly. In the future, this tagging can be done completely in the configuration file to provide optimal flexibility.

Caching. Our current implementation caches previous block and step results, leading to increased memory consumption. Transitioning to calculations that rely solely on the latest result could mitigate this issue. Furthermore, keeping data consistent is not a problem as ADIOS2 does not offer updating, overwriting or deleting data for a variable. This type of data manipulation is handled at the file system level instead.

Convenience and Defaults. To make the lives of application developers and users doing post-processing easier, several functions are computed automatically by default. For one, the mean value, the sum, and the variance are computed for every block alongside the extrema that are already present in the BP formats. So, to enable the computation of these statistics, setting the engine type to **DB-engine** is sufficient meaning that the application does not need to be recompiled. If the pre-computation is not sensible for a specific scenario, the user can specify to use only the metadata offered by the native BP formats. One reason to opt out of the default to automatically compute the additional statistics may be the concern of the increased data size. For every block[3], three additional variables of the same data type as the ADIOS2 variable are stored in the database. Assuming each array has at least one element or character and the variable is not a single value, the minimal size is 143 bytes. Assuming strings with a length of 9 characters and a three-dimensional dataset, the size is already 247 bytes. This example shows that the project namespace, the file name and the variable name can contribute significantly more to the entry size than the additional statistics. Given the small storage overhead, we find storing three additional doubles to be justifiable. Especially when considering that this means the user does not have to change anything in the application and can still benefit from the pre-computation. Due to the expected performance improvement when accessing the data, the pre-computation of the three statistics is the default. To access the pre-computed values through the CDO interface for the post-processing, writing a small wrapper is sufficient to map the CDO calls to the respective JULEA DAI calls.

6 Evaluation

In the following, we present the evaluation of the JULEA engines and the DAI.

The query application representing the analysis application is run on one of the smallerNode. The JULEA servers run on the largeNode.

smallerNode is equipped with an AMD Epyc 7443 CPU (2.85 GHz, 24 cores), 128 GB of RAM, and a Western Digital 500 GB WD5000AAKS HDD, capable of 126 MB/s sustained throughput.

largeNode is equipped with an AMD Epyc 7543 CPU (2.8 GHz, 32 cores), 1024 GB of RAM, and an Intel SSDSC2KB960G8 SSD (960 GB) with a max. performance as per specification of 510 MB/s (write), 560 MB/s (read).

FS. The local FS on the cluster is ext4 with a block size of 4096 bytes. Ceph is used as the distributed FS holding, for example, the home directories. The BP engines write to Ceph, whereas the JULEA engines write to the local storage of the nodes running the JULEA servers. This setup is chosen to avoid the overhead of an additional distributed FS when using JULEA.

Software. As the JULEA benchmark revealed, LevelDB has the best overall performance for different operations, including writing and reading. MariaDB is the best available option in JULEA to store data in a database. While the performance of SQLite running in RAM is better, the data are not persisted. Therefore, MariaDB was chosen. MariaDB is run in a Singularity container that uses the current docker image. Furthermore, we use ADIOS 2.7.

---

[3] A block is the data written by one MPI process in one step.

Real-World Application. To evaluate write and read performance, we utilized the HeatTransfer application[4] from the ADIOS2 repository. It solves the 2D Poisson equation using finite differences for the temperature distribution in homogeneous media. A constant matrix size of 1024^2 was used. We introduced a second variable, precipitation, to facilitate more complex DAI queries. Furthermore, we increased the simulation steps to 100 rather than the matrix size, intentionally stressing the JULEA engines with additional metadata to rigorously test their capacity under more demanding conditions.

ADIOS2-Query and DAI-Query. Two applications were developed to assess the DAI. Both operate within a single process, as post-processing tasks like visualization or plotting scripts are often not parallelized. The first application, ADIOS2-Query, utilizes the ADIOS2 interface for data access, illustrating how the native library handles posed queries. Notably, the ADIOS2 interface cannot access the new custom metadata. The second application, DAI-Query, employs the DAI interface to query the database and custom metadata. Pre-computation time is not separately considered, as it was previously measured in the write performance of the JULEA engines.

Raw Data. All evaluation results, as well as the specifics of how we compiled and set up the system, will be published along the survey results[5].

6.1   Write and Read Performance

In the following, the I/O performance of the two ADIOS2 (BP3 and BP4) and the two JULEA engines (KV-engine and DB-engine) is measured to demonstrate the applicability of our approach across the entire application life cycle not just the analysis phase. The results shown in the following figures are the mean write and read throughput averaged over all processes performing I/O in a step for a total of 100 steps. The write and read throughput for one and six nodes are shown in Fig. 4 for BP3, BP4 and both JULEA engines. The results for 2 and 4 nodes offer no additional insight and are omitted for the sake of space. For both JULEA engines, the performance gain grows less with a higher number of processes but does not decrease. In most cases, the engine using the key-value store is only slightly better than the one using the relational database. Given the performance differences observed benchmarking the specific backends, this is surprising. We measured the performance for `put`, `get` and `delete` operations both unbatched and batched[6]. While LevelDB reached between 59,000 and 65,000 operations per second, MariaDB only achieved between 1,000 to 9,000 operations per second. Note the wide variation, especially for the BP engines. For reading, the performance ranges from 4 GB/s to about 15 GB/s when using BP3 on 6 processes on one node as shown in Fig. 4a. This is due to caching effects that we could not eliminate. Overall, the evaluation shows that our engines provide

---

[4] https://github.com/parcio/adios2/tree/master/examples/heatTransfer.
[5] https://github.com/kiraduwe/julea.
[6] The plots for these results are not included for the sake of space.

a comparable write performance while the read performance is lacking due to caching mechanisms on ADIOS2's side.

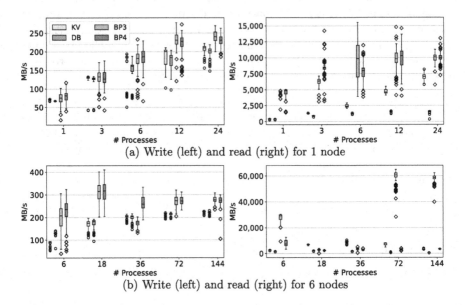

(a) Write (left) and read (right) for 1 node

(b) Write (left) and read (right) for 6 nodes

Fig. 4. HeatTransfer Results: Write and read performance for 1 node (a) and for 6 nodes (b). ADIOS2 version 2.7.1 was used.

6.2 Query Time

To evaluate the performance of the DAI and the pre-computed statistics we evaluated several queries. Four queries have been derived from the use cases we examined. They have been phrased such that they represent a real-world question. Our engines have been developed using ADIOS2 version 2.7. In order to also compare the DAI to the newer BP5 format, we included measurements for ADIOS2 version 2.8 as noted in Fig. 5.

Query 1: Find all locations where the temperature is between $-42°C$ *and* $42°C$. The first query operates on file metadata in the original BP formats. All engines, first retrieve all blocks where the temperature variable meets the condition and then the corresponding x and y coordinates are checked to deduct the location. The performance of the JULEA engines, when employing the ADIOS2 interface, is expected to be, at best, comparable to that of the BP engines, in part because they have an index for this type of metadata. This could be added to the JULEA engines as well to improve their performance in the future.

Figure 5a illustrates the results, indicating that the runtime for the JULEA engines, influenced by backend technology performance, is longest with MariaDB, reaching 2,770 ms for 96 blocks. Efforts to enhance performance involve

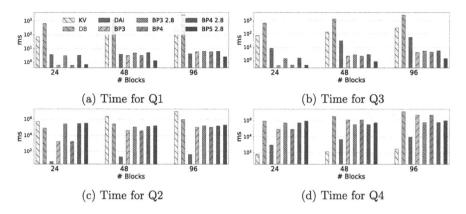

(a) Time for Q1

(b) Time for Q3

(c) Time for Q2

(d) Time for Q4

Fig. 5. The runtime for Q1 (a), Q3 (b), Q2 (c) and Q4 (d). The results were written with 1, 2 and 4 nodes with 24 processes per node which equal writing 24, 48 and 96 blocks. Note the logarithmic scale on the y-axis and the value ranges. Version 2.7 (teal) and 2.8 (blue) of ADIOS2 are used. (Color figure online)

running the database server on faster hardware, ideally in NVRAM, although this is currently unavailable in the utilized cluster. When the DAI (DAI-Query) is used instead of the ADIOS2 interface (ADIOS2-Query), the performance is improved significantly from 2,770 ms to 4 ms. This shows the advantage when the interface can access the custom data.

Query 2: What is the highest mean value of a location?
This query showcases the efficiency of pre-computed custom metadata, focusing on mean value calculations. ADIOS2-Query requires reading the variable data to compute the mean value, introducing additional time, while DAI-Query benefits from pre-computed mean values during variable writing. To optimize, additional metadata computations are combined where possible, such as reusing the sum for mean and standard deviation calculations. DAI-Query outperforms ADIOS2-Query for DB-engine, with runtimes of 41 ms 96 blocks, compared to 899,400 ms, a factor of 21,940. Notably, the DAI interface achieves impressive results even with MariaDB as the underlying technology.

Query 3: Which location had the largest increase in maximum temperature between step 1 and 100?
The third query assesses metadata querying performance across multiple steps, specifically comparing maximum temperatures for all blocks in steps 1 and 100. Unlike prior queries, this scenario avoids reliance on pre-computed features and challenges engines by requiring the reading and comparison of maximum temperatures across steps. Figure 5b illustrates consistent runtimes for JULEA engines, regardless of using one or two servers. The DB-engine engine has the longest runtime which is in part because of the lower performance of MariaDB as mentioned before. Notably, using DAI with JULEA clients achieves significantly shorter runtimes (around 31 ms for 96 blocks) compared to using MariaDB

through the ADIOS2 interface (2,720 ms), even outperforming LevelDB. The BP engines have a small runtime, for example, 5 ms for BP3. A possible reason is the number of small reads that have to be performed using the JULEA engines. Currently, there is no way to read all file metadata for a specific variable at once. In contrast, the BP engines only need to read the metadata file of their formats and all blocks' maxima are available to the application. One option to mitigate the drawbacks would be to offer an iterator to the application to allow reading multiple entries more efficiently.

Query 4: What was the highest precipitation sum for a location when the maximum temperature was above 40° C?
The chosen query for the final evaluation involves multiple variables, adding complexity compared to previous queries. In Fig. 5d, the KV-engine engine outperforms others with a runtime of approximately 270 ms. In contrast, DB-engine exhibits a longer runtime of 9060 ms for 96 blocks, a factor of about 35.

Table 1. Assessment of the query runtimes from best (1) to worst (4).

	BP	J-KV	J-DB	DAI
Query 1	1	2	3	1
Query 2	2	4	3	1
Query 3	1	3	4	2
Query 4	3	1	4	2

6.3 Discussion

When bringing the format dissection, the file metadata management in a relational database and the DAI together, they can compete with the BP engines when writing data. However, they shine when the custom metadata is relevant to the posed question and even outperform the native engines significantly by several orders of magnitude.

Query results Some interesting aspects become more evident in the condensed query results in Table 1. First, the query selection was varied and focused on different capabilities. Second, no candidate is best for each scenario. This is not surprising since previous work on access patterns for variables in self-describing data formats indicated that they could be very varied [24]. Third, the format dissection itself is not beneficial when a slow backend is used, as evidenced by the subpar results of DB-engine, whereas using a key-value store proved more efficient, largely due to the superior performance of LevelDB over MariaDB. Fourth, the pre-computation of statistics combined with the DAI is a great addition and helps improve the performance significantly even if a slow backend like MariaDB is used.
Querying across files Querying across files was not assessed due to the cluster's limited storage capacity. Despite this, the centralization of file metadata does allow for quicker cross-file queries, avoiding the need to open multiple files.

Data modelling One major challenge is that data can be modelled differently in different formats. For example, in HDF5, time can be represented using a specific array dimension, or multiple datasets can be written, each corresponding to a specific step, or even individual files for each step. This complicates the computation of additional data characteristics, e.g. for a time step. Furthermore, different scientific domains, such as climate simulations versus particle physics, introduce varied requirements for statistical analyses and grid structures.

Automatic detection of new statistics To automate the detection of statistics to precompute the following approaches could be used: 1. Machine Learning Models: Implement algorithms to predict useful statistics based on past data access patterns during analysis, enabling adaptive adjustments over time. 2. Metadata Analysis: Leverage insights from metadata such as access frequency, data size, and type to inform precomputation decisions. 3. Feedback Loops: Incorporate direct user input to tailor precomputation strategies to actual user needs. Exploring these approaches allows storage engines to dynamically adapt and optimize performance to support diverse analytical requirements of HPC data.

7 Conclusion

We present a novel data management technique utilizing ADIOS2 to boost query performance in HPC analysis applications by automating the precomputation of frequently used data characteristics that were identified in a user survey. Our Data Analysis Interface (DAI), integrated within the JULEA storage framework, demonstrates performance improvements by a factor of up to 22,000 -while supporting domain-specific features, thereby showcasing its adaptability and effectiveness across various scientific fields.

References

1. Brandt, S., Maltzahn, C., Polyzotis, N., Tan, W.C.: Fusing data management services with file systems. In: Proceedings of the 4th Annual Workshop on Petascale Data Storage, pp. 42–46 (2009). https://doi.org/10.1145/1713072.1713085
2. Breitenfeld, M.S., Pourmal, E., Byna, S., Koziol, Q.: Achieving High Performance I/O with HDF5. http://tinyurl.com/hdf5tutorial (2020). Accessed 28 Aug 2022
3. Byna, S., et al.: ExaHDF5: delivering efficient parallel I/O on exascale computing systems. J. Comput. Sci. Technol. **35**(1), 145–160 (2020). https://doi.org/10.1007/s11390-020-9822-9
4. Duwe, K., Kuhn, M.: Dissecting self-describing data formats to enable advanced querying of file metadata. In: SYSTOR, pp. 12:1–12:7. ACM (2021). https://doi.org/10.1145/3456727.3463778
5. Duwe, K., et al.: State of the art and future trends in data reduction for high-performance computing. Supercomput. Front. Innov. **7**(1), 4–36 (2020)
6. Grawinkel, M., Nagel, L., Mäsker, M., Padua, F., Brinkmann, A., Sorth, L.: Analysis of the ECMWF storage landscape. In: FAST, pp. 15–27. USENIX Association (2015)

7. Gray, J., Liu, D.T., Nieto-Santisteban, M.A., Szalay, A.S., DeWitt, D.J., Heber, G.: Scientific data management in the coming decade. SIGMOD Rec. **34**(4), 34–41 (2005)

8. Isakov, M., et al.: HPC I/O throughput bottleneck analysis with explainable local models. In: SC, p. 33. IEEE/ACM (2020)

9. Korenblum, D., Rubin, D.L., Napel, S., Rodriguez, C., Beaulieu, C.F.: Managing biomedical image metadata for search and retrieval of similar images. J. Digit. Imaging **24**(4), 739–748 (2011). https://doi.org/10.1007/s10278-010-9328-z

10. Kuhn, M., Duwe, K.: Coupling storage systems and self-describing data formats for global metadata management. In: 2020 CSCI, pp. 1224–1230 (2020). https://doi.org/10.1109/CSCI51800.2020.00229

11. Lawson, M., Gropp, W., Lofstead, J.F.: EMPRESS: accelerating scientific discovery through descriptive metadata management. ACM Trans. Storage **18**(4), 34:1–34:49 (2022)

12. Lofstead, J.F., Jimenez, I., Maltzahn, C., Koziol, Q., Bent, J., Barton, E.: DAOS and friends: a proposal for an exascale storage system. In: SC, pp. 585–596. IEEE Computer Society (2016)

13. Lofstead, J.F., Zheng, F., Klasky, S., Schwan, K.: Adaptable, metadata rich IO methods for portable high performance IO. In: 23rd IEEE IPDPS, pp. 1–10. IEEE (2009)

14. Mu, J., Soumagne, J., Byna, S., Koziol, Q., Tang, H., Warren, R.: Interfacing HDF5 with a scalable object-centric storage system on hierarchical storage. Concurr. Comput. Pract. Exp. **32**(20) (2020)

15. Niazi, S., Ismail, M., Haridi, S., Dowling, J., Grohsschmiedt, S., Ronström, M.: HopsFS: scaling hierarchical file system metadata using newSQL databases. In: FAST, pp. 89–104. USENIX Association (2017)

16. Purandare, D., Bittman, D., Miller, E.: Analysis and workload characterization of the CERN EOS storage system. In: CHEOPS@EuroSys, pp. 1–7. ACM (2022)

17. Rew, R., Davis, G., Emmerson, S., Davies, H.: NetCDF User's Guide - an interface for data access version 2.4. http://www-c4.ucsd.edu/netCDF/netcdf-guide/guide_toc.html (1996). Accessed 15 Jul 2022

18. Schulzweida, U.: CDO user guide (2022). https://doi.org/10.5281/zenodo.7112925

19. Settlemyer, B.W., Amvrosiadis, G., Carns, P.H., Ross, R.B., Mohror, K., Shalf, J.M.: It's time to talk about HPC storage: perspectives on the past and future. Comput. Sci. Eng. **23**(6), 63–68 (2021)

20. Tang, H., et al.: Tuning object-centric data management systems for large scale scientific applications. In: HiPC, pp. 103–112. IEEE (2019)

21. Tang, H., Byna, S., Dong, B., Liu, J., Koziol, Q.: SoMeta: scalable object-centric metadata management for high performance computing. In: CLUSTER, pp. 359–369. IEEE Computer Society (2017)

22. Tull, C.E., Essiari, A., Gunter, D., Li, X.S., Patton, S.J., Ramakrishnan, L.: The SPOT suite project. http://spot.nersc.gov/ (2013). Accessed 09 Oct 2020

23. Uselton, A., et al.: Parallel I/O performance: from events to ensembles. In: IPDPS, pp. 1–11. IEEE (2010)

24. Wan, L., et al.: Improving I/O performance for exascale applications through online data layout reorganization. IEEE Trans. Parallel Distributed Syst. **33**(4), 878–890 (2022)

25. Zhang, W., Byna, S., Tang, H., Williams, B., Chen, Y.: MIQS: metadata indexing and querying service for self-describing file formats. In: SC, pp. 5:1–5:24. ACM (2019)

Evolutionary Neural Architecture Search for 2D and 3D Medical Image Classification

Muhammad Junaid Ali$^{(\boxtimes)}$, Laurent Moalic, Mokhtar Essaid,
and Lhassane Idoumghar

Université de Haute-Alsace, IRIMAS UR 7499, 68093 Mulhouse, France
{muhammad-junaid.ali,laurent.moalic,mokhtar.essaid,
lhassane.idoumghar}@uha.fr

Abstract. Designing deep learning architectures is a challenging and time-consuming task. To address this problem, Neural Architecture Search (NAS) which automatically searches for a network topology is used. While existing NAS methods mainly focus on image classification tasks, particularly 2D medical images, this study presents an evolutionary NAS approach for 2D and 3D Medical image classification. We defined two different search spaces for 2D and 3D datasets and performed a comparative study of different meta-heuristics used in different NAS studies. Moreover, zero-cost proxies have been used to evaluate the performance of deep neural networks, which helps reduce the searching cost of the overall approach. Furthermore, recognizing the importance of Data Augmentation (DA) in model generalization, we propose a genetic algorithm based automatic DA strategy to find the optimal DA policy. Experiments on MedMNIST benchmark and BreakHIS dataset demonstrate the effectiveness of our approach, showcasing competitive results compared to existing AutoML approaches. The source code of our proposed approach is available at https://github.com/Junaid199f/ evo_nas_med_2d_3d.

Keywords: Evolutionary Neural Architecture Search · Medical Image Classification · AutoML · AutoDL · Automatic Data Augmentation

1 Introduction

Deep Learning (DL) algorithms have been widely used for solving real-world tasks, but designing these architectures requires domain-expert knowledge. Multiple Neural Architecture Search (NAS) approaches have been proposed to automate DL architecture design for multiple tasks [17], but they require significant searching time due to the network evaluation phase.

Multiple performance estimation strategies have been proposed to reduce the search time. This would not only reduce the search time but also assist the search algorithm in the exploration of large search space. One such strategy is the Zero

L. Franco et al. (Eds.): ICCS 2024, LNCS 14833, pp. 131–146, 2024.
https://doi.org/10.1007/978-3-031-63751-3_9

Cost (ZC) proxy, which evaluates a DL architecture on a small number of data samples to quickly estimate individual performance [10]. This approach is particularly effective in medical image classification, where traditional NAS-based methods are computationally expensive due to the large number of samples.

Data Augmentation (DA) is crucial for performance enhancement in medical image analysis tasks, as it prevents overfitting and enhances the model's generalization ability to perform well on unseen data. By creating variations of existing data, DA can help prevent overfitting. However, due to the diverse nature of medical imaging datasets, a single DA strategy may perform differently on different datasets. To address this issue, multiple automatic DA strategies have been proposed, employing different optimization strategies to search for the best DA policy. The combination of architecture components and DA expands the search space, making it challenging to find an optimal set of DA policies and the best architecture simultaneously. To tackle this issue, we divided the proposed approach into two stages: (i) architecture search and (ii) automatic DA search. At first, an architecture is searched using the proposed NAS approach, and then the best-suited pair of DA techniques is searched using the proposed automatic DA approach. The main contributions of this study are as follows:

- We have proposed an evolutionary NAS approach for both 2D and 3D medical image classification.
- The proposition of a DA technique capable of searching the best augmentation topology for a given dataset.
- Experiments on both small-scale and large-scale datasets are conducted to demonstrate the effectiveness of the proposed approach.

2   Related Work

In recent years, numerous evolutionary NAS approaches have been proposed to solve different tasks, including image classification [6] and medical image classification [3]. These approaches have adopted different performance estimation strategies to reduce the searching time [17]. For experimentation, these approaches use standard benchmark datasets like CIFAR-10 and ImageNet for image classification and MedMNIST for medical image classification.

Multiple NAS studies have been proposed for medical image classification and used MedMNIST datasets for experiments [2–4,6]. These studies have used different performance estimation strategies to reduce the searching time, such as surrogate models, ZC proxies, and One-Shot NAS approaches. The surrogate models, also known as performance predictors are machine learning models that predict the individual's fitness during evolution to reduce the search time. These machine learning models are trained on individual representations and their corresponding fitness values on the initialized population and retrained during evolution [2]. One-Shot NAS trains a supernet first, then samples sub-networks and uses a weight-sharing mechanism to save the time required for re-training. Additionally, ZC proxies use a mini-batch of data to quickly estimate the model's performance [6].

Moreover, studies have shown that DA plays a crucial role in enhancing the model generalization ability and the model performance on unseen data. These studies proposed searching for both DA and network topology simultaneously. Zhang et al. proposed a unified approach for searching both DA policy and network topology [5]. They introduced an augmentation density matching algorithm that addresses the inefficiency of density matching caused by in-domain sampling bias. They first trained a Super-Net and then used an evolutionary algorithm to search for sub-networks with optimal augmentation policy for the given dataset.

Existing studies were proposed for either 2D or 3D architectures but not both. To this aim, this study addresses both 2D and 3D NAS architectures for medical image classification. Compared to 2D NAS approaches, 3D NAS approaches are computationally expensive due to the increasing model complexity and computational costs. Incorporating ZC proxies as a performance estimation strategy could reduce the search time of the overall NAS approach. To the best of our knowledge, this is the first study to use zero-cost proxies with 3D NAS for medical image classification.

Furthermore, different meta-heuristic algorithms have been used as a search strategy in different NAS studies. Some of the famous NAS approaches have used Genetic Algorithm (GA) [9], Particle Swarm Optimization (PSO) [8], Differential Evolution (DE) [7] and other different algorithms. Unfortunately, choosing an optimal metaheuristic for a given problem is a difficult task. In this study, we performed a comparative study of famous meta-heuristics to compare their performance to choose the best-performing one.

3 Proposed Methodology

The proposed methodology is mainly divided into two primary stages: (1) the architecture search stage and (2) the DA search stage, as illustrated in Fig. 1. Initially, a population of individual neural networks is randomly generated, and then an evolutionary algorithm searches for the best-performing individual. The SynFlow ZC proxy is used for fitness evaluation, which assesses individual Neural Network (NN) performance on the validation set. The second stage involves the GA to search for the appropriate DA policy. The fitness is computed by training each model on the training set and evaluating it on the validation set. The best-performing DA strategy is used while training the final architecture. Then, the accuracy and Area Under the Curve (AUC) scores are computed on the test set. Further details regarding search space and the encoding scheme are provided in the following sections.

3.1 Search Space

In this study, we have used cell-based search space initially proposed in NAS-Net study by Zolph et al. [12] and DARTS study [11], which consists of small NN blocks called cells that can be repeated or connected in various ways to

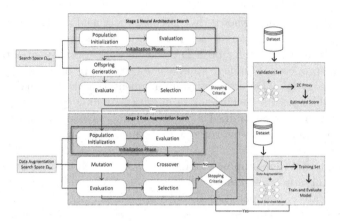

Fig. 1. Proposed methodology diagram consisting of two stages (i) neural architecture search using zero-cost proxies and (ii) data augmentation strategy using genetic algorithm

form a complete NN architecture. This search space has been widely adopted in different NAS-based studies due to its simplicity and flexibility in extending its components. These cells apply different convolution operations to get feature maps, which can be passed to other cells. It consists of two types of cells: normal and reduction cells. The normal cell computes the feature map of an input image, where convolution and pooling in the cell have a stride of 1 to keep the same resolution. In contrast, the reduction cell uses the stride of 2 to reduce the feature map dimension to down-sample the feature maps. The whole architecture is formed by stacking the cells one after another.

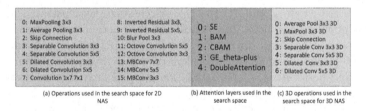

Fig. 2. Operations and attention layers for 2D architecture and 3D operations for 3D architecture used in the search space to design CNN architectures

Moreover, an attention-based search space is used which comprises 16 different convolution and pooling operations and 5 different attention layers as shown in Fig. 2. These attention layers are adopted from the Attention-DARTS [13] study. The idea behind using attention layers is to focus more on salient regions. The attention layer is used after the candidate operations. For the 3D search space, we have used seven different operations, consisting of pooling layers, dilated convolution, and separable convolution with varying kernel sizes, as shown in Fig. 2 [c].

In our 2D search space, we have used various lightweight and efficient CNN components from different architectures, including InceptionNets, OctoveNet, MobileNet, Invetable Residual Networks etc. These operations consist of mobile, dilated, separable, octove and inverted residual blocks with different kernel settings. Attention layers include Gather-Excite (GE) Attention, Squeeze and Excitation (SE), Convolution Bottleneck Attention Module (CBAM), Bottleneck Attention Module (BAM) and Double Attention (DA) blocks.

3.2   Encoding Scheme

For the representation of an individual, each candidate operation is expressed as a real value between 0 and 1. This real value is mapped to the corresponding operation by multiplying the number of operations and then applying floor operation to get the corresponding operation number. Besides, the associated attention layer is represented by a value between 1 and 5, where 1–5 means different attention types. Similarly, for 3D search space, each candidate operation in the individual is expressed by a real value between 0 and 1. An example representation of genotype and phenotype is shown in Fig. 3.

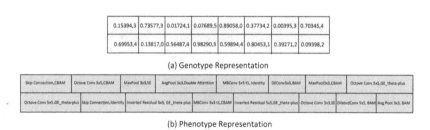

(a) Genotype Representation

(b) Phenotype Representation

Fig. 3. Genotype and phenotype representations of a sample individual consisting of 16 genes

For instance, a genotype (0.56,4),(0.42,5),(0.29,4), (0.8926,1) which is decoded into (InvRes3x3,4), (DilatedConv5x5,5), (Separable Convolution 5x5,4), (MBConv 5x5,1) such as (0.56,4) mapped into (InvRes3x3,4) and 0.56×16 where each gene is multiplied with the number of candidate operations in the search space and the floor operation is applied and corresponding operation is fetched from the list consisting of different operations.

3.3   NAS Approach

As discussed above, multiple metaheuristic algorithms have been used as search strategies in NAS. Choosing the optimal meta-heuristic for a given problem is not straightforward. Thus, we have performed a comparative analysis of different meta-heuristics and chosen the optimal one in terms of final results. Evaluating the fitness of an individual is a time-consuming task. To address this challenge, we have utilized zero-cost proxies that are based on recent pruning at initialization [27]. They use a single mini-batch of training data to compute a model score.

Abdelfattah et al. [10] proposed using existing proxies to estimate scoring in a DL architecture including SynapticFlow (SynFlow) and SNIP, which are used in this study. Recent works have shown that SynFlow assists the evolutionary algorithm in reducing the searching time while searching the optimal architecture [18]. It is noteworthy to mention that these proxies were used for both 2D and 3D datasets in this study.

3.4   Data Augmentation

DA is a series of transformations applied to the input data. It plays an essential role in DL-based medical image analysis. It increases the amount and diversity of the training data and reduces overfitting. However, manually developing a tailored DA strategy for each dataset is difficult because of the heterogeneity of medical imaging data and the different characteristics of each disease and modality. To overcome this issue, a GA-based approach is proposed to automatically search for a suitable DA policy for a given dataset.

Fig. 4. Search space consisting of different data augmentation techniques and crossover and mutation operators

Figure 4 shows the search space, crossover and mutation operators used in the proposed approach. Seven DA strategies have been used in the search space, such as horizontal and vertical flipping, random rotation, resizing, cropping, random affine, erasing, and colour jitter. Each individual consists of two different kinds of DAs that have been used. The number of DA for each dataset is set to two, as experiments with more than two were not significant in terms of improvement rate. For reproduction and mutation, a one-point crossover is used, and a random mutation strategy is adopted, replacing a bit with a randomly selected DA from the search space.

To evaluate individual fitness, we have considered AUC and accuracy scores. These measures evaluate the model's performance in different aspects: accuracy measures the correctness of the model, and AUC measures the model's ability to distinguish between positive and negative instances. The fitness function is given as:

$$Fitness = AUC + Accuracy \qquad (1)$$

4   Experimental Settings

We have performed all the experiments with the same hardware configurations for a fair comparison. All the experiments were performed on a GPU cluster with a NVIDIA A100 GPU and 96 GB of RAM. The proposed framework is implemented in Python. PyTorch library is used for DL implementation and Mealpy for implementing metaheuristics. For NN training, we have used the Adam optimizer with a MultiStep Learning Rate scheduler and a learning rate of 0.0025, a gamma rate of 0.1, and a weight decay of 0.0003. The final architecture is trained on 300 epochs; for the GA-based DA strategy, each individual is trained on 25 epochs. The accuracy and AUC scores on validation sets are used as the fitness function.

For MedMNIST implementation, we used the implementation provided by the authors and the same dataset split of train, val, and test. During the search phase, a validation set is used to evaluate the architecture and the test set is used to evaluate the final searched architecture with a given DA topology. For the Breast Cancer Histopathological Image Classification (BreakHIS) dataset, we divided the dataset into 80:10:10 ratios to create a train/validation and test set. We also used the same methodology for the MedMNIST dataset. The DA policy is only searched for MedMNIST 2D datasets, and the number of layers is set from 8 to 12 for 2D datasets and 15 to 20 for 3D datasets. For the BreakHIS dataset, a layer size of 15 is used.

Multiple trials of experiments set up the parameters of GA for DA, and PSO, DE and ACO parameters are tuned using grid-search parameter tuning. The population size and number of generations are set to 200 for all the meta-heuristics. The parameter settings for the different algorithms are given in Table 1, where the best-reported parameters of each meta-heuristic algorithm are given.

4.1   Datasets Description

We have used the datasets from the MedMNIST benchmark and the BreakHIS (Breast Histopathology) dataset for experiments. MedMNIST is a large-scale MNIST-like standard dataset benchmark for biomedical imaging. It includes 12 2D and 6 3D datasets from different organs and modalities. All images of 2D and 3D datasets are pre-processed into 28×28 resolution with corresponding labels. These datasets belong to different organs and modalities, such as histopathology, abdominal computed tomography, breast mammograms, retinal fundus, microscopy, chest X-ray, and dermoscopy images.

Apart from MedMNIST, we have also performed experiments on the BreakHis dataset, which consists of 7,904 microscopic images of breast tumors

Table 1. Parameters' settings of different meta-heuristics

Algorithm	Parameters	Values
GA (Data Augmentation)	Crossover Probability	0.95
	Mutation Probability	0.1
	Selection Technique	Tournament Selection
	Population Size	4
	Number of Generations	4
	Chromosome Length	2
PSO	c1 and c2	2
	Population Size	200
	Number of Generations	200
	Intertia weight	decreases from 0.95 to 0.4
DE	F	0.7
	Crossover Rate	0.9
	Population Size	200
	Number of Generations	200
	Mutation Strategy	DE/Current-to-best/2
LSHADE	μ_f	0.5
	Population Size	200
	Last Population Size	50
	Number of Generations	200
	μ_{CR}	0.5
ACO	Z	1.5
	q	0.7
	Population Size	200
	Number of Generations	200
	Sample Count	35

collected from 82 patients with different magnification levels (40x, 100x, 200x and 400x) and 8 different classes. As these images are high resolution, we have resized them to 224×224 for ease of training. It contains 2,480 benign and 5,091 malignant samples. This dataset includes multiple samples. The idea behind using both small and large-scale datasets is to evaluate the performance of the proposed approach.

5   Experimental Results

We have performed experiments on 2D and 3D datasets from the MedMNIST benchmark and the BreakHIS dataset. The results of the proposed approach are compared with multiple variants of ResNet architectures with two different resolutions (28 and 224), AutoML approaches (AutoSKlearn, AutoKeras and

Google AutoML Vision) and multiple NAS approaches proposed for natural images and medical images. The comparison in terms of accuracy is shown in Table 2, and the AUC is shown in Table 3.

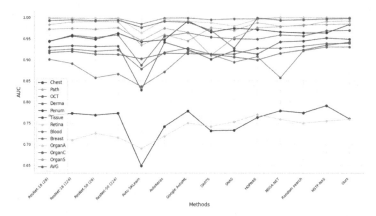

Fig. 5. Line graph representing the AUC scores of different NAS and deep learning approaches on MedMNIST 2D datasets

Figure 5 shows the AUC scores of different NAS and deep learning approaches on MedMNIST 2D datasets. Each line represents a dataset, and dotted points represent the results of different methods. These lines show better performance on NAS-based approaches than other AutoML and deep learning approaches. Furthermore, the proposed approach also has better or equal performance than other NAS-based approaches. AutoSKLearn is an AutoML tool to automate the machine learning process which includes searching for optimal classifiers, features pre-processing and data processing. AutoKeras is also an AutoML approach that uses bayesian optimization to evolve the architecture for a given problem. Google AutoML Vision is a cloud-based solution consisting of state-of-the-art NAS algorithms that design architecture automatically for the given dataset.

Existing NAS approaches to which the proposed approach is compared include Differentiable Architecture Search (DARTS), Stochastic Neural Architecture Search (SNAS), HOPNAS, Non-Dominated Sorting Genetic Algorithm (NSGA-Net), and Multi-Scale Training Free (MSTF-NAS). We have used the same results of these approaches given in the recent study named MSTF-NAS [23]. In MSTF-NAS, the authors have proposed a multi-scale, training-free NAS approach for medical image classification. They compared their proposed approach with existing NAS studies on 12 subsets from the MedMNIST benchmark.

These NAS approaches include DARTS, SNAS, and HOPNAS, which are one-shot NAS approaches based on SuperNet. Among these NAS approaches, DARTS and NSGANet are specifically designed for natural images, and SNAS and HOPNAS are designed for medical images. The results of random search on 2D MedMNIST datasets by [23] are also included in the comparison. In random

Table 2. Results comparison of the proposed approach with different NAS methods and deep learning approaches on the MedMNIST benchmark in terms of accuracy.

Methods	Chest	Path	OCT	Derma	Tissue	Retina	Blood	Breast	OrganA	OrganC	OrganS	Pneum	AVG.
ResNet-18 (28)	0.947	0.907	0.743	0.735	0.676	0.524	0.958	0.863	0.935	0.900	0.782	0.854	0.818
ResNet-18 (224)	0.947	0.909	0.763	0.754	0.681	0.493	0.963	0.833	0.951	0.920	0.778	0.864	0.821
ResNet-50 (28)	0.947	0.911	0.762	0.735	0.680	0.528	0.956	0.812	0.935	0.905	0.770	0.854	0.816
ResNet-50 (224)	0.948	0.892	0.776	0.731	0.680	0.511	0.950	0.842	0.947	0.911	0.785	0.884	0.821
Auto-SKlearn	0.779	0.716	0.601	0.719	0.532	0.515	0.878	0.803	0.762	0.829	0.672	0.855	0.721
AutoKeras	0.937	0.834	0.763	0.749	0.703	0.503	0.961	0.831	0.905	0.879	0.813	0.878	0.813
Google AutoML	0.948	0.728	0.771	0.768	0.673	0.531	0.966	0.861	0.886	0.877	0.749	**0.946**	0.809
DARTS [11]	0.934	0.872	0.712	0.749	0.648	0.510	0.953	0.832	0.926	0.791	0.808	0.874	0.800
SNAS [26]	0.938	0.850	0.708	0.737	0.708	0.515	0.946	0.811	0.918	0.891	0.778	0.871	0.805
HOPNAS [25]	0.947	0.912	0.761	0.759	0.698	0.523	0.958	0.853	0.937	0.911	0.803	0.852	0.826
NSGA-NET [24]	0.947	0.866	0.765	0.744	0.712	0.540	0.970	0.846	0.952	0.923	0.820	0.907	0.832
Random search	0.946	0.854	0.760	0.773	0.717	0.542	0.966	0.897	0.955	0.923	0.820	0.904	0.838
MSTF-NAS [23]	0.945	0.910	0.780	**0.774**	0.740	0.550	**0.976**	0.872	**0.962**	**0.936**	0.838	0.912	0.841
Ours	**0.949**	**0.920**	**0.800**	0.769	**0.760**	**0.561**	0.967	**0.911**	0.957	0.935	**0.864**	0.944	**0.861**

search, instead of using some guided search algorithm, an architecture is picked randomly till given iterations and the best architecture is selected at completion.

Table 3. Results comparison of the proposed approach with different NAS methods and deep learning approaches on the MedMNIST benchmark in terms of AUC

Methods	Chest	Path	OCT	Derma	Penum	Tissue	Retina	Blood	Breast	OrganA	OrganC	OrganS	AVG.
ResNet-18 (28)	0.768	0.983	0.943	0.917	0.944	0.930	0.717	0.998	0.901	0.997	0.992	0.972	0.922
ResNet-18 (224)	0.773	0.989	0.958	0.920	0.956	0.933	0.710	0.998	0.891	0.998	0.994	0.974	0.924
ResNet-50 (28)	0.769	0.990	0.952	0.913	0.948	0.931	0.726	0.997	0.857	0.997	0.992	0.972	0.920
ResNet-50 (224)	0.773	0.989	0.958	0.912	0.962	0.932	0.716	0.997	0.866	0.998	0.993	0.975	0.922
Auto-SKLearn	0.649	0.934	0.887	0.902	0.942	0.828	0.690	0.984	0.836	0.963	0.976	0.945	0.878
AutoKeras	0.742	0.959	0.955	0.915	0.947	0.941	0.719	0.998	0.871	0.994	0.990	0.974	0.917
Google AutoML	0.778	0.944	0.963	0.914	**0.991**	0.924	0.750	0.998	0.919	0.990	0.988	0.964	0.927
DARTS [11]	0.732	0.975	0.953	0.913	0.965	0.901	0.742	0.994	0.912	0.987	0.969	0.910	0.913
SNAS [26]	0.733	0.969	0.949	0.906	0.974	0.921	0.753	0.996	0.894	0.979	0.927	0.952	0.913
HOPNAS [25]	0.763	0.987	0.948	0.899	0.971	0.913	0.770	0.996	0.907	0.995	**0.998**	0.975	0.926
NSGA-NET [24]	0.779	0.979	0.958	0.915	0.965	0.942	**0.759**	0.999	0.857	**0.999**	0.993	0.978	0.926
Random search	0.774	0.980	0.956	0.923	0.963	0.944	0.750	0.999	0.921	**0.999**	0.994	0.982	0.932
MSTF-NAS [23]	**0.791**	0.990	0.968	0.934	0.963	0.951	0.755	0.999	0.930	0.999	0.996	0.983	0.938
Ours	0.760	**0.991**	**0.969**	**0.941**	0.983	**0.948**	**0.759**	0.999	0.930	0.999	0.997	**0.984**	**0.9383**

One of the reasons behind the good performance of random search in NAS is that, usually, NAS search spaces are high-dimensional, with a large number of possible configurations. Random sampling can effectively explore different regions of the search space in such spaces, thereby increasing the likelihood of discovering good solutions without the constraints of guided search methods.

The reason behind the performance difference between NAS approaches addressing natural images and approaches addressing medical ones is mainly the domain gap. On the one hand, natural images depict landscapes, and objects that are captured by digital means. On the other hand, medical images visualize anatomical structures and pathological conditions within the body obtained

Table 4. Results comparison of the proposed approach with different NAS methods and deep learning approaches on the 3D datasets MedMNIST benchmark in terms of AUC score.

Methods	Organ	Nodule	Fracture	Adrenal	Vessel	Synapse	Avg.
ResNet-18 + 2.5D	0.977	0.838	0.587	0.718	0.748	0.634	0.750
ResNet-18 + 3D	**0.996**	0.863	0.712	0.827	0.874	0.820	0.848
ResNet-18 + ACS	0.994	0.873	0.714	0.839	0.930	0.705	0.842
ResNet-50 + 2.5D	0.974	0.835	0.552	0.732	0.751	0.669	0.752
ResNet-50 + 3D	0.994	0.875	0.725	0.828	0.907	**0.851**	0.863
Auto-SKLearn	0.977	**0.914**	0.628	0.828	0.910	0.631	0.814
AutoKeras	0.979	0.844	0.642	0.804	0.773	0.538	0.763
Proposed	0.995	0.871	**0.728**	**0.857**	**0.940**	0.820	**0.868**

through different modalities. They also include small lesions and tumour regions. The attention mechanism highlights important regions to achieve an accurate feature extraction improving detection accuracy and interoperability. The results reveal that the proposed approach exhibits a better performance on average compared to the other studies. Regarding MSTF-NAS, the proposed approach has an advantage over DA as it helps improve model generalization ability on unseen data.

In Tables 4 and 5, the comparison in terms of the performance metrics (AUC and Accuracy) of the proposed approach with different variants of ResNet and AutoML approaches on 3D datasets from the MedMNIST benchmark is given. The conclusion from the results is that our proposed approach yields better performance in comparison with existing approaches, as the average accuracy and AUC scores are better than existing DL and AutoML approaches.

The proposed approach exhibits an overall satisfactory performance on average and less searching time than existing approaches. Following the same context, the number of parameters of searched architectures is less compared to ResNet DL architectures. The proposed approach can achieve a better exploration and exploitation of individuals thanks to the ZC proxies, which allow reduced search costs to evaluate an individual.

Similarly, experiments on the BreakHIS were conducted (the histopathology dataset, which consists of eight classes with a 200x magnification level). While prior NAS studies mainly focused on small-scale datasets, our research highlights the potential use of ZC proxies for efficiently exploring an architecture suited for large-scale datasets. A comparison of the best-performing architecture results with existing approaches is shown in Table 7. These results are based on the

Table 5. Results comparison of the proposed approach with different NAS methods and deep learning approaches on the 3D datasets MedMNIST benchmark in terms of Accuracy score.

Methods	Organ	Nodule	Fracture	Adrenal	Vessel	Synapse	Avg.
ResNet-18 + 2.5D	0.788	0.835	0.451	0.772	0.846	0.696	0.731
ResNet-18 + 3D	0.907	0.844	0.508	0.721	0.877	0.745	0.767
ResNet-18 + ACS	0.900	0.847	0.497	0.754	0.928	0.722	0.774
ResNet-50 + 2.5D	0.769	0.848	0.397	0.763	0.877	0.735	0.731
ResNet-50 + 3D	0.883	0.847	0.494	0.745	0.918	0.795	0.780
ResNet-50 + ACS	0.889	0.841	0.517	0.758	0.858	0.709	0.762
Auto-SKLearn	0.814	0.874	0.453	0.802	0.915	0.730	0.764
AutoKeras	0.804	0.834	0.458	0.705	0.894	0.724	0.736
Proposed	**0.908**	**0.877**	**0.690**	**0.805**	**0.940**	**0.846**	**0.844**

validation performance after the network training. The results reveal that the proposed approach performed better regarding multiple performance measures, taking only two hours to search an architecture with a ZC proxy.

Table 6. Comparison of Accuracy and AUC scores of different metaheuristic algorithms without data augmentation strategy.

Metaheuristics		LSHADE	ACO	PSO	DE
Path	Accuracy	**0.89**	0.79	0.80	0.86
	AUC	**0.97**	0.96	0.96	0.93
OCT	Accuracy	0.74	0.73	0.74	**0.77**
	AUC	0.95	0.94	0.95	**0.96**
OrganA	Accuracy	**0.93**	0.92	0.92	0.91
	AUC	**0.99**	**0.99**	**0.99**	0.99
OrganC	Accuracy	0.89	0.85	0.89	**0.90**
	AUC	0.98	0.98	0.98	**0.99**
OrganS	Accuracy	0.75	**0.86**	0.78	0.77
	AUC	0.96	0.75	**0.97**	**0.97**
Breast	Accuracy	**0.90**	0.80	0.89	0.89
	AUC	0.89	0.85	**0.90**	0.88
Retina	Accuracy	**0.54**	0.52	**0.54**	0.51
	AUC	**0.75**	0.74	0.74	0.72
Pneumonia	Accuracy	**0.88**	0.83	0.84	0.82
	AUC	0.95	0.95	**0.96**	**0.96**
Derma	Accuracy	0.70	0.70	**0.74**	0.65
	AUC	0.89	0.89	**0.91**	0.85
Chest	Accuracy	0.94	0.94	0.94	0.94
	AUC	0.62	**0.66**	0.65	0.65
Tissue	Accuracy	**0.67**	0.65	0.64	0.57
	AUC	**0.92**	0.90	0.91	0.47
Blood	Accuracy	0.95	0.95	**0.96**	0.93
	AUC	**0.99**	0.99	**0.99**	**0.99**
Average Value Accuracy		**0.815**	0.80	0.80	0.79
Average Value AUC		**0.90**	0.88	**0.90**	0.86

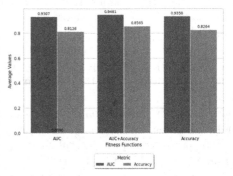

Fig. 6. Average fitness values on multiple 2D MedMNIST datasets of three different fitness functions of proposed Automatic Data Augmentation approach

5.1   Comparison of Different Meta-Heuristics

Although studies on evolutionary NAS have used different meta-heuristic algorithms, choosing a suitable meta-heuristic algorithm for a given problem remains challenging. In this study, we performed a comparative study of some of the well-known metaheuristics. It includes DE [19] alongside its variant LSHADE [22], ACO [21] and PSO algorithms [20].

The results of different meta-heuristics used to search architectures for 2D MedMNIST datasets are given in Table 6. It is worth noting that no meta-heuristic performed well on all the datasets, even if the PSO performed better on average than DE and ACO. As DE performance is sensitive towards its parameters, LSHADE performed well compared to DE, a variant of DE which is a successive history-based adoptive DE variant that keeps track of all best DE parameters with linear population size reduction. LSHADE algorithms have been widely used to solve large-scale optimization problems. It performed better despite the parameter tuning of DE due to several reasons, such as its adaptability, as it incorporates an adaptive mechanism that dynamically adjusts its parameters during the optimization process. This adaptability allows LSHADE to respond better to changes in the optimization landscape and maintain a balance between exploration and exploitation more effectively than DE.

Secondly, LSHADE explored the search space more efficiently than DE leading to the discovery of better solutions. Its methods explore diverse regions within the search space, which is crucial in finding near-optimal or optimal solutions to complex problems such as NAS.

Table 7. Results comparison of the proposed approach with existing studies on BreakHIS dataset.

	Accuracy	Precision	Recall	F1-Score
Bardou [14]	80.083	81.85	80.83	80.48
Yun Jaing [15]	92.270	90.71	92.24	91.42
Nouman et al. [16]	94.710	91.42	91.63	91.76
Proposed	95.05	93.76	93.59	93.66

Table 8. Data Augmentation strategies searched by the proposed Automatic Data Augmentation approach

Dataset	Best data augmentation searched	
OCT	RandomAffine	RandomHorizentalFlip
OrganS	RandomVerticalFlip	RandomErasing
OrganC	RandomAffine	RandomErasing
OrganA	ColorJitter	RandomAffine
Path	ColorJitter	RandomVertical
Tissue	RandomVerticalFlip	RandomRotation
Pneumonia	RandomHorizental	RandomErasing
Chest	HorizentalFlip	VerticalFlip
Breast	ColorJitter	RandomRotate30
Blood	RandomAffine	RandomErasing

6   Discussion and Conclusion

The study presents an efficient evolutionary neural architecture search method for 2D and 3D medical image classification by utilizing ZC proxies for fitness evaluation. It introduces an adaptive DA approach to address model generalization and overfitting issues and conducts a comparative study of metaheuristics to choose the optimal one for the problem at hand. The comparison of our proposed approach with existing NAS studies demonstrates its effectiveness. It outperforms existing approaches in terms of average performance. Furthermore, incorporating attention layers in the search space enables better feature extraction by prioritizing relevant regions of the image and capturing long-range dependencies.

This shows that the proposal is not only effective in the case of small-scale datasets such as MedMNIST but also in the case of large-scale datasets such as BreakHIS, requiring less searching time. DA is crucial for medical images, improving the robustness and generalization ability of DL models trained on limited datasets. Moreover, medical images often suffer from overfitting problems, which can be overcome by using DA. By searching for optimal DA sets for given datasets, an improvement in the performance is noticed. It reveals that it plays an essential role alongside NAS when searching architecture topology.

Table 9. Experimental Results on MedMNIST2D datasets on accuracy before and after data augmentation searched by the proposed Automatic Data Augmentation (ADA) approach

Data Augmentation	Path	OCT	OrganA	OrganC	OrganS	Breast	Retina	Pneumonia	Derma	Chest	Tissue	Blood
	0.900	0.780	0.930	0.900	0.841	0.900	0.540	0.926	0.970	0.740	0.940	0.670
✓	0.920	0.800	0.962	0.935	0.864	0.911	0.561	0.944	0.983	0.769	0.949	0.760

The best-reported set of DA searched by the proposed adaptive DA approach for different MedMNIST datasets is given in Table 8. This shows that the adaptive DA approach searches for various DAs. Besides, performance deterioration can emerge when using a single DA strategy. The accuracy and AUC scores before and after training the model with searched DAs are given in Tables 9 and 10. The idea behind using the combined fitness value of AUC and accuracy is to find the data augmentation set that gives better performance on both AUC and accuracy measures. We also conducted an analysis study to compare the average accuracy and AUC values using different fitness metrics (AUC, accuracy, and AUC+accuracy), as shown in Fig. 6. It clearly shows that combined fitness function leads to better individual DAs in comparison with other fitness functions.

Table 10. Experimental Results on MedMNIST2D datasets on AUC before and after data augmentation searched by the proposed Automatic Data Augmentation (ADA) approach

Data Augmentation	Path	OCT	OrganA	OrganC	OrganS	Breast	Retina	Pneumonia	Derma	Chest	Tissue	Blood
	0.970	0.960	0.990	0.990	0.970	0.925	0.750	0.970	0.910	0.941	0.670	0.920
✓	0.991	0.969	0.999	0.997	0.984	0.930	0.759	0.983	0.944	0.941	0.769	0.967

As recent studies have been conducted on small-scale datasets, e.g., the MedMNIST benchmark, we have also performed experiments on the BreakHIS dataset and demonstrated the potential ability of ZC-NAS approaches to find the best-performing neural architectures for large-scale medical image datasets. Moreover, this study demonstrates that ZC is effective on both 2D and 3D datasets, as shown for multiple 3D MedMNIST datasets. Moreover, the comparison of meta-heuristics shows that the adaptive variant of DE and PSO algorithms performed better than other metaheuristics with satisfactory accuracy and AUC scores. In the near future, we aim to extend the 3D approach for large-scale 3D medical image datasets and propose a multi-objective approach.

Acknowledgment. This work was funded by ArtIC project "Artificial Intelligence for Care" (grant ANR-20-THIA-0006-01) and co-funded by IRIMAS Institute/Université de Haute Alsace. The authors would like to thank the Mesocentre of Strasbourg for providing access to the GPU cluster

References

1. Yang, J., et al.: MedMNIST v2-A large-scale lightweight benchmark for 2D and 3D biomedical image classification. Sci. Data **10**(1), 41 (2023)
2. Ali, M.J., et al.: Designing CNNs using surrogate-assisted GA for medical image classification. In: Proceedings of Companion Conference on Genetic and Evolutionary Computation (2023)
3. Ali, M.J., et al.: Designing attention-based CNNs for medical image classification using GA with variable length-encoding. International Conference on Artificial Evolution (EA). Springer, Cham (2022). https://doi.org/10.1007/978-3-031-42616-2_13
4. Liao, P., Jin, Y., Du, W.: EMT-NAS: transferring architectural knowledge between tasks from different datasets. In: Proceedings of the IEEE/CVF Conference on Computer Vision and Pattern Recognition (2023)
5. Zhang, J., Zhang, L., Li, D.: A unified search framework for data augmentation and neural architecture on small-scale image datasets. In: IEEE Transactions on Cognitive and Developmental Systems (2023)
6. Zhang, J., et al.: An efficient multi-objective evolutionary zero-shot neural architecture search framework for image classification. Int. J. Neural Syst. **33**(05), 2350016 (2023)
7. Huang, J., et al.: EDE-NAS: an eclectic differential evolution approach to single-path neural architecture search. In: Australasian Joint Conference on Artificial Intelligence. Springer International Publishing, Cham (2022). https://doi.org/10.1007/978-3-031-22695-3_9
8. Niu, R., et al.: Neural architecture search based on particle swarm optimization. In: 2019 3rd International Conference on Data Science and Business Analytics (ICDSBA). IEEE (2019)
9. Deng, S., Sun, Y., Galvan, E.: Neural architecture search using genetic algorithm for facial expression recognition. In: Proceedings of the Genetic and Evolutionary Computation Conference Companion (2022)
10. Abdelfattah, M.S., et al.: Zero-cost proxies for lightweight NAS. arXiv preprint arXiv:2101.08134 (2021)

11. Liu, H., Simonyan, K., Yang, Y.: DARTS: differentiable architecture search. arXiv preprint arXiv:1806.09055 (2018)
12. Zoph, B., et al.: Learning transferable architectures for scalable image recognition. In: Proceedings of the IEEE Conference on Computer Vision and Pattern Recognition (2018)
13. Nakai, K., Matsubara, T., Uehara, K.: Att-DARTS: differentiable neural architecture search for attention. In: 2020 International Joint Conference on Neural Networks (IJCNN). IEEE (2020)
14. Bardou, D., Zhang, K., Ahmad, S.M.: Classification of breast cancer based on histology images using convolutional neural networks. IEEE Access 6, 24680–24693 (2018)
15. Jiang, Y., et al.: Breast cancer histopathological image classification using convolutional neural networks with small SE-ResNet module. PLoS ONE 14(3), e0214587 (2019)
16. Ahmad, N., Asghar, S., Gillani, S.A.: Transfer learning-assisted multi-resolution breast cancer histopathological images classification. Vis. Comput. 38(8), 2751–2770 (2022)
17. Ren, P., et al.: A comprehensive survey of neural architecture search: challenges and solutions. ACM Comput. Surv. (CSUR) 54(4), 1–34 (2021)
18. Vo, A., Pham, T.N., Luong, N.H.: Lightweight multi-objective and many-objective problem formulations for evolutionary neural architecture search with the training-free performance metric synaptic flow. Int. J. Comput. Inform. 47(3) (2023)
19. Feoktistov, V.: Differential Evolution. Springer, US (2006). https://doi.org/10.1007/978-0-387-36896-2
20. Kennedy, J., Eberhart, R.: Particle swarm optimization (PSO). In: Proceedings of International Conference on Neural Networks, Perth, Australia, vol. 4, no. 1, pp. 1942–1948 (1995)
21. Dorigo, M., Birattari, M., Stutzle, T.: Ant colony optimization. IEEE Comput. Intell. Mag. 1(4), 28–39 (2006)
22. Tanabe, R., Fukunaga, A.S.: Improving the search performance of SHADE using linear population size reduction. In: 2014 IEEE Congress on Evolutionary Computation (CEC), pp. 1658–1665. IEEE (2014)
23. Wang, Y., et al.: MedNAS: multi-scale training-free neural architecture search for medical image analysis. IEEE Trans. Evol. Comput. (2024)
24. Lu, Z., et al.: NSGA-Net: a multi-objective genetic algorithm for neural architecture search (2018)
25. Zhang, J., et al.: One-shot neural architecture search by dynamically pruning supernet in hierarchical order. Int. J. Neural Syst. 31(07), 2150029 (2021)
26. Xie, S., et al.: SNAS: stochastic neural architecture search. arXiv preprint arXiv:1812.09926 (2018)
27. Wang, H., et al.: Recent advances on neural network pruning at initialization (2021)

Solving Multi-connected BVPs
with Uncertainly Defined Complex Shapes

Andrzej Kużelewski[(✉)] [iD], Eugeniusz Zieniuk[iD], and Marta Czupryna[iD]

Faculty of Computer Science, University of Bialystok, Ciolkowskiego 1M,
15-245 Bialystok, Poland
{a.kuzelewski,e.zieniuk,m.czupryna}@uwb.edu.pl

Abstract. The paper presents the interval fast parametric integral
equations system (IFPIES) applied to solve multi-connected boundary
value problems (BVPs) with uncertainly defined complex shapes of a
boundary. The method is similar to the fast PIES, which uses the fast
multipole method to speed up solving BVPs and reduce RAM utilization.
However, modelling uncertainty in the IFPIES uses interval numbers
and directed interval arithmetic. Segments created the boundary have
the form of the interval Bézier curves of the third degree (curvilinear
segments) or the first degree (linear segment). The curves also required
some modifications connected with applied directed interval arithmetic.
In the presented paper, the reliability and efficiency of the IFPIES solutions were verified on multi-connected BVPs with uncertainly defined
complex linear and curvilinear domain shapes. The solutions were compared with the ones obtained by the interval PIES only due to the lack
of examples of solving uncertainly defined BVPs in the literature. All
presented tests confirm the high efficiency of the IFPIES method.

Keywords: Interval fast parametric integral equations system ·
Interval numbers · Directed interval arithmetic · Uncertainty

1 Introduction

For many years, our team has worked on developing and applying a parametric
integral equations system (PIES) to solve boundary value problems (BVPs).
The multidirectional research covers problems described by different equations,
such as Laplace's, Helmholtz or Navier-Lamé (i.e. [1]). On the other hand, some
enhancements of the method are also considered - the authors of this paper are
focused on two of them: application of uncertainty of data in the PIES [2] and
accelerating performance and reducing memory utilization of the PIES [3].

Traditional modelling and solving BVPs assumes that boundary conditions,
the boundary's shape, and the domain's parameters must be defined by real
numbers, i.e., precisely. However, the practice indicates that to obtain the mentioned data, some measurements should be carried out, which are always affected
by, e.g. gauge reading error or inaccuracy of measurement instruments. Sometimes, the approximation of the model used in the analysis of measurements

© The Author(s), under exclusive license to Springer Nature Switzerland AG 2024
L. Franco et al. (Eds.): ICCS 2024, LNCS 14833, pp. 147–158, 2024.
https://doi.org/10.1007/978-3-031-63751-3_10

may also cause errors. On the other hand, we should assume that the modelled components will be manufactured later with a certain margin of error, which was not considered during the modelling process. Therefore, considering data uncertainty in modelling and solving BVPs becomes a critical problem.

We must emphasize that classical mathematical models require exact input data values. Therefore, it is not possible to apply uncertainty directly. However, many authors modified known methods to consider uncertainty (e.g. [4–6]). The efficient way of using uncertain data in modelling and solving BVPs is the application of interval arithmetic and interval numbers. It has resulted in obtaining the interval finite element method (IFEM) [7], the interval boundary element method (IBEM) [8], and also the interval version of the PIES (IPIES) [9]. However, both the IFEM and the IBEM considered only the uncertainty of boundary conditions or material parameters. Only in a few papers describing 1D problems were the boundary shape parameters, such as beam length, uncertainly defined. The IPIES is more complete - it was developed to solve problems with uncertainly defined boundary shapes. The opportunity to consider all the uncertainties mentioned above in the IPIES is a significant advantage.

The IPIES has advantages inherited from the PIES, such as defining the boundary shape by curves widely used in computer graphics (small amount of input data) and approximating boundary conditions separated from the approximation of boundary shape. Unfortunately, it also has disadvantages connected with the PIES coefficient matrices, which are dense and non-symmetric. The way of creating these matrices requires to compute slightly complicated integrals. That process requires a lot of CPU time, especially for problems with a considerable number of segments describing the shape of the boundary (complex or large-scale problems). Unfortunately, applying interval arithmetic and interval numbers also negatively affects the computational speed and utilizes more memory (RAM).

Another problem is the method of solving the final system of algebraic equations. In traditional PIES, classical Gaussian elimination was first applied. The authors of this paper adapted the specialized libraries (such as LAPACK) to solve the system more efficiently. Also, parallelization of the PIES by OpenMP and CUDA to reduce the time of computations was proposed in our previous papers (e.g. [10,11]. However, we must still store all coefficient matrices in the operating memory. Therefore, RAM consumption is at a high level.

In the mid-1980s of the 20th century, Rokhlin and Greengard developed the fast multipole method (FMM) [12]. This compression technique reduces the utilization of RAM and computational time that is well-documented in solving potential BVPs (i.e. [13]), also in the BEM [14]. Applying the FMM to the PIES required a new approach to computing matrices coefficients and the iterative method for solving the system of algebraic equations. Also, we had to make additional changes in the FMM tree structure [15]. However, obtained that way, the fast PIES (FPIES) is an efficient method for solving BVPs [3].

The main goal of this paper is to present the interval fast Parametric Integral Equation System (IFPIES) applied for numerical solving of 2D uncertainly defined multi-connected potential complex BVPs. The IFPIES previously proposed in [16] for single-connected polygonal domains is based on both the IPIES and the FPIES. In the presented paper, the efficiency and accuracy of the IFPIES are tested on the examples of multi-connected complex uncertainly defined domains described by linear and curvilinear segments.

2 Boundary Shape and Boundary Conditions Uncertainty

The application of classical [17] and directed [18] interval arithmetic into the PIES was troublesome, as is described in previous papers (e.g. [9]). Therefore, we proposed mapping arithmetic operators to the positive semi-axis (clearly described in [9]) while applying the directed interval arithmetic to obtain the IPIES. The same strategy was also applied in the IFPIES.

The general formula of the IFPIES [3] has the following form:

$$
\frac{1}{2}u_l(\widehat{\tau}) = \sum_{j=1}^{n} \mathbb{R}\left\{ \int_{s_{j-1}}^{s_j} \widehat{U}_{lj}^{*(c)}(\widehat{\tau},\tau)p_j(s)J_j^{(c)}(s)ds \right\} -
$$
$$
\sum_{j=1}^{n} \mathbb{R}\left\{ \int_{s_{j-1}}^{s_j} \widehat{P}_{lj}^{*(c)}(\widehat{\tau},\tau)u_j(s)J_j^{(c)}(s)ds \right\},
$$
$$
l = 1, 2, ..., n, \ \ s_{l-1} \leq \widehat{s} \leq s_l, \ \ s_{j-1} \leq s \leq s_j,
$$

(1)

where: $\widehat{s}$ and s are defined in the parametric coordinate system, s_{j-1} (s_{l-1}) correspond to the beginning while s_j (s_l) to the end of interval segment S_j (S_l), n is the number of parametric segments that creates a boundary of the domain in 2D, $J_j^{(c)}(s)$ is the interval Jacobian, $u_j(s)$ and $p_j(s)$ are interval parametric boundary functions on individual segments S_j of the interval boundary, $\mathbb{R}$ is the real part of complex function.

The interval kernels modified to complex functions $\widehat{U}_{lj}^{*(c)}(\widehat{\tau},\tau)$ and $\widehat{P}_{lj}^{*(c)}(\widehat{\tau},\tau)$ have the following form [3]:

$$
\widehat{U}_{lj}^{*(c)}(\widehat{\tau},\tau) = -\frac{1}{2\pi}\ln(\widehat{\tau}-\tau),
$$
$$
\widehat{P}_{lj}^{*(c)}(\widehat{\tau},\tau) = \frac{1}{2\pi}\frac{n_j^{(c)}}{\widehat{\tau}-\tau}.
$$

(2)

where $n_j^{(c)}$ is the complex notation of normal vector to the interval curve, which creates segment j. Expressions $\widehat{\tau}, \tau$ are the complex version of parametric functions describing the boundary, which have the following interval form:

$$
\widehat{\tau} = S_l^{(1)}(\widehat{s}) + iS_l^{(2)}(\widehat{s}),
$$
$$
\tau = S_j^{(1)}(s) + iS_j^{(2)}(s),
$$

(3)

where the interval components connected with the direction of coordinates in a 2D Cartesian reference system: $S_j^{(1)} = [\underline{S}_j^{(1)}, \overline{S}_j^{(1)}], S_j^{(2)} = [\underline{S}_j^{(2)}, \overline{S}_j^{(2)}], S_l^{(1)} = [\underline{S}_l^{(1)}, \overline{S}_l^{(1)}]$ and $S_l^{(2)} = [\underline{S}_l^{(2)}, \overline{S}_l^{(2)}]$. These components have the form of directed intervals [18].

The boundary is modelled by interval Bézier curves of the first degree (linear):

$$S_k(s) = a_k^{(f)} s + b_k^{(f)}, \quad 0 \le s \le 1, \quad (f) - \text{first degree}, \tag{4}$$

and the third degree (curvilinear):

$$S_k(s) = a_k^{(t)} s^3 + b_k^{(t)} s^2 + c_k^{(t)} s + d_k^{(t)}, \quad 0 \le s \le 1, \quad (t) - \text{third degree} \tag{5}$$

where vector $S_k(s) = [S_k^{(1)}(s), S_k^{(2)}(s)]^T$, $k = \{l, j\}$ and s is a variable in the parametric reference system. Coefficients $a_k^{(f)}, b_k^{(f)}, a_k^{(t)}, b_k^{(t)}, c_k^{(t)}, d_k^{(t)}$ have also form of vectors composed of two interval components (similarly to $S_k(s)$). They are computed using interval points describing particular segments of the boundary as presented in Fig. 1 (the graphical example assumes $k = j$):

$$a_j^{(f)} = P_{e(j+2)} - P_{b(j+2)}, \quad b_j^{(f)} = P_{b(j+2)},$$

$$a_j^{(t)} = P_{e(j)} - 3P_{i2(j)} + 3P_{i1(j)} - P_{b(j)}, \quad b_j^{(t)} = 3(P_{i2(j)} - 2P_{i1(j)} + P_{b(j)}),$$

$$c_j^{(t)} = 3(P_{i1(j)} - P_{b(j)}), \quad d_j^{(t)} = P_{b(j)},$$

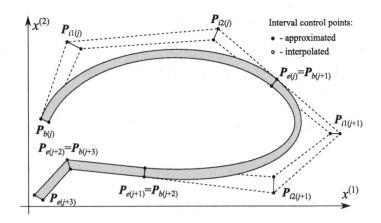

Fig. 1. The interval Bézier curves of the first and third degree used to define segments of the boundary

where coordinates of all points P, regardless of their subscript, have the form of a vector of intervals:

$$P = [P^{(1)}, P^{(2)}]^T = \left[[\underline{P}^{(1)}, \overline{P}^{(1)}], [\underline{P}^{(2)}, \overline{P}^{(2)}] \right]^T.$$

The interval boundary functions $u_j(s)$ and $p_j(s)$ in (1) present boundary conditions and the following series approximate them:

$$u_j(s) = \sum_{k=0}^{N} u_j^{(k)} L_j^{(k)}(s), \quad p_j(s) = \sum_{k=0}^{N} p_j^{(k)} L_j^{(k)}(s), \tag{6}$$

where $u_j^{(k)} = [\underline{u}_j^{(k)}, \overline{u}_j^{(k)}]$ and $p_j^{(k)} = [\underline{p}_j^{(k)}, \overline{p}_j^{(k)}]$ are unknown or given interval values of boundary functions in defined points of the segment j, N - is the number of terms in approximating series (6) and $L_j^{(k)}(s)$ – the base functions (Lagrange polynomials) on segment j.

3   Process of Solving the IFPIES

Applying the FMM into the PIES is the first step in solving the IFPIES. The FMM is based on the tree structure, which transforms interactions between the individual PIES boundary segments into interactions between some groups of segments called cells. In the IFPIES, we applied a modified version of the tree [15] (presented in Fig. 2). Unlike the classical binary tree, we had to join the beginning and end of each level to consider that a parametric 1D system describes the PIES for 2D issues. Also, the PIES's modified kernels (complex form) are expanded using the Taylor series. It allows the calculation of complex integrals to be converted into approximate sums. All process of applying the FMM into the PIES is clearly described in [15].

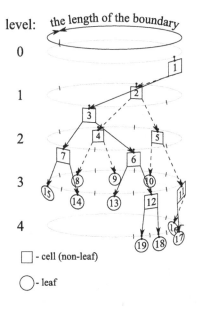

Fig. 2. The example of modified binary tree in the IFPIES for 2D problem

The process of applying the FMM into the IPIES is very similar, and as a result, we obtained the following form of integrals in (1) [16]:

$$\int_{s_{j-1}}^{s_j} \widehat{U}_{lj}^{*(c)}(\widehat{\tau}, \tau)\, p_j(s)\, J_j^{(c)}(s)\, ds = \frac{1}{2\pi} \sum_{l=0}^{N_T}(-1)^l \cdot$$

$$\left\{ \sum_{k=0}^{N_T} \sum_{m=l}^{N_T} \frac{(k+m-1)! \cdot M_k(\tau_c)}{(\tau_{el} - \tau_c)^{k+m}} \cdot \frac{(\tau'_{el} - \tau_{el})^{m-l}}{(m-l)!} \right\} \frac{(\widehat{\tau} - \tau'_{el})^l}{l!},$$

$$\int_{s_{j-1}}^{s_j} \widehat{P}_{lj}^{*(c)}(\widehat{s}, s)\, u_j(s)\, J_j^{(c)}(s)\, ds = \frac{1}{2\pi} \sum_{l=0}^{N_T}(-1)^l \cdot$$

$$\left\{ \sum_{k=1}^{N_T} \sum_{m=l}^{N_T} \frac{(k+m-1)! \cdot N_k(\tau_c)}{(\tau_{el} - \tau_c)^{k+m}} \cdot \frac{(\tau'_{el} - \tau_{el})^{m-l}}{(m-l)!} \right\} \frac{(\widehat{\tau} - \tau'_{el})^l}{l!}.$$

(7)

where: N_T is the number of terms in the Taylor expansion, $\widehat{\tau} = S_l^{(1)}(\widehat{s}) + iS_l^{(2)}(\widehat{s})$, $\tau = S_j^{(1)}(s) + iS_j^{(2)}(s)$, complex interval points τ_c, τ_{el}, τ'_c, τ'_{el} are midpoints of leaves obtained while tracing the tree structure (see [3]). Expressions $M_k(\tau_c)$ and $N_k(\tau_c)$ are called moments (and they are computed twice only) and have the form [16]:

$$M_k(\tau_c) = \int_{s_{j-1}}^{s_j} \frac{(\tau - \tau_c)^k}{k!} p_j(s) J_j^{(c)}(s)\, ds,$$

$$N_k(\tau_c) = \int_{s_{j-1}}^{s_j} \frac{(\tau - \tau_c)^{k-1}}{(k-1)!} n_j^{(c)} u_j(s) J_j^{(c)}(s)\, ds.$$

(8)

where $n_j^{(c)} = n_j^{(1)} + i n_j^{(2)}$ the complex interval normal vector to the curve created segment j.

Similarly to the original PIES, the IFPIES are written at collocation points whose number corresponds to the number of unknowns. However, during solving the IFPIES, the system of algebraic equations $A \cdot x = b$ is produced implicitly, contrary to the original PIES. It means that only the result of multiplication of the matrix A by the vector of unknowns x is used by applied iterative GMRES solver [19]. The solver is modified by applying directed interval arithmetic and directly integrated with the IFPIES. The authors also applied the same GMRES solver to the IPIES to prepare a more reliable comparison.

4   Numerical Results

All tests are performed on a PC based on Intel Core i5-4590S with 32 GB RAM. Application of the IPIES and the IFPIES are compiled by g++ 7.5.0 (-O2 optimization) on 64-bit Linux OS (kernel 6.2.0). Two multi-connected problems with linear and mixed (linear and curvilinear) segments created the shape of the boundary are considered.

4.1 L-Shaped Problem with Randomly Placed Holes

The domain boundary in the first example is composed of linear segments only, as presented in Fig. 3. Laplace's equation describes the problem. Interval boundary conditions are also presented in Fig. 3 (where u - Dirichlet and p - Neumann boundary conditions).

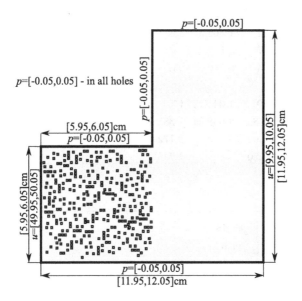

Fig. 3. L-shaped problem with randomly placed holes

The research focused on the CPU time, RAM utilization and accuracy of the IFPIES compared to the IPIES. Due to the lack of literature on solving problems with uncertainly defined boundary shapes and boundary conditions, comparison to others is impossible. The number of terms in the Taylor series is set to 25, and the value of GMRES tolerance equals 10^{-8}. The number of collocation points in all segments is constant and is changed from 2 to 8 in subsequent research.

For the L-shaped problem, we solved three examples with different numbers of holes: 400, 900 and 1600, placed in the part of the domain presented in Fig. 3. Therefore, we solved the systems from 3 212 to 51 248 equations.

As can be seen from Table 1, the IFPIES is as accurate as the IPIES, which is proved by the result of computation of the mean square error (MSE) between the lower and upper bound (infimum and supremum) of the IFPIES and the IPIES solutions. The MSE between both methods is very low and does not exceed 10^{-8}.

Table 1. Comparison between the IFPIES and the IPIES for L-shaped problem

Number of		CPU time [s]		RAM utilization [MB]		MSE	
col. pts	eqs	*IFPIES*	*IPIES*	*IFPIES*	*IPIES*	inf	sup
400 holes							
2	3 212	34.13	66.52	57.14	242	$6.29 \cdot 10^{-12}$	$5.60 \cdot 10^{-11}$
4	6 424	84.95	290.74	99	962	$3.15 \cdot 10^{-14}$	$4.67 \cdot 10^{-11}$
6	9 636	148.33	706.18	113	2 158	$8.62 \cdot 10^{-10}$	$3.47 \cdot 10^{-11}$
8	12 848	202.80	1 345.22	170	3 831	$1.11 \cdot 10^{-11}$	$1.46 \cdot 10^{-10}$
900 holes							
2	7 212	98.38	333.55	140	1 221	$8.62 \cdot 10^{-10}$	$3.47 \cdot 10^{-11}$
4	14 424	286.54	1 465.37	339	4 804	$5.14 \cdot 10^{-10}$	$5.62 \cdot 10^{-11}$
6	21 636	563.77	3 552.72	585	10 700	$1.17 \cdot 10^{-10}$	$3.65 \cdot 10^{-11}$
8	28 848	896.74	6 782.20	957	19 100	$6.95 \cdot 10^{-10}$	$6.14 \cdot 10^{-11}$
1600 holes							
2	12 812	202.80	1 345.22	170	3 831	$7.18 \cdot 10^{-8}$	$3.28 \cdot 10^{-11}$
4	25 624	1 090.83	4 658.04	1 381	15 100	$9.57 \cdot 10^{-8}$	$5.31 \cdot 10^{-11}$
6	38 436	2 343.02	—	2 772	—	—	—
8	51 248	4 130.20	—	4 745	—	—	—

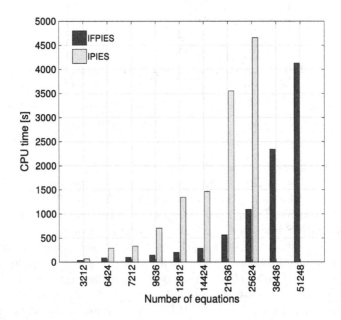

Fig. 4. Comparison of computation time of the IFPIES and the IPIES for different numbers of equations

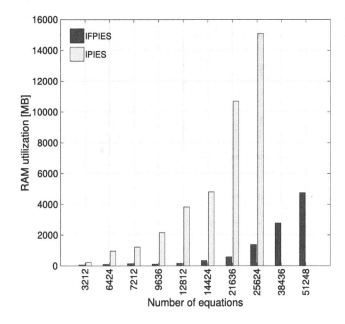

Fig. 5. Comparison of RAM utilization of the IFPIES and the IPIES for different numbers of equations

It is worth emphasising that solving the IPIES for 38 436 and 51 248 equations is not possible due to utilization of all computer memory, while the IFPIES consumed about 15% of available RAM. It means that we can solve up to about 35 000 equations using the IPIES due to exhaustion of RAM, while the IFPIES uses less than 2 GB of memory. It presents limitations of the classical IPIES contrary to contemporary one.

Also, from Figs. 4 and 5, it can be seen that the speedup of the IFPIES in relation to the IPIES grows with an increasing number of equations, whilst RAM utilization of the IFPIES is smaller and grows much slower than the IPIES. Also, for a smaller number of equations, the FMM overhead gives us a considerably smaller gain in CPU time and RAM utilization.

4.2   Current Flow Through the Plate with Holes

The domain boundary in the second example is composed of mixed linear and curvilinear segments, as presented in Fig. 6. Laplace's equation also describes the problem. Interval boundary conditions are presented in Fig. 6 (where V - potential and $\frac{\partial V}{\partial n}$ - flux).

The number of terms in the Taylor series and the value of GMRES tolerance is the same as in the previous problem, equal to 25 and 10^{-8}. The number of collocation points in all segments is also constant and has changed from 2 to 8. At last, we solved the systems from 32 048 to 128 192 equations.

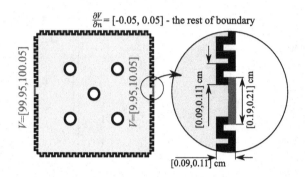

Fig. 6. Current flow through the plate with holes

Table 2. Results of solving current flow problem by the IPIES

Number of		CPU time	RAM utilization
col. pts	eqs	[s]	[MB]
2	32048	312.30	374
4	64096	623.43	892
6	96144	1 151.06	1624
8	128192	1 860.20	2473

We are able to solve only examples with the smallest number of equations using the IPIES due to exhaustion of RAM. However, the IPIES for 32 048 equations uses 23.7 GB of RAM and requires 7 031.32 s (almost 2 h) of CPU time. It is incomparably more than in the IPIES. The MSE between solutions of both methods is very small and equal to $2.22 \cdot 10^{-8}$ for lower and $1.89 \cdot 10^{-9}$ for upper bound.

As seen from Table 2, solving the example with over 128 000 equations requires less than 2.5 GB of memory and about a half hour. The obtained results confirm the very high efficiency of the IFPIES and allow us to solve large-scale examples on a standard PC in a reasonable time.

5   Conclusions

The paper presents the IFPIES in solving 2D potential curvilinear multi-connected boundary value problems with uncertainly defined boundary shapes and conditions. The method gives us the opportunity to include measurement errors (the uncertainty of measurement data) of the boundary shape and boundary conditions in calculations, which is impossible in classic practical design.

Application of the fast multipole technique in the IFPIES also allows for the highly efficient solving of complex (large-scale) engineering problems on a standard PC in a reasonable time. The real power of the IFPIES is very low RAM utilization. The IPIES is unable to cope with the solution of over 35 000 equations for 32 GB of RAM, while the IFPIES easily solves the examples with over 128 000 equations. Also, the CPU time of the IFPIES is significantly shorter than the IPIES.

The obtained results suggest that the direction of research should be continued. Our further research should cover problems modelled by other than Laplace's equations.

References

1. Kapturczak, M., Zieniuk, E., Kużelewski, A.: NURBS curves in parametric integral equations system for modeling and solving boundary value problems in elasticity. In: Krzhizhanovskaya, V.V., et al. (eds.) ICCS 2020. LNCS, vol. 12138, pp. 116–123. Springer, Cham (2020). https://doi.org/10.1007/978-3-030-50417-5_9
2. Zieniuk, E., Kużelewski, A.: Concept of the interval modelling the boundary shape using interval bézier curves in boundary problems solved by PIES. In: Simos, T.E., et al. (eds.) 12th International Conference of Numerical Analysis and Applied Mathematics ICNAAM 2014, AIP Conference Proceedings, vol. 1648, 590002. AIP Publishing LLC., Melville (2015). https://doi.org/10.1063/1.4912829
3. Kużelewski, A., Zieniuk, E.: Solving of multi-connected curvilinear boundary value problems by the fast PIES. Comput. Methods Appl. Mech. Eng. **391**, 114618 (2022)
4. Fu, C., Zhan, Q., Liu, W.: Evidential reasoning based ensemble classifier for uncertain imbalanced data. Inf. Sci. **578**, 378–400 (2021)
5. Wang, C., Matthies, H.G.: Dual-stage uncertainty modeling and evaluation for transient temperature effect on structural vibration property. Comput. Mech. **63**(2), 323–333 (2019)
6. Gouyandeh, Z., Allahviranloo, T., Abbasbandy, S., Armand, A.: A fuzzy solution of heat equation under generalized Hukuhara differentiability by fuzzy Fourier transform. Fuzzy Sets Syst. **309**, 81–97 (2017)
7. Ni, B.Y., Jiang, C.: Interval field model and interval finite element analysis. Comput. Methods Appl. Mech. Eng. **360**, 112713 (2020)
8. Zalewski, B., Mullen, R., Muhanna, R.: Interval boundary element method in the presence of uncertain boundary conditions, integration errors, and truncation errors. Eng. Anal. Boundary Elem. **33**(4), 508–513 (2009)
9. Zieniuk, E., Kapturczak, M., Kużelewski, A.: Modification of interval arithmetic for modelling and solving uncertainly defined problems by interval parametric integral equations system. In: Shi, Y., et al. (eds.) ICCS 2018. LNCS, vol. 10862, pp. 231–240. Springer, Cham (2018). https://doi.org/10.1007/978-3-319-93713-7_19
10. Kużelewski, A., Zieniuk, E.: OpenMP for 3D potential boundary value problems solved by PIES. In: Simos, T.E., et al. (eds.) 13th International Conference of Numerical Analysis and Applied Mathematics ICNAAM 2015, AIP Conference Proceedings, vol. 1738, 480098. AIP Publishing LLC., Melville (2016). https://doi.org/10.1063/1.4952334
11. Kuzelewski, A., Zieniuk, E., Boltuc, A.: Application of CUDA for Acceleration of Calculations in Boundary Value Problems Solving Using PIES. In: Wyrzykowski, R., Dongarra, J., Karczewski, K., Waśniewski, J. (eds.) PPAM 2013. LNCS, vol. 8385, pp. 322–331. Springer, Heidelberg (2014). https://doi.org/10.1007/978-3-642-55195-6_30
12. Greengard, L.F., Rokhlin, V.: A fast algorithm for particle simulations. J. Comput. Phys. **73**(2), 325–348 (1987)

13. Huang, T., Zhu, Y.X., Ha, Y.J., Wang, X., Qiu, M.K.: A hardware pipeline with high energy and resource efficiency for FMM acceleration. ACM Trans. Embed. Comput. Syst. **17**(2), 51 (2018)
14. Barbarino, M., Bianco, D.: A BEM-FMM approach applied to the combined convected Helmholtz integral formulation for the solution of aeroacoustic problems. Comput. Methods Appl. Mech. Eng. **342**, 585–603 (2018)
15. Kużelewski, A., Zieniuk, E.: The FMM accelerated PIES with the modified binary tree in solving potential problems for the domains with curvilinear boundaries. Numer. Algorithms **88**(3), 1025–1050 (2021)
16. Kużelewski, A., Zieniuk, E., Czupryna, M.: Interval modifications of the fast PIES in solving 2D potential BVPs with uncertainly defined polygonal boundary shape. In: Groen, D., et al. (eds.) Computational Science - ICCS 2022, LNCS, vol. 13351, pp. 18–25. Springer, Cham (2022). https://doi.org/10.1007/978-3-031-08754-7_3
17. Moore, R.E.: Interval Analysis. Prentice-Hall, Englewood Cliffs, New York (1966)
18. Markov, S.M.: On directed interval arithmetic and its applications. J. Univ. Comput. Sci. **1**(7), 514–526 (1995)
19. Saad, Y., Schultz, M.H.: GMRES: a generalized minimal residual algorithm for solving non-symmetric linear systems. SIAM J. Sci. Stat. Comput. **7**, 856–869 (1986)

XLTU: A Cross-Lingual Model in Temporal Expression Extraction for Uyghur

Yifei Liang[1,2], Lanying Li[3], Rui Liu[1,2], Ahtam Ahmat[1,2], and Lei Jiang[1,2(✉)]

[1] Institute of Information Engineering, Chinese Academy of Sciences,
Beijing, China
{liangyifei,liurui3221,aihetanmuaihemaiti,jianglei}@iie.ac.cn
[2] School of Cyber Security, University of Chinese Academy of Sciences,
Beijing, China
[3] Civil Aviation Flight University of China Xinjin Flight College, Chengdu, China
lly010214@163.com

Abstract. Temporal expression extraction (TEE) plays a crucial role in natural language processing (NLP) tasks, enabling the capture of temporal information for downstream tasks such as logical reasoning and information retrieval. However, current TEE research mainly focuses on resource-rich languages like English, leaving a gap for minor languages (e.g. Uyghur) in research. To address these issues, we create an English-Uyghur cross-lingual dataset specifically for the task of temporal expression extraction in Uyghur. Besides, considering the unique characteristics of Uyghur, we propose XLTU, a **Cross-L**ingual model in **T**emporal expression extraction for **U**yghur, and utilize multi-task learning to help transfer the knowledge from English to Uyghur. We compare XLTU with different models on our dataset, and the results demonstrate that our model XLTU achieves the SOTA results on various evaluation metrics. We make our code and dataset publicly available (https://github.com/lyfcsdo2011/XLTU).

Keywords: temporal expression extraction · Uyghur · cross-lingual · multi-task learning

1 Introduction

Currently, temporal expression extraction (TEE) is an important NLP task [1], which specifically refers to detecting expressions about time such as date, duration, etc. This task has wide-ranging applications in downstream tasks, including question answering [2], information retrieval [3], and causal reasoning [4]. In the past, the work of TEE mainly relies on rule-based approaches [5,6], while the current focus has shifted towards leveraging deep learning techniques [7–9]. However, the field of TEE for minor languages still lacks sufficient research and development, indicating a noticeable scarcity in this area. Due to the scarcity of annotated datasets for minor languages, it shows the suboptimal performance of deep learning methods in these languages.

L. Franco et al. (Eds.): ICCS 2024, LNCS 14833, pp. 159–173, 2024.
https://doi.org/10.1007/978-3-031-63751-3_11

In this study, we focus on addressing TEE in Uyghur, a language with distinctive characteristics that set it apart from more widely used languages. Most languages, used for pretraining (e.g. mBERT [10], XLM-R [11]), are read and written from left to right. However, Uyghur is read and written in the opposite direction, as depicted in Fig. 1. Additionally, the vocabulary of Uyghur significantly differs from that of European languages. When applying pre-trained cross-lingual models to the Uyghur language, these differences will lead to substantial deviations in feature learning and knowledge transfer, because the models cannot obtain Uyghur language features well.

Fig. 1. The order difference between English and Uyghur. English follows a left-to-right pattern, while Uyghur follows a right-to-left pattern.

In order to address these challenges, it is crucial to expand the high-quality datasets and improve the performance of TEE methods in Uyghur. In our study, we create a high-quality English-Uyghur cross-lingual dataset specifically for TEE in Uyghur. This dataset allows us to transfer the knowledge from English to Uyghur. Besides, considering the unique characteristics of Uyghur, we propose XLTU: a **Cross-L**ingual model in **T**emporal expression extraction for **U**yghur, a method based on a pre-trained model, and utilize multi-task learning (MTL) [12] to facilitate transfer of English knowledge to Uyghur in TEE (as shown in Fig. 2).

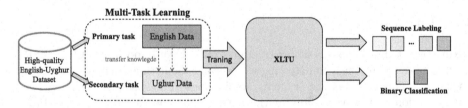

Fig. 2. Overview diagram of our work. We create a high-quality English-Uyghur dataset for TEE (the left). Besides, we propose XLTU, and utilize multi-task learning to train the model. The primary task is formulated as a sequence labeling task, and the secondary task is formulated as a binary classification task.

Our model involves two tasks: a primary task and a secondary task. In the primary task, we train the model using existing annotated English TEE data. This helps the model learn the explicit knowledge and understand the structure of temporal expressions in English. In the secondary task, we map the annotated English TEE data samples to Uyghur. This process allows us to obtain sentence-level labels (containing one or more time expressions) based on the original token-level labels. In a weakly supervised manner, we transfer the implicit knowledge learned in the target language by teaching the model to detect whether the target language contains temporal expressions.

The main contributions of this paper are: 1) We create a high-quality English-Uyghur cross-lingual dataset for TEE multi-task in minor language Uyghur. 2) We propose XLTU utilizing multi-task learning for TEE in Uyghur. 3) We show that XLTU can effectively promote the learning of Uyghur language in TEE, and achieves SOTA results on our dataset.

2   Related Work

Although TEE is very important in NLP, there are limited studies on this task, particularly for languages with limited data resources. Most existing research in this field has primarily focused on resource-rich languages like English. Currently, there are two main types of technologies used for TEE.

One is a rule/pattern-based method. HeidelTime [5] is the best-performing method so far and covers more than ten languages. It is driven by a carefully tuned set of rules. This approach is later extended to additional languages via HeidelTime-auto [13], which exploits language-independent processing and rules. Other methods, such as SynTime [6], SUTIME [14], and PTime [15], utilize heuristic rule-based approaches and pattern-learning techniques.

The second type of approach for TEE involves deep learning methods, and this is also a current major research direction. For instance, [16] proposes an RNN-based model, while [7] utilizes BERT with linear classifiers. [8] feeds mBERT embeddings into BiLSTM with CRF layers and outperforms HeidelTime-auto in four languages. [9] proposes a framework based on pre-trained models and learns in a multi-task manner. However, compared to other tasks, the performances of the deep learning-based methods reported are inferior in cross-lingual TEE. This is highly attributed to the lack of annotated datasets for minority languages. In our work, we propose XLTU to make the model learn cross-lingual features much better. Besides, we create a high-quality cross-lingual dataset to make up for the insufficient data available for minor languages.

Moreover, applying the label projection method can better solve the problem of lack of data in TEE. TMP [17] is originally proposed for cross-lingual named entity recognition (NER) [18], projecting English data in IOB (which means Inside Outside Beginning) tagging format [19] using machine translation, orthographic and phonetic similarities to other languages of the package. [9] proposes a MTL framework to transfer temporal knowledge of source languages into target languages.

In the early stage, an important motivation for MTL is to alleviate the problem of data sparsity in machine learning. When the *big data* era emerges, multi-task learning is more effective which utilizes more data from different learning tasks than single-task learning. [12] proposes a model called MT-DNN which combines multi-task learning and language model pre-training for language representation learning. MulT [20] is an end-to-end multitask learning Transformer [21] framework to simultaneously learn multiple high-level vision tasks. DeMT [22] is a novel MTL model that combines both merits of deformable CNN and query-based Transformer for multi-task learning of dense prediction.

3 English-Uyghur Cross-Lingual TEE Dataset

3.1 Temporal Expression Types

ISO-TimeML [23] has already presented the TEE dataset annotation guideline, there are four types of temporal expressions, i.e., *Date, Time, Duration,* and *Set*. *Date* refers to a calendar date, usually a day or a larger unit of time. *Time* refers to a time of day, with a granularity smaller than a day. *Duration* refers to an expression that clearly describes a period of time. *Set* refers to a regular set of time of recurrence. An intuitive representation can be seen in Table 1.

Table 1. Temporal expressions of four types. Definitions of types Seeing 3.2.

Please pay attention, I will see you next Friday, have a good rest
Date

The warrants may be exercised until 90 days after their issue date
Duration

I persist in exercising every day after work to keep a healthy body
Set

I have a stomach ache, I need to go to hospital on Friday morning.
Time

3.2 Dataset Structure

For the English dataset, following [9], we collect TE3 [1], Wikiwars [24] and Tweets [6]. As for the Uyghur datasets, a part of them is obtained through machine translation of English TE3 [1] and Tweets [6]. We also employ web crawling techniques to collect additional data, which is then carefully cleaned and filtered to ensure high-quality data for manual labeling. According to the multi-tasks we have designed, the primary task takes the form of cross-lingual sequence labeling, which includes the Named Entity Recognition (NER) [18] task. Meanwhile, the secondary task is designed as a binary classification task.

For the primary task, the training data consists of the whole English dataset in the NER format. The test data consists of Uyghur data that has been manually annotated in the NER format and is used to evaluate the cross-lingual capabilities of our model by predicting temporal expressions in Uyghur. In total, we annotate 22,726 pieces of Uyghur data for this task, including 330 pieces labeled as *Date*, 100 pieces as *Duration*, 33 pieces as *Set*, and 40 pieces as *Time* (as shown in Table 2).

Regarding the secondary task, the training data comprises Uyghur sentences obtained through machine translation from English. These sentences are manually labeled for classification.

Table 2. The statistics of the English-Uyghur cross-lingual datasets.

Language	Dataset	Domain	Expressions	Dates	Times	Durations	Sets	Tokens
English	TE3 [1]	News	1830	1471	34	291	34	124592
	Wikiwar [24]	Narrative	2634	2634	0	0	0	
	Tweet [6]	Utterance	1128	717	173	200	38	
Uyghur	Our work	Websites	503	330	40	100	33	22726

4   Proposed Model

TEE is formalized as a sequence labeling task. Inspired by [8,9,12,25], the architecture of our model is shown in Fig. 3.

4.1   Pre-trained Multilingual Model

Considering the limited availability of resources for the Uyghur language, we utilize the base XLM-Roberta model (XLM-R) [11] as the backbone. XLM-R is a state-of-the-art multilingual model and outperforms other models in various cross-lingual tasks. One of the main advantages of XLM-R is its extensive training on a wide range of languages and datasets, this gives XLM-R a larger vocabulary to learn and adapt to the characteristics of Uyghur words. It has also introduced three training targets to further enhance its performance in cross-lingual tasks. The pre-trained multilingual model consists of lexicon and Transformer encoder layers, as shown in Fig. 3. The backbone of the model is shared across all the MTL tasks during both the training and testing phases.

4.2   TextCNN

TextCNN [25] has already demonstrated that Convolutional Neural Networks (CNN) [26] can be effectively applied to text processing tasks, yielding impressive results. CNNs can be combined with pre-trained language models to further extract informative features for downstream tasks (e.g. classification), as

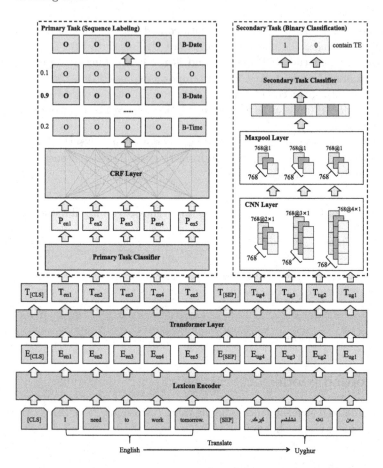

Fig. 3. Model structure of XLTU. It shows how our model transfers knowledge from English to Uyghur through the primary and the secondary tasks.

seen in models like BERT-CNN [27]. Considering the specific characteristics of the Uyghur language, we have introduced a CNN neural network after the pre-trained language model in our architecture. This allows us to leverage the last hidden state of the language model to extract additional and right-to-left features, which are then utilized in the secondary task. This CNN component enhances the model's ability to capture relevant information and improve performance on the given task, as shown in Fig. 3.

4.3   Conditional Random Fields

Conditional Random Field (CRF) [28] is widely used in sequence labeling tasks. It has been proven to enhance the performance of sequence labeling models and address issues such as mismatched predicted labels or labels that do not start with the 'B' label, such as 'O O B-Date I-Time' or 'O O I-Date I-Date'.

Figure 3 demonstrates that the undirected graphical structure of CRF enables the model to learn the context relationship between each token in both directions. Therefore, CRF helps our model better capture the contextual and right-to-left information and make more accurate predictions.

4.4 Cross-Lingual Knowledge Transfer Based on MTL

Our model is capable of transferring knowledge from English to Uyghur. To facilitate explicit and implicit knowledge transfer, we have designed the primary and secondary tasks on top of the backbone. The primary task focuses on explicitly encoded time expressions in English. It is formulated as a sequence labeling task and utilizes the training data of English to train the backbone network, which includes the primary task classifier and CRF layer. The architecture is illustrated in the top of left corner of Fig. 3. In the primary task, We incorporate two different loss functions $\mathcal{L}_t$ and $\mathcal{L}_{crf}$:

$$\mathcal{L}_t = -\sum_{i=1}^{b}\sum_{j=1}^{m_i} \mathbb{1}(y_{ij}, c) log(softmax(W_1 \cdot x)), \tag{1}$$

$$\mathcal{L}_{crf} = -log P(Y|X;\theta), \tag{2}$$

where $\mathcal{L}_t$ represents the loss between the labels directly predicted by the backbone outputs and the ground-truth labels. $\mathcal{L}_{crf}$ represents the loss between the predicted labels after the backbone outputs pass through CRF and the ground-truth labels. b is the total number of input sequences and m_i is the length of the i_{th} sequence. $x \in \mathbb{R}^d$ is the embedding of the j_{th} token in the i_{th} sequence of output by the backbone model. d is its dimension. $c = argmax(W_1 \cdot x)$ is the predicted label for each token while y_{ij} is the ground-truth label of each token. $W_1 \in \mathbb{R}^{|c| \times d}$ is the classifier parameters of the primary task. $|c|$ is the total number of unique ground-truth labels. $\mathbb{1}(,)$ is 1 if two are equal, 0 otherwise. Y is the set of ground-truth labels while X it the set of predicted labels, θ is the parameters of backbone model. $P(Y|X;\theta)$ represents the probability of labeling sequence Y under the condition of given input sequence X, which can be formulated as:

$$P(Y|X;\theta) = \frac{1}{Z(X;\theta)} \exp(\sum_{i=1}^{n}\sum_{j=1}^{k} \theta_j f_j(y_{i-1}, y_i, x_i)), \tag{3}$$

where $Z(X;\theta)$ is the normalization factor, $f_j(y_{i-1}, y_i, x_i)$ is the characteristic function of CRF, θ_j is the weight corresponding to the characteristic function. And the final target of the primary task is to minimize the $\mathcal{L}_{sl}$:

$$\mathcal{L}_{sl} = \alpha \cdot \mathcal{L}_{crf} + \beta \cdot \mathcal{L}_t, \tag{4}$$

where α and β are the weight ratios corresponding to the two losses.

After finishing the primary task, our backbone has already learned the explicit knowledge of TEE. So the secondary task implicitly captures the linguistic features of temporal expressions in Uyghur with the explicit knowledge learned in the primary task. It is formulated as a binary classification task. The input for this task is the Uyghur sequences, and the labels are sentence-level classification labels (as mentioned earlier). In this task, the language features of the last hidden state of the model are further extracted using a CNN, which helps in classifying the sequences. The secondary task enables the model to learn the features of temporal expressions in the target Uyghur language, implicitly. This is a weakly-supervised task and requires no token-level labels for each Uyghur token. The manually annotated token-level labels from the Uyghur datasets are used to evaluate the cross-lingual capability of the model after training. The ultimate objective of the secondary task is to minimize the $\mathcal{L}_{bc}$:

$$\mathcal{L}_{bc} = -\sum_{i=1}^{b} \mathbb{1}(y_i', c_i') log(softmax(W_2 \cdot x')), \tag{5}$$

where $x' \in \mathbb{R}^d$ is the sequence embedding output of CNN by passing the outputs of the backbone model to it. $c' = argmax(W_2 \cdot x')$ is the predicted sequence label of ith sequence while y_i' is the ground-truth sequence label. $W_2 \in \mathbb{R}^{2 \times d}$ is the classifier parameters of secondary task. Then We train our model concurrently by multi-task learning.

4.5   Data Format

We provide an illustrative example in Fig. 4 to demonstrate how knowledge is transferred from English to Uyghur in our model. In the primary task, the model extracts the explicit features of temporal expressions in English through a sequence labeling task. In the secondary task, the model takes the English translations of X_1 and X_2 as input. Y_3 and Y_4 indicate whether the sequences contain temporal expressions. The value of 1 indicates the presence of temporal expressions, while the value of 0 indicates their absence. These labels can be inferred from the labels Y_1 and Y_2 obtained in the primary task.

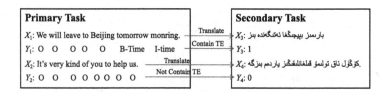

Fig. 4. An illustrative training example. In primary task, X_1 and X_2 are the inputs, while Y_1 and Y_2 are the corresponding output labels. In secondary task, X_3 and X_4 are the Uyghur translation of X_1 and X_2, while the outputs Y_3 and Y_4 can be inferred from Y_1 and Y_2 whether the sequences contain temporal expressions.

5   Experiments

5.1   Dataset

We utilize our English-Uyghur cross-lingual dataset. The dataset statistics are presented in Table 2. For the Uyghur language, we utilize the entire Uyghur dataset for test. For the English language, we utilize the entire English dataset including three separate datasets for training.

5.2   Baselines

To evaluate the performance of our model, we compare it with several popular deep learning methods, specifically focusing on cross-lingual models. We compare our model to the following models:

- mBERT [10]: This model is based on the multilingual BERT architecture.
- XLM-R [11]: We compare our model with the base and large versions of the vanilla XLM-Roberta model. XLM-R is a transformer-based language model specifically designed for cross-lingual tasks.
- XLTime [9]: We compare our model with three different variations of XLTime, which is a cross-lingual temporal expression extraction model. XLTime has shown promising results in capturing temporal expressions across multiple languages by using the MTL method.

5.3   Evaluation Approaches and Metrics

Following the previous research [1,9], we evaluate our model using two different approaches and measure the performance using F1-score, precision, and recall. The first approach is in *strict match* [1] evaluation, where all tokens of a temporal expression must be correctly identified for it to be considered as correctly extracted. This means that the predicted labels should match the ground-truth labels exactly in terms of both the sequence and the type of the temporal expression. For example, if the ground-truth labels are 'O O B-Set', any other prediction, such as 'O O B-Date', would be considered completely wrong. This evaluation approach is referred to as *with type*.

The second approach called *without type*, takes a more lenient approach. In this evaluation, as long as the labels of a temporal expression are predicted, regardless of whether the types match, it will be considered as correct. For example, if the ground-truth labels are 'O O B-Set', a prediction of labels 'O O B-Date' would be counted as correct. This approach focuses on capturing the presence of temporal expressions rather than matching their specific types.

5.4   Experiment Details

We adopt the base of the XLM-Roberta model (XLM-R) as our backbone which consists of 12 layers, 12 heads, and 270M parameters. We set the embedding dimension d as 768 to be consistent. We set batch size as 4 and dropout ratio as 0.2. We employ the AdamW as our optimizer with a learning rate of $7e^{-6}$ and a warm-up proportion of 0.5. We set the values of α and β as 0.2 and 0.8 in (4). For the CNN layer, we use a filter size of (2, 3, 4), and the kernel size is determined by the dimensions (k, d), with k corresponding to the filter size. we include a ReLU layer as an activation function following the backbone. We train all models for 50 epochs and select the best model for prediction. In order to meet the setting requirements of sequence labeling, the dataset is annotated and designed in the IOB2 format. We train all models on 8×NVIDIA Geforce RTX 3090 GPU.

5.5   Experiment Results

We evaluate our model and baselines on our dataset, employing two evaluation approaches as shown in Table 3. We observe that:

1) In both approaches, XLTU outperforms other models in terms of F1-score, recall, and precision, and achieves the SOTA results.

2) mBERT and XLTime-mBERT perform poorly in both approaches on our dataset. This is probably because their structures are not suitable for extracting features from the Uyghur language. Unlike XLM [29], mBERT simply replaces the training corpus of BERT with multilingual datasets. Although it provides shallow transfer [30] benefits for languages with vocabulary overlap, it may not be helpful for Uyghur with a completely different vocabulary.

3) Comparing XLTime-mBERT with mBERT or XLTime-XLMRbase with XLMR-base, we know that MTL does help the model to transfer knowledge. However, for XLMR-large and XLTime-XLMRlarge, MTL may have a negative impact. The large number of parameters in XLMR-large, combined with the relatively small size of our English-Uyghur dataset, may lead to overfitting during training when MTL is introduced.

4) Comparing to XLTime-XLMRbase, it shows that the introduction of CRF and CNN improves the model's perception of temporal expressions, as well as specific temporal expression label categories, enabling more accurate recognition.

5) We note that the performance of all models is not particularly high (the best F1-score is 0.66). This could be attributed to the characteristics of Uyghur itself and limited dataset. As the language characteristics of English and Uyghur are quite different, the model cannot fully capture the language characteristics of Uyghur through knowledge transfer from English. Nevertheless, compared to other models, our model still demonstrates superior cross-lingual capabilities.

Table 4 shows that XLTU performs better in predicting *Date* and *Set* labels. This discrepancy can be attributed to the complexity of the labeled data and the unequal number of labels. Most English data in these labels are simple, so does the corresponding Uyghur data. For example, 'tomorrow' is labeled as 'B-Date', while more complex expression like 'March 15' is labeled as 'B-Date I-Date'. On the other hand, the English data structure for *Duration* and *Time* labels is more complicated. These labels often consist of more than three words. In contrast, the corresponding Uyghur data typically consists of a few words, and the number

Table 3. Results of multilingual TEE on English-Uyghur cross-lingual dataset for two approaches. Number with bold is the optimal result, number with underline is the suboptimal result.

Model	w/type			w/o type		
	F1-score	Precision	Recall	F1-score	Precision	Recall
mBERT	0.14	0.39	0.08	0.14	0.53	0.08
XLMR-base	0.34	0.41	0.30	0.42	0.51	0.36
XLMR-large	<u>0.52</u>	<u>0.52</u>	<u>0.54</u>	0.56	0.55	0.58
XLTime-mBERT	0.23	0.23	0.25	0.32	0.35	0.29
XLTime-XLMRbase	0.50	**0.53**	0.52	<u>0.62</u>	<u>0.59</u>	<u>0.66</u>
XLTime-XLMRlarge	0.42	0.40	0.53	0.57	0.51	0.65
XLTU(Ours)	**0.54**	**0.53**	**0.59**	**0.66**	**0.64**	**0.67**

Table 4. Evaluation details of our model for all labels.

w/type of XLTU				
Label	F1	Precision	Recall	Support
Date	0.63	0.56	0.72	330
Duration	0.28	0.40	0.21	100
Set	0.59	0.64	0.55	33
Time	0.46	0.50	0.43	40

of labeled words does not always match the English labeled data (as shown in Fig. 5).

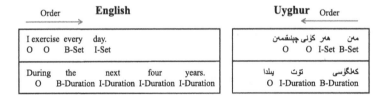

Fig. 5. Examples of labeled data in our English-Uyghur dataset. Not all labels are aligned one by one like *Duration* labels.

5.6 Ablation Study

To examine the effectiveness of the components in our model, we conducted an ablation study by removing the CRF and CNN layers. Table 5 illustrates that our model without CRF layer or CNN layer will degrade the performance on TEE task. According to the results, we know that CRF helps our model to learn contextual and right-to-left features of TEE and make more accurate predictions, while CNN can extract additional and right-to-left features of Uyghur.

Table 5. Results of Ablation Study. '-CRF' means our model without CRF layer. '-CNN' means our model without CNN layer.

Model	w/type			w/o type		
	F1	Precision	Recall	F1	Precision	Recall
XLTU	**0.54**	**0.53**	**0.59**	**0.66**	**0.64**	**0.67**
-CRF	<u>0.38</u>	0.39	<u>0.43</u>	0.53	<u>0.52</u>	0.53
-CNN	0.35	<u>0.46</u>	0.32	<u>0.56</u>	<u>0.52</u>	<u>0.60</u>
-CRF & -CNN	0.34	0.41	0.30	0.42	0.51	0.36

Table 6. Results of Comparison Experiments. 'G_i' means XLTU with ith parameter group of (α, β), such as 'G_4' represents 4th group of $(\alpha = 0.4, \beta = 0.6)$.

Group	w/type of XLTU			w/o type of XLTU		
	F1	Precision	Recall	F1	Precision	Recall
G_0	0.38	0.39	0.43	0.53	0.52	0.53
G_1	0.40	0.44	0.49	0.58	0.56	0.60
G_2	**0.54**	**0.53**	**0.59**	**0.66**	**0.64**	**0.67**
G_3	0.45	0.46	0.51	<u>0.65</u>	**0.64**	<u>0.65</u>
G_4	0.42	0.46	0.48	0.57	0.57	0.57
G_5	0.49	0.49	<u>0.54</u>	0.60	0.58	0.61
G_6	0.44	<u>0.50</u>	0.46	0.59	<u>0.63</u>	0.55
G_7	0.47	<u>0.50</u>	0.46	0.64	<u>0.63</u>	<u>0.65</u>
G_8	0.50	0.49	0.53	0.62	0.61	0.62
G_9	0.43	0.44	0.47	0.59	0.60	0.57
G_{10}	0.45	0.47	0.48	0.55	0.55	0.54

Based on these results, CRF layer plays a more significant role in the *without type* evaluation approach which helps our model better learn the English TEE features in the primary task and then be transferred in the secondary task, while the CNN layer has a greater impact in the *with type* approach. This enables the model to effectively identify the types of labels and avoid mistaking them for other types. On the other hand, the CNN layer aids in extracting Uyghur-specific features that are important for classification which enables the model to effectively identify the presence of temporal expressions.

5.7   Comparison Experiment

To investigate the impact of the two loss weight ratios, α and β, in the model in (4), We conduct additional comparison experiments. We perform 11 sets of experiments, varying the values of (α, β) from $(0, 1.0)$, $(0.1, 0.9)$, ..., to $(0.9, 0.1)$, $(1.0, 0)$. We group these experiments into G_0 to G_{10} for easy reference. We evaluate the results separately using the *with type* and *without type* evaluation approaches and visualize the experimental results based on both individual metrics and grouped results, as shown in Fig. 6.

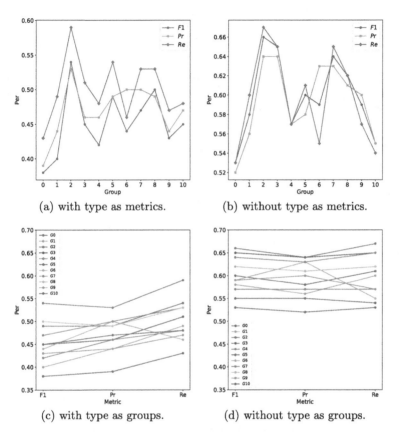

(a) with type as metrics.

(b) without type as metrics.

(c) with type as groups.

(d) without type as groups.

Fig. 6. Visualization of Comparison Experimental Results. We visualize them separately by metrics and by group for the two evaluation approaches, all of the Y-axis represent the scores. We can see that $G_2(\alpha = 0.2, \beta = 0.8)$ performs the best in terms of results.

From Table 6 we observe that $G_2(\alpha = 0.2, \beta = 0.8)$ as mentioned in Sect. 5.4, achieves the SOTA result in both the *with type* and *without type* evaluation approaches. Analyzing Fig. 6(a) and 6(b), we notice that the model performs better when the weight ratio of the CRF loss, α, is either larger or smaller (e.g., 0.2 or 0.7), but its performance is relatively poor when the ratio is close to half (e.g., 0.4 or 0.6). These findings validate our choice of setting α and β as 0.2 and 0.8 in the experiment.

6   Conclusion

We create an English-Uyghur cross-lingual dataset for temporal expression extraction tasks in Uyghur. By carefully considering the unique characteristics of Uyghur, we propose XLTU and utilize multi-task learning to help transfer the knowledge from English to Uyghur in TEE. We compare XLTU with different models, and the results demonstrate that our model XLTU achieves the SOTA results on various evaluation metrics.

In our future work, we will seek an effective method for data augmentation to expand our high-quality dataset. And we will also try to apply it to Uyghur social platforms for public opinion analysis or others.

Acknowledgement. This work is supported by Xinjiang Uygur Autonomous Region Key Research and Development Program (No.2022B03010).

References

1. UzZaman, N., Llorens, H., Derczynski, L., Allen, L., Verhagen, M., Pustejovsky, J.: Semeval-2013 task 1: Tempeval-3: evaluating time expressions, events, and temporal relations. In: Second Joint Conference on Lexical and Computational Semantics, vol. 2, Proceedings of SemEval 2013, pp. 1–9 (2013)
2. Choi, E., et al.: QuAC: question answering in context. In: Proceedings of EMNLP (2021)
3. Mitra, B., Craswell, N., et al.: An introduction to neural information retrieval. Now Foundations and Trends (2018)
4. Feder, A., et al.: Causal inference in natural language processing: estimation, prediction, interpretation and beyond (2021)
5. Strötgen, J., Gertz, M.: Multilingual and cross-domain temporal tagging. Lang. Resour. Eval. **47**(2), 269–298 (2013)
6. Zhong, X., Sun, A., Cambria, E.: Time expression analysis and recognition using syntactic token types and general heuristic rules. Proc. ACL **2017**, 420–429 (2017)
7. Chen, S., Wang, G., Karlsson, B.: Exploring word representations on time expression recognition. Tech. rep., Microsoft Research Asia (2019)
8. Lange, L., Iurshina, A., Adel, H., Strötgen, J.: Adversarial alignment of multilingual models for extracting temporal expressions from text. In: Proceedings of Workshop on Representation Learning for NLP at ACL vol. 2020, pp. 103–109 (2020)
9. Cao, Y., et al.: XLTime: a cross-lingual knowledge transfer framework for temporal expression extraction. In: Findings of NAACL 2022 (2022)
10. Devlin, J., Chang, M.-W., Lee, K., Toutanova, K.: BERT: pre-training of deep bidirectional transformers for language understanding. Proc. NAACL-HLT **2019**, 4171–4186 (2019)
11. Conneau, A., et al.: Unsupervised cross-lingual representation learning at scale. Proc. ACL **2020**, 8440–8451 (2020)
12. Liu, X., He, P., Chen, W., Gao, J.: Multi-task deep neural networks for natural language understanding. Proc. ACL **2019**, 4487–4496 (2019)
13. Strötgen, J., Gertz, M.: A baseline temporal tagger for all languages. Proc. EMNLP **2015**, 541–547 (2015)

14. Chang, A.X., Manning, C.D.: SUTIME: a library for recognizing and normalizing time expressions. In: LREC, vol. 3735, p. 3740 (2012)
15. Wentao Ding, Guanji Gao, Linfeng Shi, and Yuzhong Qu. 2019. A pattern-based approach to recognizing time expressions. In Proceedings of AAAI 2019, volume 33, pages 6335-6342
16. Laparra, E., Dongfang, X., Bethard, S.: From characters to time intervals: new paradigms for evaluation and neural parsing of time normalizations. Trans. Assoc. Comput. Linguist. **6**, 343–356 (2018)
17. Jain, A., Paranjape, B., Lipton, Z.C.: Entity projection via machine translation for cross-lingual NER. In: Proceedings of the 2019 Conference on Empirical Methods in Natural Language Processing and the 9th International Joint Conference on Natural Language Processing (EMNLP-IJCNLP), pp. 1083–1092, Hong Kong, China. Association for Computational Linguistics (2019)
18. Lample, G., Ballesteros, M., Subramanian, S., Kawakami, K., Dyer, C.: Neural architectures for named entity recognition. In: Proceedings of NAACL-HLT 2016, p. 260270 (2016)
19. Lance A Ramshaw and Mitchell P Marcus. 1999. Text chunking using transformation-based learning. In Natural language processing using very large corpora, pages 157-176. Springer
20. Bhattacharjee, D., Zhang, T., Susstrunk, S., Salzmann, M.: MulT: an end-to-end multitask learning transformer. CVPR2022 (2022)
21. Vaswani, A., et al.: Attention is all you need. In: NeurIPS (2017)
22. Xu, Y., Yang, Y., Zhang, L.: Deformable mixer transformer for multi-task learning of dense prediction. In: AAAI 2023 (2023)
23. Pustejovsky, J., Lee, K., Bunt, H., Romary, L.: ISO-TimeML: an international standard for semantic annotation. In: LREC, vol. 10, pp. 394–397 (2010)
24. Mazur, P., Dale, R.: Wikiwars: a new corpus for research on temporal expressions. Proc. EMNLP **2010**, 913–922 (2010)
25. Kim, Y.: Convolutional neural networks for sentence classification. In: EMNLP (2014)
26. LeCun, Y., Bottou, L., Bengio, Y., Haffner, P.: Gradient-based learning applied to document recognition. In: Proceeding of the IEEE (1998)
27. Lu, X., Ni, B.: BERT-CNN: a hierarchical patent classifier based on a pre-trained language model (2019). https://doi.org/10.48550/arXiv.1911.06241
28. Lafferty, J., Mccallum, A., Pereira, F.C.N.: Conditional random fields: probabilistic models for segmenting and labeling sequence data. Proc. ICML (2002). https://doi.org/10.1109/ICIP.2012.6466940
29. Conneau, A., Lample, G.: Cross-lingual language model pretraining. In: 33rd Conference on Neural Information Processing Systems of NeurIPS (2019)
30. Ben-David, S., Blitzer, J., Crammer, K., Kulesza, A., Pereira, F., Vaughan, J.W.: A theory of learning from different domains. Mach. Learn. **79**(1-2), 151–175 (2010). https://doi.org/10.1007/s10994-009-5152-4

From Sound to Map: Predicting Geographic Origin in Traditional Music Works

Daniel Kostrzewa$^{(\boxtimes)}$ ⓘ and Paweł Grabczyński

Department of Applied Informatics, Silesian University of Technology, Gliwice,
Poland
daniel.kostrzewa@polsl.pl

Abstract. Music is a ubiquitous phenomenon. In today's world, no one
can imagine life without its presence, and no one questions its significance
in human life. This is not a new phenomenon but has been prevalent for
hundreds of years. Therefore, an automated approach to understanding
music plays a nontrivial role in science. One of the many tasks in Music
Information Retrieval is the categorization of musical compositions. In
this paper, the authors address the rarely explored topic of classifying
traditional musical compositions from different cultures into regions (con-
tinents), subregions, and countries. A newly created dataset is presented,
along with preliminary classification results using well-known classifiers.
The presented work marks the beginning of a long and fascinating scientific journey.

Keywords: Traditional Music · Music Information Retrieval ·
Machine Learning · Classification

1 Introduction

Living on Earth, we are surrounded by the experience of "multiple worlds". We
perceive nuances in songs, dances, instruments, and languages, all indicative of
the complexity and richness that envelops us. Diversity manifests itself, among
other aspects, in the traditional music of virtually all ethnic groups inhabiting
the Earth.

In the public domain, terms such as traditional music, ethnic music, and folk
music are often used interchangeably. To understand the topic addressed in this
work, it is essential to introduce some distinctions. Traditional and ethnic music
refers to the music inherent to a specific ethnic group, passed down from generation to generation, and performed on traditional instruments. Folk music, while
inspired by ethnic music, is just its stylization [13]. Contemporary instruments
unrelated to a particular region are often employed in folk compositions.

It is challenging to generalize the sound of traditional music. The term refers
to various musical forms and styles, dependent on culture and region. In some

This work is partially funded by statutory research funds of Department of Applied
Informatics, Silesian University of Technology, Poland.

L. Franco et al. (Eds.): ICCS 2024, LNCS 14833, pp. 174–188, 2024.
https://doi.org/10.1007/978-3-031-63751-3_12

cases, even geographically distant areas may share similar musical cultures; for example, certain traditions in the United States and Canada trace their roots back to Great Britain and Ireland [2].

Nowadays, people performing traditional music are often touring artists who share their musical heritage with others [9]. However, the noticeable increase in interest among listeners is counterbalanced by the number of people abandoning their musical traditions in favor of imitating other styles. Urbanization serves as the primary cause of this phenomenon. Even if individuals migrating from rural to urban areas initially maintain their cultural identity, ultimately, Western lifestyle may displace native traditions [1].

Each ethnic group has developed its distinctive musical style, reflecting diversity in the use of specific instruments, musical scales, or the application of unconventional rhythms. The production and propagation of sound waves are physical phenomena that can be described using numbers and mathematical formulas. In machine learning, numerical features contribute to creating a model capable of capturing subtle differences between the music of different ethnic groups. However, the classifiers can serve not only to recognize the origin of composition but also to analyze songs, offering a chance for a deeper understanding of diversity and the relationships between the various worlds on Earth.

In this context, the main goal and primary contribution of this work is predicting the geographic origin of traditional musical compositions based on sound analysis with the use of classic machine learning methods. The collateral goals are to describe the newly created dataset and perform preliminary results obtained by standard classification methods.

The remainder of the paper is as follows. Section 2 presents the related work and state-of-the-art of the task undertaken in this paper. The dataset prepared for the experiments is thoroughly described in Sect. 3. Section 4 outlines the methods used for the classification, while the outcomes of the experiments are shown in Sect. 5. Section 6 concludes the work and describes future work.

2 Related Work

The prediction of the geographic origin of musical compositions remains a niche topic, with only two papers presenting a comprehensive approach to the issue [7, 21]. The direct influence on the development of this work was an article by Fang Zhou et al. [21]. The researchers addressed the problem of predicting the geographic origin of musical compositions and utilized machine learning methods for this purpose. They mentioned the use of booklets accompanying CDs for data labeling. It can be assumed that the researchers created a dataset (published on UCI Machine Learning Repository, [3]) based on songs from their own collection. The dataset consists of 1059 examples from 33 countries, with each instance having geographical coordinates corresponding to the capital city of the country from which the composition originates. Unfortunately, determining the recording region is somewhat generalized in cases where clear information is lacking.

Upon analyzing the data, it was discovered that the classes are not balanced; for instance, traditional music from Belize is represented by 11 examples, while

India has 69. Additionally, the class sizes are relatively small. Feature extraction was performed using the MARSYAS program, creating vectors of 68 features for each composition while maintaining the program's default settings. Statistics were based on the entire composition (rather than its parts), and the model's behavior was not verified on feature subsets.

The identification of these issues and the desire to tackle this challenging classification task were the reasons for delving into the topic of predicting the geographic origin of musical compositions.

Access to musical examples is also provided by some online archives. In this manner, Kedyte et al. [7] acquired compositions to create their dataset. Unfortunately, the source of the recordings is no longer available. The researchers focused solely on the area of the United Kingdom and, after preprocessing, obtained 10055 examples. The method of determining the geographical coordinates of points was not discussed in the paper.

The next two papers [6,18] were built upon [21] and [7]. However, topics that combine music and machine learning are gaining popularity. Research includes determining the musical style of a composition [16] or distinguishing "Western" music from that influenced by other cultures [5,10].

The application of different classifiers, hyperparameter optimization, and result analysis are standard stages in the machine learning process. In the experiment by Zhou et al. [21], the best result was achieved using the Random Forest classifier, while Kedyte et al. [7] built a neural network. Schedl et al. [18] propose a different hybrid approach. Two separate methods were used for prediction, and their combined results yielded higher effectiveness. The first method was based on typical feature analysis extracted from compositions using the KNN classifier. The second method involved data mining for data obtained from the Internet. The highest-rated websites related to queries about the song's name, biography, and origin were retrieved using the Bing Search API. The content of these pages was combined into one document for each composition. Subsequently, based on the list of country names, the frequency of each country's occurrence in the document was determined, and the one with the highest score was selected.

Although the prediction of data geolocation appears in various contexts, such as determining the origin of a photo, the topic of traditional music has not been developed for almost a decade. The lack of adequate datasets and the difficulty in acquiring materials may contribute to the limited interest in this area. Furthermore, no work has been found that frames the issue as a classification problem. Researchers approach the task in the context of regression, attempting to determine the geographical coordinates of a musical composition. The model's effectiveness is measured by the average distance error between points on the sphere (orthodrome). The following results were achieved: 3113 km [21], 1825 km [18], and 114 km for the United Kingdom [7]. Assessing the results of the mentioned models is challenging. In the case of [21] and [18], the average distance between all points in the dataset could serve as a reference point, but such information was not provided in the respective studies.

3 Dataset

The work on this issue can be divided into two stages: data acquisition and feature extraction and the selection of classifiers and their optimization. While several datasets are available online, they mostly focus on the traditional music of individual countries. The dataset prepared as part of the work [21] is likely the only publicly available collection presenting data from multiple regions worldwide. Its drawbacks are small size (1059 compositions), significant class imbalances, and a "mechanical" approach to feature extraction. For this reason, a decision has been made to create a new dataset addressing the described problems.

The process of data acquisition began with gathering information about the traditional music of each country. Analyzing the List of Intangible Cultural Heritage maintained by UNESCO served as a good starting point. The elements on the list represent culturally significant phenomena for each country/region, with music being a crucial part. Another knowledge repository on traditional music is the Naxos Music Library platform [14], enabling the search for albums and recordings from around the world. The platform also provides insights into some booklets accompanying the records. However, the most substantial information was obtained directly from the websites of record labels specializing in traditional music. Among the notable labels are Smithsonian Folkways Recordings [19], Ocora [15], VDE-Gallo Records [20], and Maison Des Cultures Du Monde [12]. Most of the recordings are available on streaming services. Selected tracks were used to create the dataset following the permissible use defined in the Copyright and Related Rights Act.

The collected data (Fig. 1) underwent the following verification: checking the validity of including a recording in the dataset (e.g., distinguishing traditional from folk music) and ensuring adherence to the specified duration limits for each track, ranging from 30 s to 15 min. Longer soundtracks were omitted to avoid excessive memory load. The resulting database comprises 12,860 recordings from 44 countries worldwide. The highest number of examples were gathered from the Democratic Republic of the Congo and Siberia (both with 334 examples), while the fewest were from Scotland (227 examples). Additionally, each track received a label indicating its region (continent) and subregion of origin. The division into regions and subregions was done according to the United Nations geoscheme. There are more significant differences in class sizes, especially at the regional level.

Despite the dataset being created, it still requires thorough verification from ethnomusicologists. Therefore, it is not yet publicly available. Undoubtedly, work on it will continue, and ethnomusicologists have expressed interest in contributing to the construction of this dataset. This marks the first part of developing a comprehensive dataset. The second equally crucial element is the extraction of numerical features.

Using audio files directly in analytical tasks would be impractical due to their size. Therefore, input data undergo transformations. Values calculated in this way describe characteristic elements of music, such as timbre, melody, and

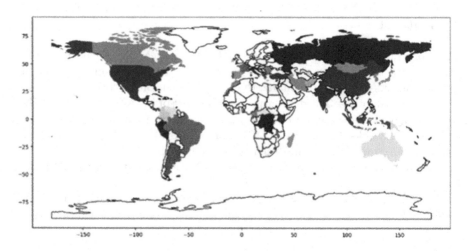

Fig. 1. A map illustrating the content of the dataset created for the purposes of this study (countries); the darker the color is, the more examples are present.

rhythm. Furthermore, feature extraction helps eliminate irrelevant information, improving the efficiency of machine learning algorithms.

When discussing feature extraction, it is essential to consider the duration of examples. Analyzing the entire composition is a rare practice, with only [21] mentioning it. Some papers propose cutting a portion of the recording, e.g., a 10-s [7] or a 30-s snippet [5,16]. Typically, the beginning of the track is analyzed. However, an algorithm can be employed to choose a suitable segment, such as one with the highest spectrum energy (the culmination point of the composition). In this study, the feature extraction process from an audio file began with cutting a 30-s segment of the composition, possessing the highest energy. The assumption was made that the strongest signal is synonymous with the most characteristic excerpt.

A total of 323 features were extracted from each audio recording, which can be categorized into the following groups: timbre (219 features), melody (72 features), and rhythm (25 features). The next two attributes (album and title) serve to identify the track during dataset creation. The last 5 categories are labels: latitude, longitude, region (continent), subregion, and country of origin. The geographical coordinates point to the capital city of the country. Exceptions are the coordinates of Korea and Siberia. Due to data acquisition from both North and South Korea, coordinates of the Korean Demilitarized Zone were used. In the case of Siberia, coordinates indicate the central part of the area.

Attributes were calculated using pre-existing functions from the Librosa library. To better understand the data distribution, the following statistics were applied for most features: mean, median, standard deviation, maximum value, minimum value, skewness, and kurtosis. Extracting a large number of features aims to minimize the impact of values that might inaccurately characterize a specific composition.

The dataset employs various methods for signal description in both the time and frequency domains. Features belonging to the first group can be directly derived from the audio file, providing a straightforward way to analyze the signal. To obtain attributes from the frequency domain, it is necessary to transform the audio signal, for example, into a Mel spectrogram.

One of the parameters computed in the time domain is the Zero-Crossing Rate (ZCR). This indicator defines the frequency of changes in the signal value from positive to negative and vice versa. Additionally, ZCR can indicate the noise in the recording-usually, it achieves much higher values if the signal is noisy. The standard deviation [4] is a particularly useful statistic describing ZCR. Different musical instruments (and human voices) produce sound in their own distinctive ways, allowing the prediction of the likely distribution of this value.

Next time domain features are connected with the rhythm. The *librosa.onset_ detect* function detects the beginning of the sound, called onsets, by analyzing the envelope of the signal. A sudden increase in amplitude can indicate the occurrence of a sound. The number of detected onsets was used to calculate the frequency of events per second. Another valuable parameter is the onset_intervals. This array contains the time differences between consecutive onsets. This information helps determine if there are rhythmic patterns in the composition or if the rhythm undergoes frequent changes. Kurtosis was used to assess rhythm variability. This statistic provides insights into the frequency of extreme values and the ratio of the value distribution to the normal distribution. In addition to detected onsets, there is the onset_strength function. This is an example of a method for signal description in the frequency domain. The function calculates the spectrum difference between adjacent frames. Information about the strength of successive sounds can be helpful in classifying a recording. Instruments characterized by a sharp sound achieve higher onset_strength values. An example is the djembe, one of the most popular African drums. The sound produced by the djembe depends on the way it is hit and the part of the membrane that is hit. The onset_strength value is usually higher for percussion instruments. In contrast, Japanese court music, gagaku, exhibits a slow tempo and the occurrence of long tones, often performed in unison.

Traditional music is often based on musical scales, which are sequences of sounds arranged according to fixed patterns, defining the distances between consecutive tones. One of the earliest scales confined within an octave was the pentatonic scale, employed in ancient Chinese, Greek, and Peruvian music, known for its smooth sound, free of dissonances. "Western" music has been based on the seven-tone scale (the eighth tone is a repetition of the first, an octave higher) for several centuries. This relationship is reflected in Fig. 2. The sound material of traditional Chinese and Peruvian music is based on the pentatonic scale, as evidenced by the dominance of tones belonging to this scale. Argentine tango is constructed on a seven-note minor scale, where the G tone predominates in this example due to the composition being in the key of G minor. The last example features a solo performed on the didgeridoo. The sound produced by the instrument resembles a buzzing, oscillating between adjacent tones (in this case, D

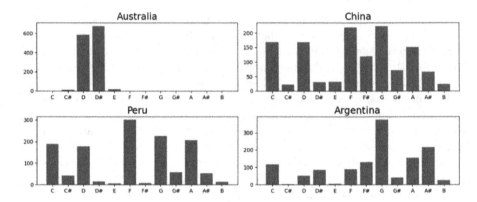

Fig. 2. Frequency of occurrence of individual pitch classes. Sources: Aboriginal Songs of the Northern Territory - Devilman (Australia), Chinese Masterpieces of the Pipa and Qin - Wild Geese Descending (China), Traditional Music of Peru, Vol. 5 - Cusqueño (Peru), Tango de Buenos Aires - Balada para Mi Muerte (Argentina).

and D#). The obtained frequencies were transformed into tones and normalized to one octave. The sorted array of results was also used to determine the chroma contrast coefficient. This ratio represents the sum of the six most frequently occurring pitch classes divided by the sum of the remaining pitch classes [10]. In principle, "Western" compositions should have a lower chroma contrast value compared to compositions based, for example, on the pentatonic scale, as the seven-tone scale includes more naturally occurring tones. The extreme case is represented by didgeridoo recordings, where typically only 4 pitch classes are detected. Calculating chroma contrast would involve division by 0, leading to the automatic setting of the value to 1000.

In addition to the described features, essential information regarding timbre includes Mel-Frequency Cepstral Coefficients (MFCC, Fig. 3) [11] and harmonics. It is worth discussing these parameters more precisely, constituting as much as 44% of all attributes in the created dataset. MFCC coefficients are obtained by analyzing the Mel spectrogram. They undergo further transformations, including Discrete Cosine Transform (DCT), resulting in a simplified representation of sound. Several processing stages follow the creation of the Mel spectrogram, including pre-emphasis and windowing (differentiation and segmentation), fast Fourier transform (processing the segment into a frequency spectrum and modification according to the Mel scale), Mel-scale filtering (using triangular filters), normalization (subtracting the mean and normalizing the variance of each coefficient). Using the Mel scale, which concentrates higher frequencies, MFCC is calculated using DCT.

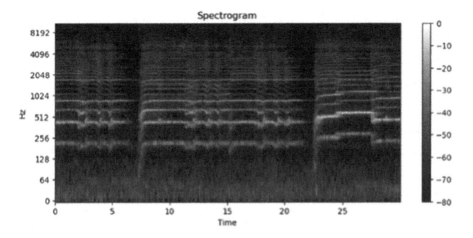

Fig. 3. Spectrogram of the song on the Mel scale. Source: The Gagaku - Kashin.

MFCC parameters play a significant role in identifying musical genres [8]. By extracting characteristic audio features, they enable the differentiation between music styles such as jazz, rock, or classical music. They are also employed in streaming services for automatic music sorting and serve as input data for neural networks used in music genre classification.

4 Methods

The modest size of the feature-extracted CSV file (approximately 55 MB) contributed to the acceptable (in minutes) computation time required for evaluating results across various models. In the pursuit of identifying the optimal classifier, a diverse set of algorithms was systematically tested. While some approaches were discarded based on insights from literature and personal observations, the primary objective was to assess numerous classifiers and compare their performance comprehensively. The following algorithms were employed in the experiment.

Dummy Classifier - serves as a reference point for other classifiers. Predictions are made based on a selected strategy, bypassing the analysis of input data. The parameter *strategy* is set to *uniform*, allowing each class to be chosen with equal probability.

Gaussian Naive Bayes Classifier - a method based on Bayes' theorem. For each class, the probability of the occurrence of individual features is calculated. The class with the highest score is chosen as the model's prediction. Gaussian Naive Bayes assumes that data are independent of each other, meaning the occurrence of one feature has no impact on the presence of another. Additionally, the algorithm assumes that values for each class follow a normal distribution. While this method may be suitable for specific tasks such as text classification or spam filtering, its inability to detect correlations between features limits its effectiveness in complex classification tasks.

Decision Tree Classifier - a classifier employing a decision tree-based approach. The model's structure resembles a tree, with nodes representing individual features, branches denoting decisions made by the algorithm, and leaves presenting the selected classes. Based on successive attributes, the algorithm divides the data into progressively smaller subsets. The order of features chosen for set partitioning is determined by measures such as the Gini index or entropy. The advantage of using a decision tree lies in the ease of interpreting results. The model allows for tree visualization and analysis of the most significant features identified by the algorithm. Moreover, it performs well with both numerical and categorical features. However, decision trees are inherently unstable, a small change in the data can lead to an entirely different tree structure. For complex classification tasks, it might be necessary to utilize advanced algorithms like Random Forest or Gradient Boosting, which are also based on decision trees.

Stochastic Gradient Descent Classifier - the classifier's name does not directly refer to the model but to the training method. The stochastic gradient descent (SGD) method is applied using a linear classifier as an example. SGD continuously updates the model's parameters in the direction where the values of the loss function decrease. In contrast to the Gradient Descent method, where updating a parameter in a given iteration requires processing the entire training set, Stochastic Gradient Descent allows the use of randomly selected samples. The method is efficient for large datasets as it does not necessitate processing all data at once.

Logistic Regression Classifier - despite the term regression in the model's name, it is more commonly used for classification than regression. The algorithm operates by employing the logistic (sigmoidal) function, transforming any value into a range between (0, 1). Each feature carries a weight determining its influence on the prediction. Various optimization methods are used to minimize the difference between expected and predicted classes. Although logistic regression is mainly associated with binary classification, choosing the parameter $multi\_class='multinomial'$ enables its application to multiclass classification. The ease of interpreting the model's results is a significant advantage of logistic regression (weights of individual features). The method is effective in simple tasks, especially when a linear relationship exists between variables. However, for more complex data, the application of logistic regression may be ineffective.

Gradient Boosting Classifier - a technique involving the sequential training of simple models (such as decision trees) and their optimization through the evaluation of the previous predictor. The residual error, i.e., the difference between the expected and predicted values, is calculated for each decision tree. In successive iterations, the algorithm creates new decision trees that consider the errors made in previous iterations. The direction of changes is determined by the gradient of the loss function. The Gradient Boosting Classifier often achieves very good results in many cases, frequently outperforming classifiers like Random Forest [17]. It handles complex data and outliers well. However, the downsides of this technique include high computational complexity and the risk of overfitting when inappropriate parameters are used.

K-Nearest Neighbors Classifier - the algorithm is based on the principle of similarity among data points located closely in the feature space. The model determines the distance between a given instance and the remaining points in the dataset. Subsequently, a specific number of nearest points (neighbors) are selected, and through voting, the most frequently occurring class is identified. KNN requires feature scaling to prevent large values from dominating computations. The model performs well in tasks with multiple classes and nonlinear decision boundaries. On the other hand, the algorithm is highly sensitive to imbalanced datasets, where the dominance of one or a few classes can impact prediction accuracy.

Random Forest Classifier - a method classified under ensemble learning, which involves combining multiple simple models to obtain a more accurate final model. RF is based on decision trees, and the algorithm creates a "forest" consisting of a specified number of these trees. Each tree is independent and trains on different subsets of data (using bootstrap sampling), allowing for greater diversity in models and preventing overfitting. Then, similar to KNN, the most frequently occurring class selected by individual decision trees is chosen. The algorithm handles outliers effectively. In multiclass classification, RF achieves significantly better results than a single decision tree. The use of a large number of trees helps minimize the negative impact of outliers.

Support Vector Machine Classifier - the classifier's task is to find a suitable hyperplane that separates individual classes. The optimal hyperplane is maximally distant from all classes. Support vectors, points in the feature space closest to the hyperplane, play a crucial role in determining its position. The removal of a specific support vector changes the position of the hyperplane. If the data is not linearly separable, it undergoes transformation into a higher dimension where linear separation is achieved. SVM performs well with data featuring a large number of features. The ability to apply different kernel functions (for transforming data into higher dimensions) allows the model to be used for various purposes. However, interpreting the classifier's operation is generally more challenging than in the case of, for example, logistic regression.

Linear Support Vector Machine Classifier - this classifier is also based on the support vector method and employs a linear kernel function. Unlike SVM, the kernel function cannot be changed. The method is effective for data that can be separated using a line or hyperplane, especially when dealing with a higher number of features. Linear SVC is a faster classifier but struggles more with complex patterns and feature relationships than SVM. It should be assumed that the application of the linear version of the SVM algorithm may be doomed to failure due to the characteristics of the data.

Neural Network - neural networks process data in a manner similar to the human brain, primarily manifested in their structure. The human brain is composed of neurons connected by synapses, and communication between neurons occurs through electrical signals controlling synapses. Artificial neural networks feature interconnected neurons as well, and each connection has a weight adjusted during the machine learning process.

For constructing the network, the Keras library was utilized. A simple architecture was implemented, comprising an input layer (128 neurons, ReLU activation function), two hidden layers (128 and 64 neurons, ReLU activation function), and an output layer (the number of neurons dynamically determined based on the number of classes, softmax activation function). Compilation parameters were set as: optimizer - Adam, loss function - sparse categorical crossentropy. To prevent overfitting and enhance the model's performance, an early stopping callback was employed. This callback monitored changes in the validation accuracy metric with a patience parameter set to 10. At the end of each epoch, the model was compared with the best and, if necessary, overwritten.

5   Experiments

At the initial stage of the experiment, a preliminary assessment of classifier results was conducted. The task involved determining the origin of a musical piece, categorized into three levels: region (continent), subregion, and country. Calculations were performed for three categories: region, subregion, and country. All features were utilized in the analysis, having been previously scaled. Standardization was necessary as the value ranges of certain attributes significantly varied, which could adversely affect the performance of classifiers like SVM. The dataset was divided into training and testing sets in a 70:30 ratio.

For evaluating the model's quality, cross-validation was employed. The dataset was divided into 10 subsets. After evaluating each subset, the average and standard deviation of the accuracy metric were computed. Table 1 shows the quantitative results (average accuracy and its standard deviation) for the selected classifiers with the default parameters.

The Dummy Classifier achieved a result close to the probability $P = 1/n$, where n represents the number of classes. The Dummy Classifier was applied as a reference point for more complex algorithms. Gaussian Naive Bayes, Decision Tree, and Stochastic Gradient Descent yielded average results, placing them towards the lower end of the classification effectiveness ranking. Logistic Regression and Linear SVC performed the task with very similar effectiveness. Both classifiers are linear models that seek a hyperplane to separate classes. Gradient Boosting and Random Forest Classifier required the most time to find a solution. Random Forest achieved better results than Gradient Boosting, especially in country classification. However, the Gradient Boosting result is below expectations, and overfitting could be a reason, as this classifier is less resistant to it than Random Forest.

K-Nearest Neighbors and Support Vector Machine Classifier achieved good results, especially SVM (the best result in country classification). The relatively high effectiveness of KNN suggests that data points close to each other in the feature space may belong to the same class. The best-performing model in the experiment was the neural network, which achieved the highest score for region and subregion classification.

At the second stage of the experiment, classifiers with the highest accuracy metric were selected for further optimization: K-Nearest Neighbors, Random

Table 1. Comparison of the average accuracy and its standard deviation of the classifiers.

Classifier	Region (continent)	Subregion	Country
Dummy Classifier	16.12% (1.10%)	5.38% (0.57%)	2.51% (0.56%)
Gaussian Naive Bayes	36.40% (1.19%)	20.78% (1.23%)	21.97% (1.40%)
Decision Tree	40.24% (1.68%)	24.23% (0.89%)	18.14% (0.54%)
Stochastic Gradient Descent	49.86% (1.77%)	32.64% (1.61%)	28.76% (1.32%)
Logistic Regression	57.23% (1.67%)	40.31% (1.78%)	36.84% (1.10%)
Linear SVC	57.37% (1.27%)	39.76% (1.64%)	36.29% (1.23%)
Gradient Boosting	59.55% (1.44%)	38.21% (1.12%)	33.35% (0.81%)
Random Forest	60.10% (0.81%)	47.38% (1.64%)	49.47% (1.81%)
K-Nearest Neighbors	64.06% (1.47%)	50.29% (1.51%)	45.82% (1.11%)
Support Vector Machine	66.60% (1.11%)	52.93% (1.42%)	**51.39%** (2.05%)
Neural Network	**70.41%** (0.83%)	**54.77%** (1.36%)	49.47% (0.99%)

Forest, Support Vector Machine, and Neural Network. Suboptimal sets of hyperparameters were identified using grid search. The search included values for the following parameters: KNN (*n_neighbors* - the number of neighbors, *weights* - the weight function, *algorithm* - the algorithm used to compute the nearest neighbors), Random Forest (*n_estimators* - the number of trees in the forest, *max_features* - the number of features considered when finding the best split, *bootstrap* - specifies whether the tree is built from the whole dataset or a sample of data), SVM (*C* - regularization parameter, *kernel* - the type of kernel used in the algorithm, *gamma* - kernel coefficient), Neural Network (*optimizer* - the algorithm adjusting model parameters during training, *activation* - activation function, *batch_size* - the number of samples processed before updating the model, *neurons* - the number of neurons). Experiments also involved changes in network architecture, such as adding hidden layers or dropouts.

All classifiers improved their results by approximately 3–4% (Table 2). The most significant changes in effectiveness increased by about 10%. Each model presents a different approach to the classification problem. KNN makes decisions based on local similarity in the feature space, Random Forest utilizes decision trees, SVM finds the optimal hyperplane separating classes, and neural networks search for complex patterns using neurons. Despite their differences, all classifiers achieved satisfactory results.

Table 2 shows average accuracy before and after hyperparameter optimization for the selected classifiers as well as a set of found hyperparameter values. The best values are in bold.

The presented experiment allowed for the selection of the best classifiers. In the category of region (continent) and subregion, the SVM algorithm emerged as the winner, achieving average accuracies of 72.42% and 60.26%, respectively. For the task of country prediction, Random Forest performed the best, achieving an average accuracy of 60.03%.

Table 2. Comparison of the average accuracy after hyperparameters optimization.

Classifier	Parameters	Accuracy Before Opt.	Accuracy After Opt.
Region (continent)			
Random Forest	n_estimators = 500, max_features = 40, bootstrap = False	60.10%	66.43%
K-Nearest Neighbors	n_neighbors=4, weights= 'distance', algorithm = 'auto'	64.06%	68.12%
Support Vector Machine	C = 0.1, gamma = 1, kernel = 'poly'	66.60%	**72.42%**
Neural Network	optimizer = 'adam', activation = 'relu', batch_size = 32, neurons = 512, 256, epochs = 50	70.41%	71.02%
Subregion			
Random Forest	n_estimators = 500, max_features = 60, bootstrap = False	47.38%	57.67%
K-Nearest Neighbors	n_neighbors=3, weights= 'distance', algorithm = 'auto'	50.29%	58.04%
Support Vector Machine	C = 0.1, gamma = 1, kernel = 'poly'	52.93%	**60.26%**
Neural Network	optimizer = 'adam', activation = 'relu', batch_size = 32, neurons = 512, 256, epochs = 50	54.77%	60.06%
Country			
Random Forest	n_estimators = 500, max_features = 10, bootstrap = False	49.47%	**60.03%**
K-Nearest Neighbors	n_neighbors=3, weights= 'distance', algorithm = 'auto'	45.82%	54.28%
Support Vector Machine	C = 0.1, gamma = 1, kernel = 'poly'	51.39%	57.62%
Neural Network	optimizer = 'adam', activation = 'relu', batch_size = 32, neurons = 512, 256, epochs = 50	49.47%	55.78%

6   Conclusions and Future Work

The prediction of the geographic origin of musical compositions is an area that is still waiting to be thoroughly explored. This paper presents a comprehensive approach to the problem, from acquiring musical compositions and creating a dataset to applying various classifiers.

The obtained results can be considered satisfactory, especially the effectiveness of predicting the country of origin of a composition at a level of 60% for 44 classes. A holistic approach is needed to characterize the sound, involving musical features such as timbre, melody, rhythm, and ethnomusicological knowledge.

Collecting the necessary data proved to be a tedious task, requiring a thorough examination of each album and track. The effectiveness of classifiers depended on the quality of the acquired data and extracted features. On average, each country is represented by 292 compositions, ten times more than in the dataset of the previous work [21]. In some regions of the world, traditional music is highly diverse, so even the representation of several hundred recordings may be insufficient to capture this diversity. A possible solution is to divide countries into smaller areas, such as separating the northern and southern parts of India.

The created dataset was tested using ten different classifiers. The best results, after hyperparameter optimization, were achieved by K-Nearest Neighbors, Random Forest (best in the country category: 60.03%), Support Vector Machine (best in the region (continent): 72.42%, and subregion category: 60.26%), and Neural Network.

To improve results for region and subregion, it would be necessary to balance the data. For this purpose, new compositions from minority classes could be added to the dataset, or oversampling methods, such as the Synthetic Minority Over-sampling Technique (SMOTE), could be employed.

In further research, extracting information about the most significant features from the mentioned estimators would be valuable. This involves examining the paths created in the decision trees of the Random Forest classifier and identifying the most frequently repeated attributes that lead to correct predictions. Moreover, combining different methods, as demonstrated in [18], could contribute to performance improvement.

The attractiveness of the topic of this study arises from its interdisciplinary nature. Predicting the geographic origin of compositions combines machine learning, ethnomusicology, and geography. Such topics demonstrate that the only limit in tasks posed to machine learning is imagination.

References

1. Elbourne, R.: The study of change in traditional music. Folklore **86**(3–4), 181–189 (1975)
2. Frank, A.: That's the way I've always learned: the transmission of traditional music in higher education. Ph.D. thesis, East Tennessee State University (2014)

3. Geographical Original of Music dataset website: (2014). https://www.archive.ics. uci.edu/dataset/315/geographical%2Boriginal%2Bof%2Bmusic. Accessed 10 Feb 2024
4. Giannakopoulos, T., Pikrakis, A.: Introduction to audio analysis: a MATLAB® approach. Academic Press (2014)
5. Gómez, E., Haro, M., Herrera, P.: Music and geography: content description of musical audio from different parts of the world. In: ISMIR, pp. 753–758 (2009)
6. Hamoen, J., Engbers, S.: Explaining the geographic origin of music (2021). www. jellehamoen.nl/wp-content/uploads/2021/07/Research-paper-REDI-Sharon-Engbers-Jelle-Hamoen.pdf
7. Kedyte, V., Panteli, M., Weyde, T., Dixon, S.: Geographical origin prediction of folk music recordings from the united kingdom. In: 18th International Society for Music Information Retrieval Conference (2017)
8. Kostrzewa, D., Kaminski, P., Brzeski, R.: Music genre classification: looking for the perfect network. In: Paszynski, M., Kranzlmüller, D., Krzhizhanovskaya, V.V., Dongarra, J.J., Sloot, P.M.A. (eds.) ICCS 2021. LNCS, vol. 12742, pp. 55–67. Springer, Cham (2021). https://doi.org/10.1007/978-3-030-77961-0_6
9. Littlefield, M.D.: Folk music in new England: a living tradition (2020)
10. Liu, Y., Xiang, Q., Wang, Y., Cai, L.: Cultural style based music classification of audio signals. In: 2009 IEEE International Conference on Acoustics, Speech and Signal Processing, pp. 57–60. IEEE (2009)
11. Logan, B., et al.: Mel frequency cepstral coefficients for music modeling. In: ISMIR, vol. 270, p. 11. Plymouth, MA (2000)
12. Maison Des Cultures Du Monde website: http://www.boutiqueenligne. maisondesculturesdumonde.org/. Accessed 10 Feb 2024
13. Metzig, C., Gould, M., Noronha, R., Abbey, R., Sandler, M., Colijn, C.: Classification of origin with feature selection and network construction for folk tunes. Pattern Recogn. Lett. **133**, 356–364 (2020)
14. Naxos Music Library World website: www.naxosmusiclibrary.com/world. Accessed 10 Feb 2024
15. Ocora website: https://www.radiofrance.com/les-editions/collections/ocora. Accessed 10 Feb 2024
16. Patil, N.M., Nemade, M.U.: Music genre classification using MFCC, K-NN and SVM classifier. Int. J. Comput. Eng. Res. Trends **4**(2), 43–47 (2017)
17. Piryonesi, S.M., El-Diraby, T.E.: Data analytics in asset management: cost-effective prediction of the pavement condition index. J. Infrastruct. Syst. **26**(1), 04019036 (2020)
18. Schedl, M., Zhou, F.: Fusing web and audio predictors to localize the origin of music pieces for geospatial retrieval. In: Ferro, N., et al. (eds.) ECIR 2016. LNCS, vol. 9626, pp. 322–334. Springer, Cham (2016). https://doi.org/10.1007/978-3-319-30671-1_24
19. Smithsonian Folkway Recordings website: https://folkways.si.edu/. Accessed 10 Feb 2024
20. VDE-Gallo Records website: https://vdegallo.com/. Accessed 10 Feb 2024
21. Zhou, F., Claire, Q., King, R.D.: Predicting the geographical origin of music. In: 2014 IEEE International Conference on Data Mining, pp. 1115–1120. IEEE (2014)

A Novel Multi-Criteria Temporal Decision Support Method - Sustainability Evaluation Case Study

Aleksandra Bączkiewicz[1]([⊠])(ID), Jarosław Wątróbski[1,2](ID),
and Artur Karczmarczyk[3](ID)

[1] Institute of Management, University of Szczecin, ul. Cukrowa 8, 71-004 Szczecin,
Poland
{aleksandra.baczkiewicz,jaroslaw.watrobski}@usz.edu.pl
[2] National Institute of Telecommunications, ul. Szachowa 1, 04-894 Warsaw, Poland
[3] Faculty of Computer Science and Information Technology, West Pomeranian
University of Technology in Szczecin, ul. Żołnierska 49, 71-210 Szczecin, Poland
artur.karczmarczyk@zut.edu.pl

Abstract. Moving toward a sustainable society requires the development of reliable indices, indicators, and computational methods that supply the tools, such as decision support systems used in assessing the achievement of sustainable development goals. The aim of this paper is to present an intelligent decision support system that enables multi-criteria evaluation, taking into account the temporal variability of the performance of the assessed alternatives. The framework of this DSS is based on the method called Data vARIability Assessment - Measurement of Alternatives and Ranking according to COmpromise Solution (DARIA-MARCOS). The proposed method was used for an exemplary multi-criteria analysis problem concerning the implementation of the sustainable development goals included in Sustainable Development Goal 11 (SDG 11), focused on sustainable cities and communities. SDG 11 aims to develop toward making cities and human settlements inclusive, safe, resilient, and sustainable. The methodical framework implemented in the demonstrated DSS ensures an efficient, automatized, and objective assessment of a multi-criteria temporal decision-making problem and gives an unequivocal, clear outcome. The results proved the usability of the developed DSS in the multi-criteria temporal evaluation of sustainable development focused on sustainable cities and communities.

Keywords: decision support system · sustainability assessment · multi-criteria temporal assessment · sustainable society · DARIA-MARCOS

1 Introduction

Innovative models, algorithms, and tools involving the implementation of computational methods provide important contributions to the field of sustainability development assessment [2]. The implementation of sustainable development,

L. Franco et al. (Eds.): ICCS 2024, LNCS 14833, pp. 189–203, 2024.
https://doi.org/10.1007/978-3-031-63751-3_13

evaluation of this process, and monitoring progress require the development of indicators, indexes, and measurement instruments [3]. Indicators for assessing the implementation of Sustainable Development Goals (SDGs) are a reliable data source. Hence, their use is recommended [10]. Indicators for assessing the sustainable development of the SDGs were officially developed by the United Nations in 2015 with a group of experts [8]. As a result, the development of analytical tools for assessing the achievement of sustainable development goals is essential. From a technical point of view, multiple frameworks, interpretability, and selection of indicators justify the need to use decision support systems (DSSs) powered by objective computational methods for this purpose [5]. The multitude of indicators that need to be considered in sustainability assessment justifies using multi-criteria decision analysis methods (MCDA) in DSS [9]. MCDA methods evaluate compromises between several quantitative and qualitative criteria and facilitate complex decisions [7]. However, MCDA methods evaluate alternatives against a set of criteria for a situation at a single moment in time [12]. This justifies the need to develop methods that take into account the dynamics of results over time [13].

The aim of this research is to present an intelligent decision support system that allows multi-criteria evaluation, considering the temporal variability of the performance of the assessed alternatives. The purpose of the authors is to develop a multi-criteria approach that provides the opportunity to simultaneously include in evaluating the performance of a given time interval in subsequent years, along with the dynamics of fluctuation. The framework of this DSS is based on the method named Data vARIability Assessment - Measurement of Alternatives and Ranking according to COmpromise Solution (DARIA-MARCOS). The classic MARCOS (Measurement of Alternatives and Ranking according to COmpromise Solution) method is employed in the developed DARIA-MARCOS method as a module for the annual assessment of alternatives. Based on the obtained annual utility function values in the DARIA-MARCOS method, parameters such as the value and direction of the variability of annual scores of alternatives used in the next stages of DARIA-MARCOS are then determined. The MARCOS method is based on defining the relationship between alternatives and reference solutions, which are ideal and anti-ideal solutions [28]. Based on the defined relationships, the utility functions of the alternatives are determined, and a ranking of the compromises with respect to the ideal and anti-ideal solutions is created. Utility functions represent the position of an alternative concerning the ideal and anti-ideal solution. The best alternative is the one that is closest to the ideal and, at the same time, furthest from the anti-ideal reference point [21].

The DARIA-MARCOS method gives the possibility of the evaluation of alternatives with simultaneous consideration of multiple assessment criteria and the temporal aggregation of the achieved scores into a single unambiguous score presented as utility function values and rankings. Therefore, the proposed DSS is developed to provide a complete automated assessment, not requiring engagement analysts to assess the variability of performances over time. The demonstrated tool automatically considers the temporality of annual scores. The automatic incorporation of variability in the final stage gives reliable results that would be difficult in the case of subjective analysis of individual scores from following years by analysts without the support of supplementary computational methods.

The application of the proposed DSS was shown in this research through the example of a multi-criteria temporal assessment of Sustainable Development Goal 11 (SDG 11) implementation by selected European countries. SDG 11 was introduced by the United Nations (UN) in the 2030 Agenda for Sustainable Development. It was approved by all United Nations Member States in 2015 [22,24]. SDG 11 encourages making cities and human settlements safe, stable, sustainable, and inclusive [18]. According to the assumptions of SDG 11, cities should develop towards the creation of safe, stable, and sustainable inclusive settlements [15]. At the same time, countries must improve resource efficiency, strive to reduce pollution, and counter poverty [11]. One such example is improving municipal waste management. In the future, cities should provide equal opportunities for all people and access to basic services, energy, housing, transportation, and more [4].

The proposed system powered by the DARIA-MARCOS method can find application in assessing the achievement of the targets contained in SDG 11 by European countries in any chosen time frame, giving a view of the trend of progress or regression in development with respect to sustainable cities and communities, compared to other countries.

The rest of the paper is organized as follows. Section 2 provides the literature review including related works focused on existing multi-criteria methods considering temporality in assessment. Section 3 presents the methodology of the research performed. Research results are demonstrated and discussed in Sect. 4. In the end, in Sect. 5, conclusions are stated, and directions for further work are indicated.

2 Related Works

The necessity of applying a temporal approach when evaluating multi-criteria problems requiring consideration of multiple periods is emphasized in several research papers. Martins and Garcez conducted a multidimensional and multi-period assessment of road safety using an aggregation of various road safety indicators recorded for different periods [16]. The authors used the Multi-criteria Multi-Period Outranking Method (MUPOM) proposed by Frini and Amor [12]. The MUPOM method belongs to the outranking methods, thus it considers the sustainability requirements [25]. The application of the proposed method is presented using the example of selecting the most favorable scenario for sustainable forest management. Sustainability assessment problems require consideration of complexity involving multiple criteria and periods. This challenge was successfully addressed by Urli, Frini, and Amor, who, in their research work, proposed the PROMETHEE-MP (PROMETHEE (Preference Ranking Organization Method for Enrichment Evaluations) for Multi-Period) method based on a double aggregation involving multi-criteria aggregation and temporal aggregation [23]. The practical application of the proposed approach was demonstrated for the problem of evaluating sustainable forest management. A temporal extension of the PROMETHEE II method was used to create a ranking of emerging economies in terms of HDI (Human Development Index) in the work of Banamar and Smet [6]. The method is based on aggregating scores over time using an arithmetic mean. Another outranking method, considering both temporality and uncertainty, is the SMAA-TRI generalization based on the ELECTRE TRI (ELimination and Choice Expressing the Reality TRI) method proposed by Mouhib and Frini presented for a problem of temporal assessment of sustainable development in forest management [17]. In addition to outranking methods, another approach that takes into account temporality in multi-criteria evaluation is the multi-period single synthesizing criterion approach based on the Technique for Order of Preference by Similarity to Ideal Solution (TOPSIS) method proposed by Frini and Benamor [13]. The presented temporal extension of the TOPSIS method was applied for multi-period assessment of forest management.

However, the methods discussed have some limitations. Temporal extensions of PROMETHEE, such as PROMETHEE-MP and MUPOM, require a complex computational procedure, including multiple aggregations, which makes these algorithms complicated to apply [12,23]. The SMAA-TRI procedure also requires repetition [17]. The temporal extension of TOPSIS is a more straightforward method than the cited temporal outranking approaches. However, it requires performing a double TOPSIS procedure, first for periods, then for the results achieved for periods. It also has the disadvantage of needing clear rules for determining the weights of periods [13]. Therefore, the authors in the present work aimed to develop an approach that considers a multiplicity of criteria and periods in sustainability assessment that will be simple and, at the same time, automatically take into account the variability of results over time using clear rules.

3   Methodology

This section presents methodical aspects of the proposed DSS, including a novel methodology called DARIA-MARCOS and a multi-criteria assessment model regarding targets incorporated in SDG 11. The proposed DARIA-MARCOS method was applied in this research to evaluate 27 selected European countries against implementing the goals included in the Sustainable Development Goal 11 (SDG 11) framework focused on sustainable cities and communities. SDG 11 is among one of the seventeen SDGs aimed at promoting sustainable development in various aspects set in the 2030 Agenda by the United Nations in 2015. SDG 11 aims to develop toward making cities and human settlements inclusive, safe, resilient, and sustainable. Since the world's population is constantly increasing, it is essential to accommodate everyone and build modern, sustainable cities. Intelligent urban planning creates safe, affordable, and resilient cities with green and culturally inspiring living conditions. The particular nine targets included in SDG 11 are listed in Table 1.

Table 1. Multi-criteria model of assessment in relation to the achievement of SDG 11 goals.

Criterion	Description	Unit	Aim
C_1	Severe housing deprivation rate	Percentage	↓
C_2	Population living in households considering that they suffer from noise	Percentage	↓
C_3	Settlement area per capita	Square metres per capita	↑
C_4	Road traffic deaths - considering total type of roads	Rate	↓
C_5	Premature deaths due to exposure to fine particulate matter (PM2.5)	Rate	↓
C_6	Recycling rate of municipal waste	Percentage	↑
C_7	Population connected to at least secondary wastewater treatment	Percentage	↑
C_8	Share of buses and trains in inland passenger transport	Percentage	↑
C_9	Population reporting occurrence of crime, violence or vandalism in their area	Percentage	↓

3.1   The DARIA-MARCOS Method

This section provides the basics and assumptions of the newly developed multi-criteria temporal DARIA-MARCOS. Software developed for the proposed DSS, including, among other items, a DARIA class providing five methods for determining variability, including the entropy and datasets used in this study, is available in an open GitHub repository at link https://github.com/energyinpython/DARIA-MARCOS. Measurement of Alternatives and Ranking according to COmpromise Solution (MARCOS) method included in stages of the DARIA-MARCOS is described based on [21].

Step 1. Create a decision matrix defined by $X^p = [x_{ij}^p]_{m \times n}$ with performance values of m alternatives concerning n criteria for each evaluated period of time, where following periods are represented by $p = 1, 2, \ldots, t$ and t denotes number of time periods evaluated. Single decision matrix for a single period of time is presented in Eq. (1).

$$X^p = [x_{ij}^p]_{m \times n} = \begin{bmatrix} x_{11}^p & x_{12}^p & \cdots & x_{1n}^p \\ x_{21}^p & x_{22}^p & \cdots & x_{2n}^p \\ \vdots & \vdots & \vdots & \vdots \\ x_{m1}^p & x_{m2}^p & \cdots & x_{mn}^p \end{bmatrix} \tag{1}$$

Step 2. Extend each decision matrix created for partical period of time p by ideal (AI^p) and anti-ideal (AAI^p) solutions as shown in Eq. (2)

$$X^p = [x_{ij}^p]_{m+2 \times n} = \begin{bmatrix} x_{aa1}^p & x_{aa2}^p & \cdots & x_{aan}^p \\ x_{11}^p & x_{12}^p & \cdots & x_{1n}^p \\ x_{21}^p & x_{22}^p & \cdots & x_{2n}^p \\ \vdots & \vdots & \vdots & \vdots \\ x_{m1}^p & x_{m2}^p & \cdots & x_{mn}^p \\ x_{ai1}^p & x_{ai2}^p & \cdots & x_{ain}^p \end{bmatrix} \tag{2}$$

The anti-ideal solution (AAI^p) is the worst alternative and the ideal solution (AI^p) is the best alternative. AAI^p is determined according to Eq. (3) and AI^p is established with Eq. (4), where B denotes profit criteria and C represents cost criteria.

$$AAI^p = x_j^{p\ min} \ if \ j \in B \ and \ x_j^{p\ max} \ if \ j \in C \tag{3}$$

$$AI^p = x_j^{p\ max} \ if \ j \in B \ and \ x_j^{p\ min} \ if \ j \in C \tag{4}$$

Step 3. Normalize the extended initial matrix X^p. Normalized matrix $N^p = [n_{ij}^p]_{m+2 \times n}$ are calculated using Eqs. (5) for cost criteria and (6) for profit criteria, where x_{ij} and x_{ai} are elements of extended initial matrix X.

$$n_{ij}^p = \frac{x_{ai}^p}{x_{ij}^p} \ if \ j \in C \tag{5}$$

$$n_{ij}^p = \frac{x_{ij}^p}{x_{ai}^p} \ if \ j \in B \tag{6}$$

Step 4. Calculate the weighted matrix $V^p = [v_{ij}^p]_{m+2 \times n}$ by multiplying the normalized matrix N by criteria weight values w_j^p for j-th criterion, according to Eq. (7). Criteria weights can be determined subjectively by decision-makers or by using objective weighting methods that determine weights based on a decision matrix. In this research, criteria weights were determined using the objective weighting method called CRITIC (Criteria Importance Through Inter-criteria Correlation) method for objective determination of criteria weights [1].

$$v_{ij}^p = n_{ij}^p w_j^p \tag{7}$$

Step 5. Calculate the utility degree of alternatives K_i^p with Eqs. (8) and (9), where S_i^p $(i = 1, 2, \ldots, m)$ denotes the sum of the elements in the weighted matrix V^p calculated by Eq. (10).

$$K_i^{p-} = \frac{S_i^p}{S_{aai}^p} \tag{8}$$

$$K_i^{p+} = \frac{S_i^p}{S_{ai}^p} \tag{9}$$

$$S_i^p = \sum_{j=1}^{n} v_{ij}^p \tag{10}$$

Step 6. Determine the utility function of alternatives $f(K_i^p)$. The utility function is the compromise of a given alternative in relation to the ideal and anti-ideal solution. The utility function of alternatives is represented by Eq. (11)

$$f(K_i^p) = \frac{K_i^{p+} + K_i^{p-}}{1 + \frac{1-f(K_i^{p+})}{f(K_i^{p+})} + \frac{1-f(K_i^{p-})}{f(K_i^{p-})}} \tag{11}$$

where $f(K_i^{p-})$ denotes the utility function in relation to the anti-ideal solution. On the other hand, $f(K_i^{p+})$ denotes the utility function in relation to the ideal solution. Utility functions in relation to the ideal and anti-ideal solutions are established using Eqs. (12) and (13)

$$f(K_i^{p-}) = \frac{K_i^{p+}}{K_i^{p+} + K_i^{p-}} \tag{12}$$

$$f(K_i^{p+}) = \frac{K_i^{p-}}{K_i^{p+} + K_i^{p-}} \tag{13}$$

Step 7. Construct the matrix $S = [s_{pi}]_{t \times m}$ shown in Eq. (14) containing annual MARCOS utility function values of alternatives s_{pi} (for MARCOS method they are represented by K_i^p) collected for t periods in rows, where following periods are numbered by $p = 1, 2, \ldots, t$ and m alternatives a in columns, where subsequent alternatives are numbered by $i = 1, 2, \ldots, m$. Subsequent periods are represented by $y_1, \ldots, y_p, \ldots, y_t$.

$$S = \begin{array}{c|ccccc} & a_1 & \ldots & a_i & \ldots & a_m \\ \hline y_1 & s_{11} & \ldots & s_{1i} & \ldots & s_{1m} \\ \vdots & \vdots & \ldots & \vdots & \ldots & \vdots \\ y_p & s_{p1} & \ldots & s_{pi} & \ldots & s_{pm} \\ \vdots & \vdots & \ldots & \vdots & \ldots & \vdots \\ y_t & s_{t1} & \ldots & s_{ti} & \ldots & s_{tm} \end{array} \tag{14}$$

Step 8. Calculate the variability of obtained scores in matrix S received using the MARCOS method for each assessed period. The variability value is calculated using the entropy method [27] provided in steps 8.1–8.3. Entropy was selected for measuring variability as the most common objective method. Entropy measures uncertainty and provides a quantitative measure of information content.

Step 8.1. Normalize matrix S using sum normalization method to get normalized matrix $K = [k_{pi}]_{t \times m}$ where $p = 1, 2, \ldots, t$ and $i = 1, 2, \ldots, m$, t represents periods number and m denotes alternatives number.

$$k_{pi} = \frac{s_{pi}}{\sum_{p=1}^{t} s_{pi}} \tag{15}$$

Step 8.2. Calculate the entropy value E_i for each ith alternative according to Eq. (16) [27].

$$E_i = -\frac{\sum_{p=1}^{t} k_{pi} ln(k_{pi})}{ln(t)} \qquad (16)$$

Step 8.3. Calculate the variability value represented by d_i as Eq. (17) shows.

$$d_i = 1 - E_i \qquad (17)$$

Step 9. Determine the direction of score variability. The threshold value provided in Eq. (19) with Eq. (18) is employed to calculate the variability direction for each ith alternative.

$$thresh_i = \sum_{p=2}^{t} s_p - s_{p-1} \qquad (18)$$

$$dir_i = \begin{cases} 1 & if\ thresh_i > 0 \\ -1 & if\ thresh_i < 0 \\ 0 & if\ thresh_i = 0 \end{cases} \qquad (19)$$

Step 10. The MARCOS utility function values for alternatives received for the most recent period t is updated with the value of the variability of scores d_i in all investigated periods according to its direction using Eq. (20),

$$S_i = S_i^t + d_i \cdot dir_i \qquad (20)$$

where S_i defines the score achieved by given alternative a_i updated by adding variability values multiplied by variability direction, S_i^t represents the score of given alternative a_i reached in the most recent period t investigated, d_i represents values of the variability of alternative's a_i scores over all analyzed periods $p = 1, 2, \ldots, t$ calculated using entropy method, and dir_i defines directions of variability d_i, which may be equal to 1 for increasing scores, -1 for decreasing scores or 0 for stable scores. Alternatives are defined by a_i ($i = 1, 2, \ldots, m$).

Step 11. The purpose of the final step is to rank the alternatives according to the final scores S following the descending order as for the MARCOS method.

4   Results

This section provides results given by the DARIA-MARCOS method. In the first stage, individual evaluation of each year was carried out. The list of European countries evaluated in this research is provided in Table 2. The selection of just these 27 countries is justified by the availability of data against all criteria of the SDG 11 framework. The analysis considers the most recent seven years (2015–2022) for which data is available in the Eurostat database. The data were accessed on 23 January 2024. Table 2 presents sample performance values of evaluated countries collected for 2022. The datasets for the other years included in the analysis are made available in an open GitHub repository.

Table 2. Sample dataset with performances regarding implementation of targets included in SDG 11 collected for 2022.

Country	C_1	C_2	C_3	C_4	C_5	C_6	C_7	C_8	C_9
Belgium	2.3	14.5	583.5	4.6	44	55.5	84.03	14.4	10.8
Bulgaria	8.6	8.8	623.4	8.2	158	28.2	65.05	10.1	19.1
Czechia	2	13.3	634.4	5.1	81	43.3	84.7	15	6.1
Denmark	2.8	18.2	1053.8	2.6	21	57.6	97.8	13.3	7.3
Germany	1.2	21.6	586.7	3.3	39	67.8	96.32	11.2	8.2
Estonia	2.1	8	1484.4	3.6	7	30.3	82	10.7	5.5
Ireland	1.4	10.3	972.7	2.7	9	40.8	62.3	14.3	11.3
Greece	5.8	20.1	710.2	5.9	95	21	94.7	12.9	18.1
Spain	3.4	21.9	577.5	3.2	30	36.7	86.93	12.6	14.1
France	3.8	20.7	845.1	4.8	30	43.8	79.85	14	17.7
Croatia	5.1	8.1	722.5	7.4	96	31.4	31.39	11.2	2.4
Italy	6.1	14.3	484.3	5.4	79	51.9	59.6	17.2	8.4
Cyprus	1.6	14	939	5	70	15.3	83.48	12.7	10.4
Latvia	11.5	12.5	1276.1	7.8	75	44.1	76.48	11.5	5.3
Lithuania	5.4	14.7	1090.5	4.2	77	44.3	76.94	5.3	3.3
Luxembourg	1.6	19.7	565.2	5.5	12	55.3	97	13.7	11
Hungary	7.6	9.3	811.5	5.6	107	34.9	84.23	20.7	5.3
Malta	1	30.8	201.4	1.7	37	13.6	7.4	14.1	11.4
Netherlands	1.5	25.5	456.9	2.9	32	57.8	99.52	10.6	15.7
Austria	3	16.8	740.1	4.1	36	62.5	99.1	18.8	5.7
Poland	7.9	12.6	633.7	5.9	125	40.3	75.2	13.7	4.4
Portugal	3.9	25.1	689.1	5.4	20	30.4	55.8	8.7	6.6
Romania	14.3	16.1	528.4	9.3	103	11.3	52.6	17	8.8
Slovenia	3.1	15	625.1	5.4	56	60.8	67.61	10	7.3
Slovakia	3.2	9.9	631.8	4.9	98	48.9	69.9	16.3	4.3
Finland	1	14.1	2447.6	4.1	3	39	85	12.3	7
Sweden	2.5	17.3	2223	2	6	39.5	96	15.9	13.8

Results of DARIA-MARCOS comprise variability value of annual scores, direction of variability, DARIA-MARCOS utility function value and final ranking. Mentioned results are included in Table 4. It can be noted that Malta achieved the highest variability of annual performance toward improvement (0.00298). As a result, even though Malta was ranked 26th in 2015–2019, the improvement in 2020–2022 resulted in the country moving up to rank 24 in 2020 and up to rank 23 in 2021–2022 and achieving rank 23 in the temporal ranking of DARIA-MARCOS. The DARIA-MARCOS gives greater relevance to these most recently evaluated years because it is the most important from the perspective of policymakers and stakeholders. Results involving utility function values of alternatives for each year and annual rankings created based on utility function values are displayed in Table 3.

Table 3. Annual MARCOS utility function values and ranks.

Country	Utility function values								Ranks							
	2015	2016	2017	2018	2019	2020	2021	2022	2015	2016	2017	2018	2019	2020	2021	2022
Belgium	0.455	0.458	0.445	0.463	0.462	0.454	0.459	0.457	9	10	12	15	16	14	14	14
Bulgaria	0.361	0.357	0.365	0.371	0.384	0.372	0.351	0.350	22	23	23	23	23	23	24	24
Czechia	0.407	0.426	0.448	0.456	0.472	0.482	0.467	0.467	18	14	11	16	12	10	11	11
Denmark	0.522	0.525	0.511	0.536	0.500	0.486	0.533	0.519	4	4	5	5	6	9	5	7
Germany	0.464	0.485	0.474	0.480	0.476	0.520	0.524	0.520	7	7	8	10	11	5	6	6
Estonia	0.510	0.521	0.583	0.564	0.595	0.551	0.564	0.571	5	5	4	4	4	3	3	3
Ireland	0.573	0.593	0.621	0.656	0.602	0.515	0.539	0.538	3	3	3	2	3	6	4	4
Greece	0.341	0.342	0.342	0.362	0.350	0.342	0.334	0.334	24	24	24	24	24	25	26	26
Spain	0.449	0.457	0.483	0.473	0.466	0.397	0.395	0.394	10	11	7	11	15	21	22	22
France	0.423	0.420	0.426	0.440	0.433	0.395	0.400	0.395	11	15	15	18	21	22	21	21
Croatia	0.413	0.416	0.419	0.464	0.477	0.472	0.461	0.460	14	17	17	14	10	11	13	13
Italy	0.359	0.377	0.396	0.438	0.427	0.406	0.411	0.407	23	21	21	19	22	20	19	20
Cyprus	0.407	0.377	0.387	0.435	0.433	0.412	0.408	0.408	17	22	22	20	20	19	20	19
Latvia	0.389	0.393	0.397	0.400	0.446	0.429	0.432	0.432	20	20	20	22	17	17	16	16
Lithuania	0.392	0.460	0.407	0.464	0.469	0.434	0.440	0.448	19	9	19	13	13	15	15	15
Luxembourg	0.416	0.446	0.443	0.468	0.483	0.469	0.497	0.481	13	12	13	12	9	12	9	10
Hungary	0.421	0.440	0.454	0.527	0.501	0.490	0.482	0.482	12	13	10	6	5	8	10	9
Malta	0.318	0.267	0.290	0.325	0.314	0.354	0.375	0.373	26	26	26	26	26	24	23	23
Netherlands	0.480	0.487	0.499	0.491	0.468	0.455	0.466	0.465	6	6	6	9	14	13	12	12
Austria	0.459	0.481	0.471	0.509	0.498	0.520	0.523	0.522	8	8	9	7	7	4	7	5
Poland	0.408	0.408	0.409	0.420	0.435	0.415	0.426	0.425	16	19	18	21	19	18	18	18
Portugal	0.318	0.337	0.323	0.358	0.337	0.316	0.338	0.337	25	25	25	25	25	26	25	25
Romania	0.255	0.260	0.271	0.285	0.297	0.305	0.294	0.293	27	27	27	27	27	27	27	27
Slovenia	0.410	0.419	0.432	0.455	0.437	0.430	0.426	0.426	15	16	14	17	18	16	17	17
Slovakia	0.381	0.414	0.421	0.496	0.496	0.510	0.500	0.496	21	18	16	8	8	7	8	8
Finland	0.691	0.738	0.740	0.708	0.727	0.695	0.684	0.686	1	1	1	1	1	1	1	1
Sweden	0.647	0.640	0.632	0.640	0.638	0.580	0.597	0.597	2	2	2	3	2	2	2	2

It can be observed that Finland is the leader of all annual rankings for the years included in this analysis. This confirms Finland's strong and stable position over the seven years investigated. On the other hand, the last rank in all the annual rankings received in the research is held by Romania, which indicates the country's poor implementation of the goals of SDG 11 compared to the other countries analyzed.

The cases presented above show that temporal analysis of the performance of countries such as Finland and Romania, whose performance in all years is maintained at constant positions, is easy to carry out using classical MCDA methods that give results for single moments in examined time. However, it can be observed that a high variability of results over the time range studied is evident in the significant majority of the countries considered. This variability is expressed in variable utility function values achieved by countries in subsequent years, as reflected in shifts in rankings. In such situations, in order to obtain a single unambiguous and easy-to-interpret result that takes into account the full range of time under study, additional computational methods are needed to complement classical MCDA methods.

Table 4. Results of temporal assessment performed with DARIA-MARCOS.

Country	Variability	Direction	Score	Rank	Country	Variability	Direction	Score	Rank
Belgium	0.00003	↑	0.4574	14	Lithuania	0.00084	↑	0.4488	15
Bulgaria	0.00021	↓	0.3502	24	Luxembourg	0.00067	↑	0.4815	10
Czechia	0.00065	↑	0.4673	11	Hungary	0.00111	↑	0.4826	9
Denmark	0.00022	↓	0.5190	7	Malta	0.00298	↑	0.3762	23
Germany	0.00051	↑	0.5207	6	Netherlands	0.00021	↓	0.4647	12
Estonia	0.00059	↑	0.5714	3	Austria	0.00053	↑	0.5220	5
Ireland	0.00141	↓	0.5370	4	Poland	0.00012	↑	0.4251	18
Greece	0.00015	↓	0.3339	26	Portugal	0.00035	↑	0.3378	25
Spain	0.00158	↓	0.3922	22	Romania	0.00088	↑	0.2940	27
France	0.00037	↓	0.3947	21	Slovenia	0.00020	↑	0.4262	17
Croatia	0.00077	↑	0.4612	13	Slovakia	0.00252	↑	0.4982	8
Italy	0.00086	↑	0.4079	20	Finland	0.00023	↓	0.6860	1
Cyprus	0.00050	↑	0.4084	19	Sweden	0.00037	↓	0.5962	2
Latvia	0.00060	↑	0.4327	16					

Slovakia (0.00252) was another country that received a large variability towards improvement. Slovakia was ranked 21st in 2015 but performed much better in subsequent years, which enabled the country to be ranked 7th-8th in 2018–2022. Slovakia's improved performance promoted the country to eighth place in the DARIA-MARCOS temporal ranking. A different situation due to high variability in performance towards worsening (0.00158) occurs for Spain. Spain ranked 7–11 in 2015–2018, dropped to 15th in 2019, 21st in 2020, and 22nd in 2021–2022. The significance of the most recent year, decreasing annual performances, and a drop in subsequent rankings caused Spain to rank only 22nd in the final DARIA-MARCOS temporal ranking. If the variability is minor, then even if it is associated with worsening, it will not result in a degradation of the DARIA-MARCOS ranking, as in the case of Sweden. For this country, there was a variability of 0.00037 toward worsening, which is nevertheless small enough that Sweden maintained its second place with annual rankings in the DARIA-MARCOS temporal ranking. Other countries can be analyzed analogously to the examples discussed.

Finally, the comparative analysis involving determining the correlation for the following annual rankings and the temporal DARIA-MARCOS ranking was performed in order to compare the convergence between them. For this aim, the Weighted Spearman correlation coefficient r_w was employed [26]. Results are visualized in the form of a heatmap in Fig. 1. It can be outlined that the DARIA-MARCOS ranking involving six analyzed years 2015–2022 demonstrates the lowest convergence for the ranking generated for the very earliest year in the investigation. On the other hand, for the successive years investigated, the correlation grows. In the end, the highest correlation is noticed when comparing the DARIA-MARCOS ranking with the annual ranking generated for the most recent year, 2022. This is proved by the fact that in the temporal DARIA-MARCOS method employed in this research, the most recent period is the most important, which is updated with the variability of the scores for the successive years analyzed. In the approach of this novel method, the highest significance of the most recent year was established, as this is the period of most importance and interest to decision-makers and stakeholders from the perspective of sustainable development.

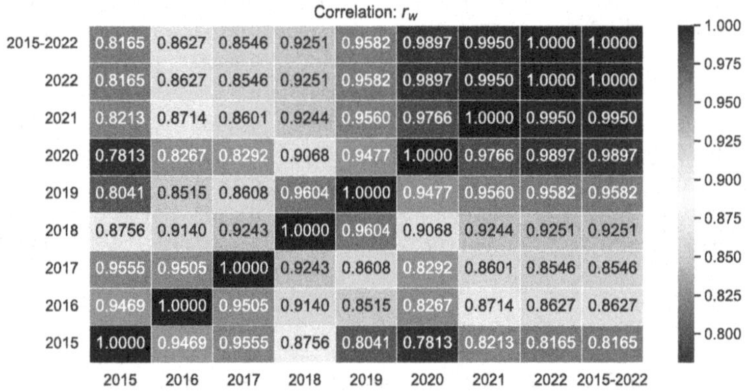

Fig. 1. r_w correlation of the DARIA-MARCOS ranking with annual rankings.

However, in accordance with individual analytical requirements for particular decision-making problems, the DARIA-MARCOS method is adaptable, and its idea enables this concept to be easily modified. Creating the final score may include updating the average score of all periods instead of the score achieved for the most recent period. Measures of variability other than entropy, such as the Gini coefficient [14], standard deviation [20], and statistical variance [19] can also be used to measure variability in performance.

5   Conclusions

The research work demonstrated in this paper a novel method DARIA-MARCOS employed in DSS for multi-criteria temporal assessment of any decision problem. The practical application of the presented DSS was shown using a practical example of evaluation of the implementation of targets included in SDG 11, which calls for making cities and human settlements inclusive, safe, resilient, and sustainable. A multi-criteria temporal assessment employing DARIA-MARCOS was performed for selected European countries in the time interval covering eight years, 2015–2022. The methodical framework implemented in the demonstrated DSS ensures an efficient, automatized, and objective assessment of a multi-criteria temporal decision-making problem and gives an unequivocal, clear outcome. The results proved the usability of the developed DSS in the multi-criteria temporal evaluation of sustainable development focused on sustainable cities and communities. The comparative analysis demonstrated that the results delivered by the presented DSS are reliable. Therefore, it can also be employed for multi-criteria temporal evaluation in other development areas.

Advantages of the proposed method over other existing temporal approaches include low computational complexity due to the lack of need to perform multiple aggregations, the use of a measure of variability to aggregate results that reflects variability over time more adequately than simple methods such as the arithmetic mean, the lack of need to determine the significance of individual periods, and the possibility of expanding the approach in the future to include measures of variability other than entropy. The DARIA-MARCOS method can be a valuable tool for researchers and practitioners, as it reflects the impact of data variability on the final form of the model. Replacing oversimplifications that smooth out variability over time introduces new analytical possibilities into the model. Directions for further work involve developing multi-criteria temporal methods based on other multi-criteria decision analysis (MCDA) methods and measures of variability and comparative analysis of the results.

Acknowledgments. This research was partially funded by National Science Centre, Poland 2022/45/N/HS4/03050, and Co-financed by the Minister of Science under the "Regional Excellence Initiative" Program (A.B., J.W.).

 Regionalna Inicjatywa Doskonałości
 Minister of Science Republic of Poland
 Ministry of Science and Higher Education Republic of Poland

References

1. Abdel-Basset, M., Mohamed, R.: A novel plithogenic TOPSIS-CRITIC model for sustainable supply chain risk management. J. Clean. Prod. **247**, 119586 (2020). https://doi.org/10.1016/j.jclepro.2019.119586
2. Abdella, G.M., Kucukvar, M., Onat, N.C., Al-Yafay, H.M., Bulak, M.E.: Sustainability assessment and modeling based on supervised machine learning techniques: the case for food consumption. J. Clean. Prod. **251**, 119661 (2020). https://doi.org/10.1016/j.jclepro.2019.119661
3. Agyemang, P., Kwofie, E.M., Fabrice, A.: Integrating framework analysis, scenario design, and decision support system for sustainable healthy food system analysis. J. Clean. Prod. **372**, 133661 (2022). https://doi.org/10.1016/j.jclepro.2022.133661
4. Akuraju, V., Pradhan, P., Haase, D., Kropp, J.P., Rybski, D.: Relating SDG11 indicators and urban scaling-An exploratory study. Sustain. Urban Areas **52**, 101853 (2020). https://doi.org/10.1016/j.scs.2019.101853

5. Baffo, I., Leonardi, M., Bossone, B., Camarda, M.E., D'Alberti, V., Travaglioni, M.: A decision support system for measuring and evaluating solutions for sustainable development. Sustain. Futures **5**, 100109 (2023). https://doi.org/10.1016/j.sftr.2023.100109
6. Banamar, I., Smet, Y.D.: An extension of PROMETHEE II to temporal evaluations. Int. J. Multicriteria Decis. Making **7**(3–4), 298–325 (2018). https://doi.org/10.1504/IJMCDM.2018.094371
7. Dabous, S.A., Zeiada, W., Zayed, T., Al-Ruzouq, R.: Sustainability-informed multi-criteria decision support framework for ranking and prioritization of pavement sections. J. Clean. Prod. **244**, 118755 (2020). https://doi.org/10.1016/j.jclepro.2019.118755
8. Delanka-Pedige, H.M.K., Munasinghe-Arachchige, S.P., Abeysiriwardana-Arachchige, I.S.A., Nirmalakhandan, N.: Evaluating wastewater treatment infrastructure systems based on UN sustainable development goals and targets. J. Clean. Prod. **298**, 126795 (2021). https://doi.org/10.1016/j.jclepro.2021.126795
9. Deshpande, P.C., Skaar, C., Brattebø, H., Fet, A.M.: Multi-criteria decision analysis (MCDA) method for assessing the sustainability of end-of-life alternatives for waste plastics: A case study of Norway. Sci. Total Environ. **719**, 137353 (2020). https://doi.org/10.1016/j.scitotenv.2020.137353
10. D'Adamo, I., Gastaldi, M., Ioppolo, G., Morone, P.: An analysis of sustainable development goals in Italian cities: performance measurements and policy implications. Land Use Policy **120**, 106278 (2022). https://doi.org/10.1016/j.landusepol.2022.106278
11. Enoh, M.A., Njoku, R.E., Okeke, U.C.: Modeling and mapping the spatial-temporal changes in land use and land cover in Lagos: a dynamics for building a sustainable urban city. Adv. Space Res. **72**(3), 694–710 (2023). https://doi.org/10.1016/j.asr.2022.07.042
12. Frini, A., Amor, S.B.: MUPOM: a multi-criteria multi-period outranking method for decision-making in sustainable development context. Environ. Impact Assess. Rev. **76**, 10–25 (2019). https://doi.org/10.1016/j.eiar.2018.11.002
13. Frini, A., Benamor, S.: Making decisions in a sustainable development context: a state-of-the-art survey and proposal of a multi-period single synthesizing criterion approach. Comput. Econ. **52**(2), 341–385 (2018). https://doi.org/10.1007/s10614-017-9677-5
14. Luptáčik, M., Nežinskỳ, E.: Measuring income inequalities beyond the Gini coefficient. CEJOR **28**(2), 561–578 (2020). https://doi.org/10.1007/s10100-019-00662-9
15. Martínez-Córdoba, P.J., Amor-Esteban, V., Benito, B., García-Sánchez, I.M.: The commitment of Spanish local governments to sustainable development goal 11 from a multivariate perspective. Sustainability **13**(3), 1222 (2021). https://doi.org/10.3390/su13031222
16. Martins, M.A., Garcez, T.V.: A multidimensional and multi-period analysis of safety on roads. Accid. Anal. Prev. **162**, 106401 (2021). https://doi.org/10.1016/j.aap.2021.106401
17. Mouhib, Y., Frini, A.: TSMAA-TRI: a temporal multi-criteria sorting approach under uncertainty. J. Multi-Criteria Decis. Anal. **28**(3–4), 185–199 (2021). https://doi.org/10.1002/mcda.1742
18. Musavengane, R., Siakwah, P., Leonard, L.: The nexus between tourism and urban risk: towards inclusive, safe, resilient and sustainable outdoor tourism in African cities. J. Outdoor Recreat. Tour. **29**, 100254 (2020). https://doi.org/10.1016/j.jort.2019.100254

19. Rao, R.V., Patel, B.K., Parnichkun, M.: Industrial robot selection using a novel decision making method considering objective and subjective preferences. Robot. Auton. Syst. **59**(6), 367–375 (2011). https://doi.org/10.1016/j.robot.2011.01.005
20. Shi, J., et al.: Optimally estimating the sample standard deviation from the five-number summary. Res. Synth. Methods **11**(5), 641–654 (2020). https://doi.org/10.1002/jrsm.1429
21. Stević, Ž, Pamučar, D., Puška, A., Chatterjee, P.: Sustainable supplier selection in healthcare industries using a new MCDM method: measurement of alternatives and ranking according to compromise solution (MARCOS). Comput. Ind. Eng. **140**, 106231 (2020). https://doi.org/10.1016/j.cie.2019.106231
22. Tsalis, T.A., Malamateniou, K.E., Koulouriotis, D., Nikolaou, I.E.: New challenges for corporate sustainability reporting: United Nations' 2030 Agenda for sustainable development and the sustainable development goals. Corp. Soc. Responsib. Environ. Manag. **27**(4), 1617–1629 (2020). https://doi.org/10.1002/csr.1910
23. Urli, B., Frini, A., Amor, S.B.: PROMETHEE-MP: a generalisation of PROMETHEE for multi-period evaluations under uncertainty. Int. J. Multicriteria Decis. Making **8**(1), 13–37 (2019). https://doi.org/10.1504/IJMCDM.2019.098042
24. Vaidya, H., Chatterji, T.: SDG 11 sustainable cities and communities. In: Franco, I.B., Chatterji, T., Derbyshire, E., Tracey, J. (eds.) Actioning the Global Goals for Local Impact. SSS, pp. 173–185. Springer, Singapore (2020). https://doi.org/10.1007/978-981-32-9927-6_12
25. Wątróbski, J.: Temporal PROMETHEE II-New multi-criteria approach to sustainable management of alternative fuels consumption. J. Clean. Prod. **413**, 137445 (2023). https://doi.org/10.1016/j.jclepro.2023.137445
26. Wątróbski, J., Bączkiewicz, A., Ziemba, E., Sałabun, W.: Sustainable cities and communities assessment using the DARIA-TOPSIS method. Sustain. Urban Areas **83**, 103926 (2022). https://doi.org/10.1016/j.scs.2022.103926
27. Wu, R.M., et al.: A comparative analysis of the principal component analysis and entropy weight methods to establish the indexing measurement. PLoS ONE **17**(1), e0262261 (2022). https://doi.org/10.1371/journal.pone.0262261
28. Zhang, Z., et al.: A hybrid MCDM model for evaluating the market-oriented business regulatory risk of power grid enterprises based on the Bayesian best-worst method and MARCOS approach. Energies **15**(9), 2978 (2022). https://doi.org/10.3390/en15092978

Modeling Tsunami Waves at the Coastline of Valparaiso Area of Chile with Physics Informed Neural Networks

Alicja Niewiadomska[1], Paweł Maczuga[1], Albert Oliver-Serra[2],
Leszek Siwik[1], Paulina Sepulveda-Salaz[4], Anna Paszyńska[1,3],
Maciej Paszyński[1(✉)], and Keshav Pingali[5]

[1] AGH University of Krakow, Kraków, Poland
paszynsk@agh.edu.pl
[2] University of Las Palmas de Gran Canaria (ULPGC), Gran Canaria, Spain
albert.oliver@ulpgc.es
[3] Jagiellonian University, Krakow, Poland
anna.paszynska@uj.edu.pl
[4] Pontifical Catholic University of Valparaiso, Valparaíso, Chile
paulina.sepulveda@pucv.cl
[5] The Oden Institute for Computational Engineering, The University of Texas at
Austin, Austin, USA
pingali@cs.utexas.edu

Abstract. The Chilean coast is a very seismically active region. In the
21st century, the Chilean region experienced 19 earthquakes with a magnitude of 6.2 to 8.8, where 597 people were killed. The most dangerous earthquakes occur at the bottom of the ocean. The tsunamis they cause are very dangerous for residents of the surrounding coasts. In 2010, as many as 525 people died in a destructive tsunami caused by an underwater earthquake. Our research paper aims to develop a tsunami simulator based on the modern methodology of Physics Informed Neural Networks (PINN). We test our model using a tsunami caused by a hypothetical earthquake off the coast of the densely populated area of Valparaiso, Chile. We employ a longest-edge refinement algorithm expressed by graph transformation rules to generate a sequence of triangular computational meshes approximating the seabed and seashore of the Valparaiso area based on the Global Multi-Resolution Topography Data available. For the training of the PINN, we employ points from the vertices of the generated triangular mesh.

1 Introduction

Tsunami waves (or just tsunamis) are "high-energy" ocean waves caused by the sudden displacement of large masses of ocean water, originating most frequently from (high-magnitude) (underwater) earthquakes (or, less frequently, from underwater volcanic eruptions, calving of glaciers, massive landslides or meteorite impacts). Despite the high energy and moving at speeds of up to

© The Author(s), under exclusive license to Springer Nature Switzerland AG 2024
L. Franco et al. (Eds.): ICCS 2024, LNCS 14833, pp. 204–218, 2024.
https://doi.org/10.1007/978-3-031-63751-3_14

several hundred kilometers per hour, due to their low height (up to several dozen centimeters) and long length (up to several hundred kilometers), tsunami waves may even be unnoticed in the open ocean. The situation changes, however, dramatically when they reach the coastal zone, where they accumulate (often rapidly) and, reaching heights of up to several dozen meters, break onto the land, flooding and destroying everything that comes within their range of influence.

Due to their unpredictable nature, devastating power, and huge area and range of impact, we, as humankind, can still not predict their occurrence with 100% effectiveness and in a long enough advance to get a chance to evacuate people from endangered areas. Also, it is not possible to keep all the endangered areas like a greenfield or uninhabited sites, so as important as improving the methods and tools for monitoring and warning against their occurrence, an important direction of research are methods and techniques for modeling the propagation and behavior of these waves (especially while reaching the coastal zones) to make it possible and reliable to simulate the potential impact of their occurrence and, for example, to adjust evacuation routes, procedures, and coastal areas development plans appropriately to minimize their potential devastating effects.

Due to the high seismic activity, tsunami waves are (most) frequently observed in the Pacific region, where in the so-called (Pacific) Ring of Fire, or the Circum-Pacific Seismic Belt, there are most of Earth's active volcanoes located (more than 450 in total) and approximately 90% of all earthquakes in the world occur.

One of the countries particularly exposed to the devastating effects of tsunami waves due to its location in the area of the Circum-Pacific Seismic Belt, the length of the coastline (approximately 6.5 thousand kilometers), and a large number of islands and islets located near the coastline and in the open ocean (around 3,000 in total) is south-American Chile [12]. How real is that dangerous for the Chilean territory (and the coast in particular) let it be proven by the fact that on the list of the strongest earthquakes recorded in history, as many as five can be classified as located in the Chilean region (coastline or the interior), including so-called the Great Chilean Earthquake, on May 22, 1960, considered, with the magnitude of 9.5 Richter degrees, as the strongest earthquake recorded so far in the history of seismic measurements. So, that is the direct motivation for the research undertaken and discussed in this paper.

As for the particular region considered in our research, we focused on the Valparaiso area. It has been chosen for two reasons. First, it is located approximately halfway along Chile's coastline so that the results may be representative of the broader part of the Chilean coast. Second, Valparaiso (the city and the region) is one of Chili's most densely populated and urbanized regions. Also, it is one of the most popular tourist regions on the west coast of South America, so the topic may be even more important and interesting for the Chilean community, researchers, and authorities.

The goal of this paper is to develop the Physics Informed Neural Network (PINN) model of a tsunami based on non-linear wave equations and the topography of the seabed and seashore in the Valparaiso region of Chile. Prof. Karniadakis proposed the PINN [1] in 2019 as a method of solving Partial Differential Equations (PDEs) by training the neural networks. PINN has been implemented in the DeepXDE library [9]. It is an extensive library supporting TensorFlow, PyTorch, JAX, and PaddlePaddle, with huge functionality, including ODEs, PDEs, complex geometries, and different initial and boundary conditions for solving forward and inverse problems. Another library for solving the wave equation, Allan-Cahn equations, Volterra integrodifferential equations, and variational minimization problems is IDRLnet [10], which uses pytorch, numpy, and Matplotlib.

In this paper, we have implemented the PINN solver for the non-linear wave equation using our PINN-2DT library [14] implemented in PyTorch and executed in GoogleColab. It enables simple implementation and execution of the wave-equation simulations. The novelty of our computational method lies in the fact that we select the points for training PINN residuals following the adaptive mesh refinement procedure. Namely, we employ a longest-edge refinement algorithm [17] to generate an accurate triangular computational mesh approximating the seabed and seashore of the Valparaiso area. Our adaptive method employs the Global Multi-Resolution Topography Database (GMRT) [11]. We express the longest-edge refinement algorithm by graph transformation rules. For the training of the PINN, we use the points from the vertices of the generated triangular mesh. Following the ideas presented in our previous model [8,15,18], we express the triangular mesh refinement algorithm by a set of graph transformation rules.

Alternative available state-of-the-art simulators are based on finite volume method (GeoClaw [19]), finite difference method (COMCOT [20], CoulWAVE [21]), a hybrid of finite difference-finite volume methods (Celeris base [22]), or finite element method [8]. We verify the correctness of our PINN simulator by comparing it to the finite element method solver [8] executed on the model "pool" example. The PINN model's potential advantage over the finite element method solver lies mainly in its simplicity of implementation, employing space time formulations, and lack of problems with time stepping and stabilization [7].

2 Modeling of Tsunami Wave with Physics Informed Neural Networks

Our basic model for simulations of the tsunami waves is based on the formulation using the wave equation described in [8]. Let us focus on a strong form of the wave equation: Find $u \in C^2(0,1)$ for $(x,y) \in \Omega = [0,1]^2$, $t \in [0,T]$ such that:

$$\frac{\partial^2 u(x,y,t)}{\partial t^2} - \left(\frac{\partial}{\partial x}, \frac{\partial}{\partial y}\right) \cdot \left(g(u(x,y,t) - z(x,y))\left(\frac{\partial u}{\partial x}, \frac{\partial u}{\partial y}\right)\right) = 0, \quad (1)$$
$$(x,y,t) \in \Omega \times [0,T],$$

Here $z(x, y)$ stands for the bathymetry (the seabed and the seashore topography function), $g = 9,81$ is the acceleration due to the Earth's gravity, and $u(x, y, t)$ represents the water level. We start the simulation by assuming the initial tsunami wave shape:

$$u(x, y, 0) = u_0(x, y), \quad (x, y) \in \Omega. \tag{2}$$

We assume the no-reflection zero-Neumann boundary condition:

$$\frac{\partial u}{\partial n} = 0, \quad \text{on } \partial\Omega \times [0, T]. \tag{3}$$

We expand the wave equation PDE by computing the partial derivatives:

$$
\begin{aligned}
\frac{\partial^2 u(x, y, t)}{\partial t^2} &- \left(g \left(\frac{\partial u(x, y, t)}{\partial x} - \frac{\partial z(x, y)}{\partial x} \right) \frac{\partial u(x, y, t)}{\partial x} \right) \\
&- \left(g \left(u(x, y, t) - z(x, y) \right) \frac{\partial^2 u(x, y, t)}{\partial x^2} \right) \\
&- \left(g \left(\frac{\partial u(x, y, t)}{\partial y} - \frac{\partial z(x, y)}{\partial y} \right) \frac{\partial u(x, y, t)}{\partial y} \right) \\
&- \left(g \left(u(x, y, t) - z(x, y) \right) \frac{\partial^2 u(x, y, t)}{\partial y^2} \right) = 0
\end{aligned}
\tag{4}
$$

We employ the Physics Informed Neural Network approach, proposed by Karniadakis [1]. The neural network is understood there as a continuous function that represents the solution:

$$u(x, y, t) = PINN(x, y, t) = A_n\sigma\left(A_{n-1}\sigma(...\sigma(A_1 \begin{bmatrix} x \\ y \\ t \end{bmatrix} + B_1)... + B_{n-1}\right) + B_n. \tag{5}$$

Here, A_i represents the matrices of layers of the neural network, and B_i represents the bias vectors. Also, σ is the activation function.

We choose the sigmoid activation function after our analysis in [2], where we show it best fits for wave equation training with PINN. In our training, we employed three internal layers, each one with 300 neurons. The motivation was that a smaller number of neural networks (4 layers with 80 neurons) experience some problems with the training of the wave equations.

We define the loss function as the residual of the PDE.

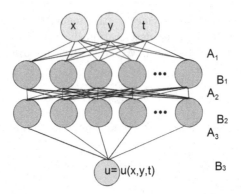

Fig. 1. Physics informed neural network as a continuous function representing a solution function. The matrix entries $A_1^{kl}, A_2^{kl}, A_3^{kl}$ represents the weight of edges connecting the neurons. The vector entries B_1^k, B_2^k, B_3^k represent the biases assigned to neurons.

$$
\begin{aligned}
LOSS_{PDE}&(x,y,t) = \\
\left(\frac{\partial^2 PINN(x,y,t)}{\partial t^2}\right. &- \left(g\left(\frac{\partial PINN(x,y,t)}{\partial x} - \frac{\partial z(x,y)}{\partial x}\right)\frac{\partial PINN(x,y,t)}{\partial x}\right) \\
&- \left(g\left(PINN(x,y,t) - z(x,y)\right)\frac{\partial^2 PINN(x,y,t)}{\partial x^2}\right) \\
&- \left(g\left(\frac{\partial PINN(x,y,t)}{\partial y} - \frac{\partial z(x,y)}{\partial y}\right)\frac{\partial PINN(x,y,t)}{\partial y}\right) \\
&\left.- \left(g\left(PINN(x,y,t) - z(x,y)\right)\frac{\partial^2 PINN(x,y,t)}{\partial y^2}\right)\right)^2
\end{aligned}
\tag{6}
$$

On top of the loss function related to the residual of the PDE, we also need to define the loss function for training the initial condition $u_0(x,y)$:

$$
LOSS_{Init}(x,y,0) = (PINN(x,y,0) - u_0(x,y))^2
\tag{7}
$$

We also define the loss of the residual of the Neumann boundary condition:

$$
LOSS_{BC}(x,y,t) = \left(\frac{\partial PINN(x,y,t)}{\partial n} - 0\right)^2.
\tag{8}
$$

During the training, we adjust the weights of the matrices representing the neural network layers. We also adjust entries of the vector representing the neural network biases. This is done by using the derivatives of the neural network with respect to particular entries. The sketch of the gradient descent procedure is the following:

The sketch of the training procedure is the following

- Repeat
  - Select points $(x, y, t) \in \Omega \times [0, T]$ from vertices of the computational mesh (excluding its boundary edges) approximating the bathymetry, update the weights: $A_{i,j}^k = A_{i,j}^k - \eta \frac{\partial LOSS_{PDE}(x,y,t)}{\partial A_{i,j}^k}$, $B_i^k = B_i^k - \eta \frac{\partial LOSS_{PDE}(x,y,t)}{\partial B_i^k}$.
  - Select point $(x, y) \in \partial\Omega$ from vertices of the computational mesh located on its boundary, approximating the seabed and the seashore, and update the weights: $A_{i,j}^k = A_{i,j}^k - \eta \frac{\partial LOSS_{BC}(x,y,t)}{\partial A_{i,j}^k}$, $B_i^k = B_i^k - \eta \frac{\partial LOSS_{BC}(x,y,t)}{\partial B_i^k}$.
  - Select point $(x, y, 0) \in \Omega \times \{0\}$ from vertices of the computational mesh located on its boundary at time moment 0, update the weights $A_{i,j}^k = A_{i,j}^k - \eta \frac{\partial LOSS_{Init}(x,y,0)}{\partial A_{i,j}^k}$, $B_i^k = B_i^k - \eta \frac{\partial LOSS_{Init}(x,y,0)}{\partial B_i^k}$.
- Until $w_{PDE} LOSS_{PDE} + w_{BC} LOSS_{BC} + w_{Init} LOSS_{Init} \leq \delta$

As for the parameters, we run our experiments with the following setup:

- Maximum number of epochs: 15000,
- $\eta = 0.00015$ (lines 4,5 and 6).

The selection of the training rate came from experimenting with smaller and larger values. The weights w_{PDE}, w_{BC}, w_{Init} are selected by the adaptive loss weighting of Neural Networks with multi-part loss functions method.

During the training procedure, we employ the ADAM algorithm [3], storing the derivatives computed in previous iterations and taking the weighted average to modify the neural network's weights. In this way, it avoids being trapped in local minima. There are also several modern modifications to the ADAM algorithm available [4–6], but ADAM seems to be the most popular one.

3   Verification of the Code

To verify the correctness of the PINN, we compare with finite element method [8] executed on the "pool" scenario, using the non-linear wave equation

$$\frac{\partial^2 u}{\partial t^2} - \nabla (g(u - z)\nabla u) = 0, \tag{9}$$

We employ a Bubnov-Galerkin [23] finite element method for spatial discretization and an explicit finite difference time-stepping scheme for temporal discretization. Namely, we introduce a finite difference approximation of the second time derivative $\frac{\partial^2 u}{\partial t^2} \approx \frac{u_t - 2u_{t-1} + u_{t-2}}{dt^2}$:

$$\underbrace{u_t}_{\text{Next state}} = \text{'} \underbrace{u_{t-1}}_{\text{Previous state}} + \underbrace{u_{t-1} - u_{t-2}}_{\text{States difference}} + \underbrace{\Delta t^2}_{(\text{Time step})^2} \underbrace{\nabla (g(u_{t-1} - z)\nabla u_{t-1})}_{\text{Physics}}. \tag{10}$$

We multiply by test functions v, integrate by parts, and apply the boundary conditions $\nabla u_t \cdot n = 0$. We solve: Find $u \in V$:

$$\begin{aligned} (u_t, v) &= (u_{t-1}, v) + C (u_{t-1} - u_{t-2}, v) \\ &\quad - \Delta t^2 (g(u_{t-1} - z)\nabla u_{t-1}, \nabla v) \quad \forall v \in V \end{aligned} \tag{11}$$

Following [8], we have introduced a damping constant C in front of the wave propagation term to stabilize the finite element formulation.

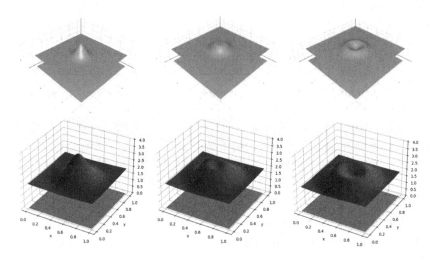

Fig. 2. Model simulation of the wave equation in a "pool" using the finite element method code (first row) and PINN solver (second row).

4 Generation of the Computational Mesh Describing the Seashore and Seabed Near Valparaiso Region

For the training of the Physics Informed Neural Network model, we select the points as vertices of mesh triangles of the most refined mesh from the adaptive procedure towards the GMRT [11] database data. As the refinement criterion we estimate the error between the triangular elements approximation of the topography and the GMRT topography data. For the adaptation, we employ the longest-edge refinement algorithm [8]. We have executed 22 iterations of the longest-edge refinement algorithm starting from a rectangle partitioned into two triangles. The adaptive mesh refinements were continued until the required topography accuracy was reached. The overview of the refinement process is presented in Fig. 3. The resulting mesh is shown in Fig. 4. The mesh is employed as a barymetry function $z(x, y)$. On top of that mesh, we introduce the sea level $u(x, y)$. Table 1 presents the number of triangles and vertices of the generated meshes. The final mesh has 91,458 triangles and 45,957 vertices. We employ these 45,957 vertices to train the PINN model.

The longest-edge refinement algorithm employed for the generation of the computational mesh approximating the seashore and seabed of the Valparaiso region of Chile has been expressed by graph transformations. Some representative transformations are presented in Figs. 5 and 6. The rules presented in this paper are the transformation of the hyper-graph grammar implementation of the longest-edge refinement algorithm described in [16] into the composite graph grammar, employed previously for hp adaptive mesh refinements [15]. They express the rules for the creation of the new triangular elements in a way

Fig. 3. The sequence of computational meshes approximating the seabed and seashore at the Valparaiso area of Chile, obtained by executing the longest edge refinement algorithm starting from the rectangular domain partitioned into two triangles.

Table 1. The number of vertices and triangles generated by an adaptive longest-edge refinement algorithm executed from two triangles.

Iteration	Number of vertices	Number of triangles
0	4	2
1	5	4
2	9	8
3	13	16
4	25	32
5	41	64
6	81	128
7	145	256
8	284	502
9	519	972
10	1008	1890
11	1834	3540
12	3443	6658
13	6050	11869
14	10126	19904
15	15492	30619
16	21144	41866
17	25205	49974
18	28103	55763
19	30909	61372
20	34409	68369
21	38808	77165
22	45957	91458

that no hanging nodes are generated in the mesh. They employ the predicates of applicability, telling when the transformation can be executed.

5 Numerical Results

After initial experiments with the ADAM optimization algorithm [3], due to the fact that we have residual, boundary and initial loss functions, thus, the multi-objective optimization problem, we switched to the SoftAdapt algorithm [13] executing ADAM algorithm with adaptive weighting of three loss functions. The convergence of these three loss functions, as well as the convergence of the total loss function, is presented in Fig. 7.

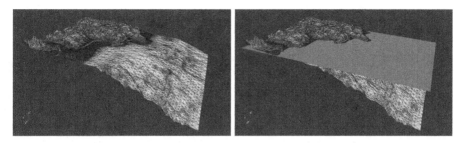

Fig. 4. The final computational mesh representing the seashore and seabed near the Valparaiso area obtained after 22 iterations of the longest-edge refinement algorithm. The topographic mesh represents the bathymetry function $z(x,y)$; the sea level represents the undisturbed water level $u(x,y,t)$.

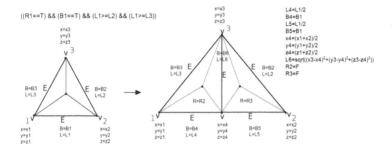

Fig. 5. Production (P1).

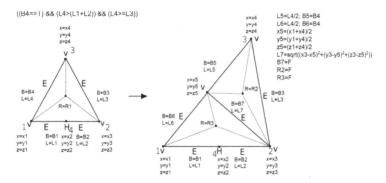

Fig. 6. Production (P2).

The residual loss first learns the trivial zero solution, but when the model learns the initial and boundary conditions, it adjusts and maintains a high accuracy of the order of 0.001. The initial loss learns the initial state with an accuracy of almost 0.003. The boundary loss is also trained with an accuracy of 0.001.

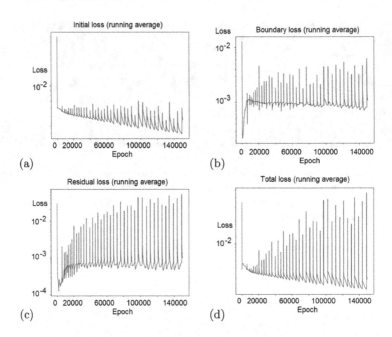

Fig. 7. The convergence of the training for the initial (a), boundary (b), residual (c), and total (d) loss.

As a result, the PINN has learned the solution with a total loss function accuracy of 0.003. The initial condition is given by:

$$u_0(x,y,0) = 0.04 \exp\left(-120\sqrt{(x-x_0)^2 + (y-y_0)^2}\right) + 0.5 \qquad (12)$$

in the relative units of $[0,1]^3$ of the computational domain. This setup represents the following interpretation. The ocean level is set at 0.5 (the middle of the horizontal height of the domain). The seashore spans between 0.5 and 0.55, so the initial tsunami wave height is set to 0.55 (see first panel in Fig. 5). The center of the initial tsunami wave is located in $1/5$ of the domain in the x direction and $1/2$ of the domain in the y direction (see the first panel in Fig. 5). The snapshots of the simulations are presented in Fig. 9. We can predict the propagation of the tsunami waves along the coastline as well as check the water elevation in the coastal area.

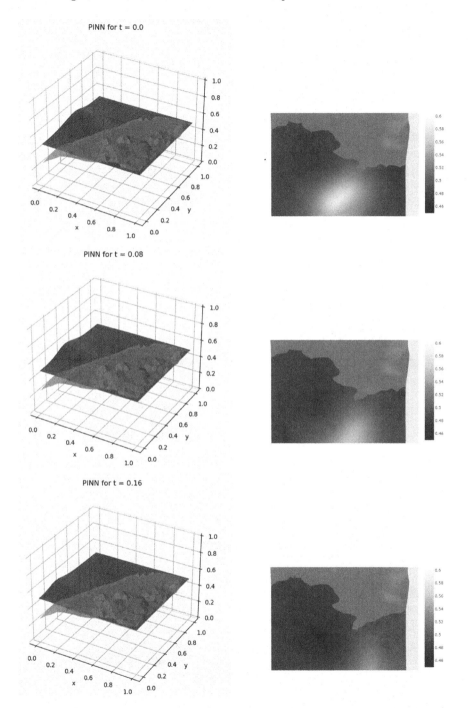

Fig. 8. Snapshots from tsunami simulation near the Valparaiso region of Chile (left panels). Changes of the coastline caused by the tsunami near the Valparaiso region of Chile (right panels).

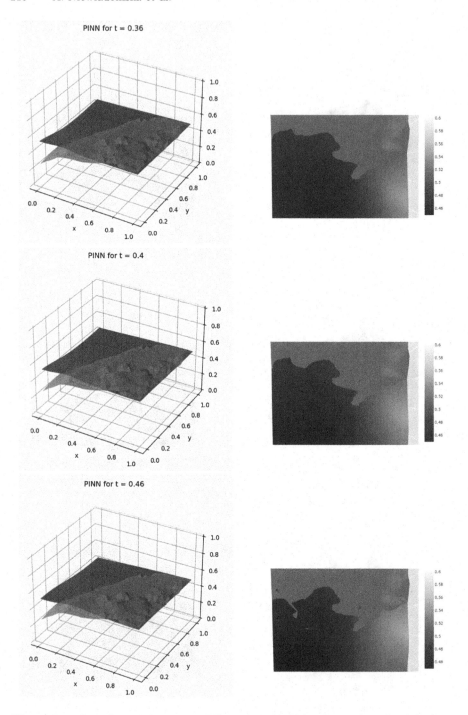

Fig. 9. Snapshots from tsunami simulation near the Valparaiso region of Chile (left panels). Changes of the coastline caused by the tsunami near the Valparaiso region of Chile (right panels).

6   Conclusions

In this paper, we showed how to use physics-informed neural networks to simulate tsunami wave propagation for an assumed seafloor and coastal topography. Our application consisted of the following building blocks. The PINN model used a nonlinear wave propagation equation [8]. Neural network learning was based on the residuum of the equation, boundary condition, and initial condition using the ADAM algorithm [3], adaptive loss function weight selection [13]. The topography of the seabed and waterfront was approximated using an adaptive generation algorithm for computational grids built from triangular elements. Adaptation was performed using the longest-edge refinement method on the basis of the Global Multi-Resolution Topography Database [11]. The points selected for learning the neural network, in particular for sampling the equation residuum, boundary, and initial condition, were chosen based on the nodes of the generated computational grid. Having the computational meshes generated for the coastline of Chile, it takes a couple of days for a master's degree student to implement and run the PINN simulator from scratch, employing the mesh points for training. Implementing the finite element method solver [8] requires weeks of a very careful development and debugging, including issues with the stabilization of the simulation, proper selection of time iteration schemes, and dealing with simulator behavior at the coastline area. The PINN solver provides similar accuracy results as a linear finite element [8] solver. The future work may involve comparison to higher-order finite element method solvers [24,25].

Acknowledgements. The work of Albert Oliver Serra was supported by "Ayudas para la recualificación del sistema universitario español" grant funded by the ULPGC, the Ministry of Universities by Order UNI/501/2021 of 26 May, and the European Union-Next Generation EU Funds The authors are grateful for support from the funds the Polish Ministry of Science and Higher Education assigned to AGH University of Krakow. The visit of Maciej Paszyński at Oden Institute was supported by J. T. Oden Research Faculty Fellowship.

References

1. Raissi, M., Perdikaris, P., Karniadakis, G.E.: Physics-informed neural networks: A deep learning framework for solving forward and inverse problems involving nonlinear partial differential equations. J. Comput. Phys. **378**, 686–707 (2019)
2. Maczuga, P., Paszyński, M.: Influence of activation functions on the convergence of physics-informed neural networks for 1D wave equation. In: Computational Science – ICCS 2023: 23rd International Conference, Prague, Czech Republic, July 3-5, 2023, Proceedings, Part I, pp. 74–88 (2023)
3. Kingma, D.P., Lei Ba, J.: ADAM: a method for stochastic optimization (2014). arXiv:1412.6980
4. Loshchilov, I., Hutter, F.: Decoupled Weight Decay Regularization arxiv.org/abs/1711.05101 (2019)
5. Chen, X., et al.: Symbolic Discovery of Optimization Algorithms, arxiv.org/abs/2302.06675 (2023)

6. Deng, Y., Hu, H., Song, Z., Weinstein, O., Zhuo, D.: Training Overparametrized Neural Networks in Sublinear Time, arxiv.org/abs/2208.04508 (2022)
7. Chen, Y., Yongfu, X., Wang, L., Li, T.: Modeling water flow in unsaturated soils through physics-informed neural network with principled loss function. Comput. Geotech. **161**, 105546 (2023)
8. Maczuga, P., Oliver-Serra, A., Paszyńska, A., Valseth, E., Paszyński, M.: Graphgrammar based algorithm for asteroid tsunami simulations. J. Comput. Sci. **64**, 101856 (2022)
9. Lu, L., Meng, X., Mao, Z.: DeepXDE: a deep learning library for solving differential equations. SIAM Rev. **63**(1), 208–228 (2021)
10. Peng, W., Zhang, J., Zhou, W., Zhao, X., Yao, W., Chen, X.: IDRLnet: A Physics-Informed Neural Network Library, arxiv2107.04320 (2021)
11. Global Multi-Resolution Topography Data Synthesis https://www.gmrt.org/
12. https://en.wikipedia.org/wiki/List_of_earthquakes_in_Chile
13. Ali Heydari, A., Thompson, C.A., Mehmood, A.: SoftAdapt: techniques for adaptive loss weighting of neural networks with multi-part loss functions. arXiv:1912.12355v1 (2019)
14. Maczuga, P., et al.: Physics Informed Neural Network Code for 2D Transient Problems (PINN-2DT) Compatible with Google Colab arXiv:2310.03755v2 (2024)
15. Paszyńska, A., Paszyński, M., Grabska, E.: Graph transformations for modeling *hp*-adaptive finite element method with triangular elements. In: Bubak, M., van Albada, G.D., Dongarra, J., Sloot, P.M.A. (eds.) ICCS 2008. LNCS, vol. 5103, pp. 604–613. Springer, Heidelberg (2008). https://doi.org/10.1007/978-3-540-69389-5_68
16. Podsiadło, K., et al.: Parallel graph-grammar-based algorithm for the longest-edge refinement of triangular meshes and the pollution simulations in Lesser Poland area. Eng. Comput. **37**, 3857–3880 (2021)
17. Rivara, M.C.: New longest-edge algorithms for the refinement and/or improvement of unstructured triangulations. Int. J. Numerical Methods Eng. **40**, 3313–3324 (1997)
18. Paszyński, M., Paszyńska, A.: Graph Transformations for Modeling Parallel hp-Adaptive Finite Element Method, Parallel Processing and Applied Mathematics: 7th International Conference, pp. 1313–1322. Gdansk, Poland (2007)
19. LeVeque, R.J., George, D.L., Berger, M.J.: Tsunami modelling with adaptively refined finite volume methods. Acta Numer **20**, 211–289 (2011)
20. Wang, X.: User manual for COMCOT version 1.7. Cornel University (2009)
21. Lynett, P., Liu, P.L.F., Sitanggang, K.I., Kim, D.: Modeling Wave Generation, Evolution, and Interaction with Depth-Integrated, Dispersive Wave Equations COULWAVE Code Manual Cornell University Long and Intermediate, Wave Modeling Package V. 2.0, Cornell University, Itacha, New York (2008)
22. Tavakkol, S., Lynett, P.: Celeris base: An interactive and immersive Boussinesq-type nearshore wave simulation software. Comput. Phys. Commun. **248** (2020) Article 106966
23. Becker, E.B., Carey, G.F., Oden, J.T.: Finite elements: an introduction, vol. 1. Prentice Hall (1981)
24. Woźniak, M., Łoś, M., Paszyński, M., Dalcin, L., Calo, V.M.: Parallel fast isogeometric solvers for explicit dynamics. Comput. Inf. **36**(2), 423–448 (2017)
25. Łoś, M., Munoz-Matute, J., Muga, I., Paszyński, M.: Isogeometric Residual Minimization Method (iGRM) with direction splitting for non-stationary advection-diffusion problems. Comput. Math. Appl. **79**(2), 213–229 (2020)

Automated Discovery of Concurrent Models of Decision-Making Systems from Data

Zbigniew Suraj$^{(\boxtimes)}$, Piotr Grochowalski , and Paweł Drygaś

Institute of Computer Science, University of Rzeszów, 1 Prof. S. Pigonia Str., 35-310 Rzeszów, Poland
{zsuraj,pgrochowalski,padrygas}@ur.edu.pl

Abstract. The paper presents a methodology for building concurrent models of decision-making systems based on knowledge extracted from empirical data. We assume that the data is represented by a decision table, while the decision-making system is represented by a Petri net. Decision tables contain conditional attribute values obtained from measurements or other sources. A Petri net is constructed using all true and acceptable rules generated from a given decision table. Rule factors and other parameters needed to build the net model are also computed from the data table. Three operators In, Trs and Out interpreted as uninorms are used to describe the dynamics of the net model. The expected behavior of the model is achieved by proper organization of its work. The theoretical basis of the methodology is the concepts, methods and algorithms derived from the theory of rough sets, fuzzy sets and Petri nets.

Keywords: Rough set · Fuzzy set · Petri net · Decision rule · Uninorm

1 Introduction

One of the challenges of modern artificial intelligence [1] and IT is building intelligent systems that can function reliably and efficiently, also in uncertain conditions. It seems that in order to satisfactorily meet this task, it is no longer enough to use the achievements of science in one field. In this paper, we try to solve the problem posed in it, using an approach based on three widely recognized theories: rough sets, fuzzy sets and Petri nets. Rough set theory is an important mathematical and artificial intelligence technique proposed by Pawlak in 1982 [2]. The main feature of this theory, from the point of view of practical applications, is the classification of empirical data, and thus decision making. Rough sets and fuzzy sets are quite closely related. The latter were introduced in 1965 by Zadeh [3]. Both of these concepts can be regarded as generalizations of a set in the classical sense. In turn, Petri nets were introduced by Petri in 1962 [4] as one of the formalisms used to model and analyze the behavior of systems of concurrent processes.

L. Franco et al. (Eds.): ICCS 2024, LNCS 14833, pp. 219–234, 2024.
https://doi.org/10.1007/978-3-031-63751-3_15

In the paper, we assume that empirical knowledge about the modeled system is presented in the form of a given decision table in the sense of Pawlak [5]. The decision table DT consists of a series of rows labeled by the names of elements from the set of objects U, represented by a vector of conditional attribute values from the set A along with the corresponding decision d. The input data for the construction of a decision-making system model are rules generated from a given decision table. The resulting system model is represented by a weighted priority uninorm Petri net (WPUP-net), which enables decision making as soon as there are enough values of conditional attributes represented by the so-called starting places in the net model, based on the knowledge encoded in DT (cf. [6]). We consider two types of rules generated from DT: true and acceptable. A rule is true in DT if and only if any object u in DT that matches the left-hand side of the rule also matches its right-hand side, and there is an object u that matches its left-hand side. A rule is acceptable in DT if the match of any object to the rule is not exact, but only to some non-zero degree (cf. [7]). A rule is active when values are specified for all attributes on its left-hand side. We assume that the net model proposed in this paper will transfer information from one attribute to another as quickly as possible. Therefore, there is a need to generate both all true and acceptable rules from DT [6]. Finally, our net model is an implementation of a set of generated rules and their parameters using WPUP-nets. The proposed net model works as follows. There are two phases in every net computation. In the first phase it is checked whether the values of any conditions are known, if so, the net tries to run decision rules, if possible, if not, then in the second phase the net tries to generate new information about the condition values and send them across the net. The entire computational process is carried out through the appropriate organization of the net operation based on the prioritization of transitions. Transitions representing conditional rules have a lower priority than transitions representing decision rules. This gives the desired effect.

It is worth emphasizing that in our approach, in addition to calculating rules from a given decision table, all rule coefficients and other parameters needed to build a Petri net model are also calculated from it, and they do not come from an expert in a given field of application. This aspect clearly distinguishes our approach from other approaches commonly available in the literature. This also means that the process of modeling decision-making systems using our methodology can be largely automated, not only modeling, but also analyzing and verifying the correctness of its operation using specialized software such as PNeS (Petri Net System), which is designed to assist users in modeling and analyzing systems of concurrent processes with different types of Petri nets [8]. In 1984, Lipp [9] published the first paper on fuzzy Petri nets, which work well in modeling systems operating in imperfect information environments. Since their introduction, they have enjoyed unflagging interest among people dealing with artificial intelligence and computer science [10]. As indicated in the review literature [11,12], these nets have some disadvantages and therefore are not fully suitable for modeling complex decision systems. For this reason, there are many alternative models in the literature that increase both the ability to represent knowledge more accurately and the ability to reason more intelligently. To the best of our knowledge, we are not aware of studies using the Petri nets based on uninorms that fit into the above-mentioned research trend. Uniforms were introduced by Yager and Rybalov in 1996 [13] as a generalization of t-norms and t-conorms, almost universally used to describe the dynamics of fuzzy Petri nets. With uninorms, logical connectives AND and OR can be modeled more adequately than using triangular norms. By manipulating the values of the neutral element in uninorms, such an effect is obtained. Since their introduction, uninorms have been used in expert systems [14,15], decision systems [16,17], and others. In this paper, uninorms are used both to represent knowledge and to describe the decision-making process implemented in WPUP-net models.

The work is theoretical in nature and contains potential applications in modeling decision-making systems operating in an uncertain environment. We believe this is an important extension of the research described in [18]. This extension applies, among others: (1) adding information about uninorms; (2) development of a new Petri net model based on uninorms, not triangular norms; (3) modified description of both the structure and behavior of the new net model; (4) a modified net representation of rule knowledge, in which the In, Trs and Out operators are now uninorms, thanks to which the logical operators And and Or appearing in the rules can be better modeled in the environment of uncertain information; (5) modification of the operating algorithm of the developed net model; (6) building and analyzing the operation of a net model of a decision-making system described by a decision table based on a new type of Petri net, from which decision and conditional rules were extracted along with the necessary parameter values needed to automatically build such a model. We consciously use similar examples in the introductory part of this work, as well as in the main example illustrating our approach to help the potential reader see the similarities and differences between the current methodology and that described in [18].

The rest of this paper is structured as follows. Section 2 recalls the basic concepts and notations for rough sets and uninorms, and illustrates them with examples. In Sect. 3, a new model of Petri nets based on uninorms is presented. Section 4 contains net representations of rules. In Sect. 5, an algorithm that describes how the WPUP-net should work is introduced. Section 6 gives an example to illustrate our methodology. In Sect. 7, conclusions and directions for further work are presented.

2   Backgrounds and Examples

2.1   Rough Sets

A pair $S = (U, A)$ is called an *information system* if U is a nonempty finite set of objects, called the *universe*, A is a nonempty finite set of *attributes* and $a : U \rightarrow V_a$ for every $a \in A$. The set V_a is called the *value set* of a, and $V = \bigcup_{a \in A} V_a$ is called the *domain* of A.

A *decision table* is a pair $DT = (U, A \cup \{d\})$, where A is a nonempty set of *conditional attributes*, $d \notin A$ is a *decision attribute* (*decision*). Any decision table $DT = (U, A \cup \{d\})$ can be represented by a table with a number of rows equal to the size of the universe U and a number of columns equal to the size of the set $A \cup \{d\}$. The value $a(u)$ appears at the position corresponding to row u and column a.

Let $S = (U, A)$ be an information system, $B, C \subseteq A$. A set C is dependent to *degree* k on B in S if $k = \gamma(B, C) = \frac{|POS_B(C)|}{|U|}$, where the set $POS_B(C)$ is the positive region of the partition U/C w.r.t. B, i.e., it is the set of all elements of U that can be uniquely classified to blocks of the partition U/C by means of B [22]. If $k = 1$ a set C is *totally* dependent on B, if $k = 0$ a set C is *totally* independent on B and otherwise C is *roughly* dependent on B.

Let $B, C \subseteq A$, and $B' \subseteq B$. We say that a set B' *is a relative reduct of* B w.r.t. C, if B' is a minimal subset of B and $\gamma(B, C) = \gamma(B', C)$.

Let $DT = (U, A \cup \{d\})$ be a decision table, $B \subseteq A$. We consider two natural coefficients of the significance based on the degree of dependency between the attribute sets B and $\{d\}$:

$$\sigma_1(B, d, a) = \gamma(B, \{d\}) - \gamma(B - \{a\}, \{d\}) = \frac{|POS_B(\{d\})| - |POS_{B-\{a\}}(\{d\})|}{|U|},$$

$$\sigma_2(B, d, a) = \frac{\gamma(B, \{d\}) - \gamma(B - \{a\}, \{d\})}{\gamma(B, \{d\})} = \frac{|POS_B(\{d\})| - |POS_{B-\{a\}}(\{d\})|}{|POS_B(\{d\})|},$$

and denoted by $\sigma_1(a)$ ($\sigma_2(a)$), when B and $\{d\}$ are understood. σ_1 measures the difference between $\gamma(B, \{d\})$ and $\gamma(B - \{a\}, \{d\})$, i.e., it determines how the value of $\gamma(B, \{d\})$ changes after removing the attribute a, whereas σ_2 normalizes this difference. In practice, the more important the attribute a, the greater the value of both coefficients. It is true that: $0 \leq \sigma_1(a) \leq \sigma_2(a) \leq 1$.

Let $DT = (U, A \cup \{d\})$ be a decision table, $B \subseteq A \cup \{d\}$, and $V = \bigcup_{a \in A} V_a \cup V_d$. Expressions of the form $a = v$, where $a \in B$ and $v \in V_a$ are called *descriptors* over B and V. By DESC(B, V), we denote the set of all descriptors over B and V which is the smallest such set and closed w.r.t. classical connectives: OR (disjunction), AND (conjunction) and NOT (negation).

Let $\tau \in$ DESC(B, V). The set of all objects in U with property τ is called the *meaning* of τ in the decision table DT and denoted by $\| \tau \|$.

Let DT be a decision table and DESC(A, V_a), $a \in A$ be the set of *conditional formulas of DT*. Any expression of the form $\tau \to d = v$, where $\tau \in$ DESC(A, V_a), $v \in V_d$ and $\| \tau \| \neq \emptyset$ is called a *decision rule* r in DT. The formula τ is called the *predecessor* and the formula $d = v$ *successor* of the decision rule r. A non-empty set of objects $\| \tau \|$ is *matching* the decision rule, and the set of objects $\| \tau \| \cap \| (d = v) \|$ is *supporting* the rule. By *accuracy factor* of the decision rule r we mean the number $acc(r) = \frac{\| \tau \| \cap \| (d=v) \|}{\| \tau \|}$, whereas by *coverage factor* of r the number $cov(r) = \frac{\| \tau \| \cap \| (d=v) \|}{\| (d=v) \|}$. The *strength factor* of the decision rule r is the number $str(r) = \frac{\| \tau \| \cap \| (d=v) \|}{|U|}$. We say that a decision rule r is *true* in DT, if $acc(r) = 1$, and it is *acceptable* in DT, if $0 < acc(r) < 1$. A decision rule r is called *minimal* if it has the minimum number of descriptors on the left-hand side. It is obvious that: $0 \leq str(r) \leq acc(r) \leq 1$ and $0 \leq str(r) \leq cov(r) \leq 1$ for every decision rule r in DT.

Remark. (1) All the terms defined above for decision rules also apply to conditional rules of the form: $\tau \rightarrow a = v$, where $\tau, a = v \in \mathrm{DESC}(A, V_a)$. (2) In this paper, $\sigma_2(a)$, $acc(r)$ and $cov(r)$ are the parameters used to characterize the rules (see Sect. 4).

Table 1. A decision table

$U / A \cup \{d\}$	H	M	T	F
u_1	no	yes	high	yes
u_2	yes	no	high	yes
u_3	yes	yes	very high	yes
u_4	no	yes	normal	no
u_5	yes	no	high	no
u_6	no	yes	very high	yes

Example 1. Consider the decision table DT from Table 1. We have: $U = \{u_1, u_2, u_3, u_4, u_5, u_6\}$; $A = \{$H, M, T$\}$, where H is headache, M - muscle pain, T - temperature; $d = $ F and means flu. The attribute d represents the expert's decision to diagnose flu based on the patient's interview. Attribute values from $A \cup \{d\}$ are presented inside the table. We can calculate: $\gamma(\{$H, M, T$\}, \{$F$\}) = \frac{2}{3}$, $\gamma(\{$T$\}, \{$F$\}) = \frac{1}{2}$, $\gamma(\{$H$\}, \{$F$\}) = \gamma(\{$M$\}, \{$F$\}) = 0$, and two relative reducts w.r.t. $\{$F$\}$, $R_1 = \{$H,T$\}$ and $R_2 = \{$M,T$\}$ of the set of conditions $\{$H,M,T$\}$. Using the formulas for σ_1 and σ_2 for Table 1, we obtain the following measures of the significance of some attributes from the set A w.r.t. the classification generated by: (1) the conditional attributes A: $\sigma_1(\mathsf{H}) = 0$, $\sigma_2(\mathsf{H}) = 0$, $\sigma_1(\mathsf{M}) = 0$, $\sigma_2(\mathsf{M}) = 0$, $\sigma_1(\mathsf{T}) = \frac{1}{2}$, $\sigma_2(\mathsf{T}) = \frac{3}{4}$; (2) the relative reduct R_1: $\sigma_1(\mathsf{H}) = \frac{1}{6}$, $\sigma_2(\mathsf{H}) = \frac{1}{4}$, $\sigma_1(\mathsf{T}) = \frac{2}{3}$, $\sigma_2(\mathsf{T}) = 1$; (3) the relative reduct R_2: $\sigma_1(\mathsf{M}) = 0$, $\sigma_2(\mathsf{M}) = \frac{1}{4}$, $\sigma_1(\mathsf{T}) = \frac{2}{3}$, $\sigma_2(\mathsf{T}) = 1$. Furthermore, using the method of generating the minimal rules in DT [6], we obtain the following rules along with a list of numerical factors corresponding to:

1. Nontrivial functional dependencies between the values of conditions T and H in the reduct R_1: $r_1 := (\mathsf{T}=$very high$) \rightarrow (\mathsf{H}=no)$; $[\sigma_2(\mathsf{T}) = 1, cov(r_1) = \frac{1}{3}$ / $str(r_1) = \frac{1}{6}$; $acc(r_1) = \frac{1}{2}]$; $r_2 := (\mathsf{T}=$very high$) \rightarrow (\mathsf{H}=yes)$; $[\sigma_2(\mathsf{T}) = 1$, $cov(r_2) = \frac{1}{3}$ / $str(r_2) = \frac{1}{6}$; $acc(r_2) = \frac{1}{2}]$; $r_3 := (\mathsf{T}=$high$) \rightarrow (\mathsf{H}=no)$; $[\sigma_2(\mathsf{T}) = 1, cov(r_3) = \frac{1}{3}$ / $str(r_3) = \frac{1}{6}$; $acc(r_3) = \frac{1}{3}]$; $r_4 := (\mathsf{T}=$high$) \rightarrow (\mathsf{H}=yes)$; $[\sigma_2(\mathsf{T}) = 1, cov(r_4) = \frac{2}{3}$ / $str(r_4) = \frac{1}{3}$; $acc(r_4) = \frac{2}{3}]$; $r_5 := (\mathsf{T}=$normal$) \rightarrow (\mathsf{H}=no)$; $[\sigma_2(\mathsf{T}) = 1, cov(r_5) = \frac{1}{3}$ / $str(r_5) = \frac{1}{6}$; $acc(r_5) = 1]$.

2. Nontrivial functional dependencies between the values of conditions from $R_1=\{H,T\}$ and the decision F: $r_6 :=$ (T=very high) $\rightarrow$ (F=yes); $[\sigma_2(T) = 1$, $cov(r_6) = \frac{1}{4}$ / $str(r_6) = \frac{1}{6}$; $acc(r_6) = 1]$; $r_7 :=$ (H=no) AND (T=high) $\rightarrow$ (F=yes); $[\sigma_2(H) = \frac{1}{4}, \sigma_2(T) = 1, cov(r_7) = \frac{1}{4}$ / $str(r_7) = \frac{1}{6}$; $acc(r_7) = 1]$; r_8 := (H=yes) AND (T=high) $\rightarrow$ (F=yes); $[\sigma_2(H) = \frac{1}{4}, \sigma_2(T) = 1, cov(r_8) = \frac{1}{4}$ / $str(r_8) = \frac{1}{6}$; $acc(r_8) = \frac{1}{2}]$; $r_9 :=$ (H=yes) AND (T=high) $\rightarrow$ (F=no); $[\sigma_2(H) = \frac{1}{4}, \sigma_2(T) = 1, cov(r_9) = \frac{1}{2}$ / $str(r_9) = \frac{1}{6}$; $acc(r_9) = \frac{1}{2}]$; r_{10} := (T=normal) $\rightarrow$ (F=no); $[\sigma_2(T) = 1, cov(r_{10}) = \frac{1}{2}$ / $str(r_{10}) = \frac{1}{6}$, $acc(r_{10}) = 1]$. Note that r_5, r_6, r_7 and r_{10} are true in Table 1, while the rest are only acceptable.

Remark. These rules can also be generated from Table 1 using e.g. PNeS [8].

2.2   Uninorms

A mapping $U: [0, 1]^2 \rightarrow [0, 1]$ is called a *uninorm* if it is commutative, associative, nondecreasing, and there exists $e \in [0, 1]$ (called neutral element) such that $U(e, x) = x$ for all $x \in [0, 1]$.

The function U becomes a t-norm when $e = 1$ and an s-norm (t-conorm) when $e = 0$. Both classes of triangular norms are commonly used in fuzzy logic [19]. It is true that for every $(x, y) \in [0, e) \times (e, 1] \cup (e, 1] \times [0, e)$, $min(x, y) \leq U(x, y) \leq max(x, y)$. Moreover, $U(0, 1) \in \{0, 1\}$ for all uninorms U [20]. If $U(0, 1) = 0$, then the uninorm U is called *andlike* (or *conjunctive*), and if $U(0, 1) = 1$, then U is called *orlike* (or *disjunctive*).

Fact. If U is a uninorm with $e \in (0, 1)$ and the functions $x \rightarrow U(x, 1)$ and $x \rightarrow U(x, 0)$ ($x \in [0, 1]$) are continuous, except perhaps $x = e$. Then U can be determined using one of the formulas below.

(a) If $U(0, 1) = 0$ then

$$U(x, y) = \begin{cases} eT_U(\frac{x}{e}, \frac{y}{e}) \text{ if } (x, y) \in [0, e]^2 \\ e + (1 - e)S_U(\frac{x-e}{1-e}, \frac{y-e}{1-e}) \text{ if } (x, y) \in [e, 1]^2 \\ min(x, y) \text{ in other cases} \end{cases} \quad (1)$$

(b) If $U(0, 1) = 1$ then

$$U(x, y) = \begin{cases} eT_U(\frac{x}{e}, \frac{y}{e}) \text{ if } (x, y) \in [0, e]^2 \\ e + (1 - e)S_U(\frac{x-e}{1-e}, \frac{y-e}{1-e}) \text{ if } (x, y) \in [e, 1]^2 \\ max(x, y) \text{ in other cases} \end{cases} \quad (2)$$

In both formulas given above, T_U is a t-norm and S_U is an s-norm. We denote the class of uninorms of the form (1) by U_{min}, and the class of uninorms of the form (2) by U_{max}. The above relationships allow us to determine general formulas for uninorms with a neutral element e. In this paper, we limit our considerations

Table 2. Uninorms with $e \in (0,1)$ corresponding to three basic t-norms [21]

$Name$	$Formula \ for \ x,y \in [0,e]$
Zadeh t-uninorm	$U_{Z_t}^e(x,y) = min(x,y)$
Goguen t-uninorm	$U_{G_t}^e(x,y) = \frac{xy}{e}$
Lukasiewicz t-uninorm	$U_{L_t}^e(x,y) = max(0, x+y-e)$

Table 3. Uninorms with $e \in (0,1)$ corresponding to three basic s-norms [21]

$Name$	$Formula \ for \ x,y \in [e,1]$
Zadeh s-uninorm	$U_{Z_s}^e(x,y) = max(x,y)$
Goguen s-uninorm	$U_{G_s}^e(x,y) = \frac{x+y-xy-e}{1-e}$
Lukasiewicz s-uninorm	$U_{L_s}^e(x,y) = min(1, x+y-e)$

to six uninorms, simple in the mathematical notation, which general formulas along with names borrowed from the names of the appropriate triangular norms used to determine them are given in Tables 2 and 3.

Example 2. Let $U \in U_{min}$ and $U' \in U_{max}$ be uninorms defined as follows:

$$U(x,y) = \begin{cases} max(0, x+y-1/2) \text{ if } (x,y) \in [0,1/2]^2 \\ min(1, x+y-1/2) \text{ if } (x,y) \in [1/2,1]^2 \\ min(x,y) \text{ in other cases} \end{cases}$$

$$U'(x,y) = \begin{cases} max(0, x+y-1/2) \text{ if } (x,y) \in [0,1/2]^2 \\ min(1, x+y-1/2) \text{ if } (x,y) \in [1/2,1]^2 \\ max(x,y) \text{ in other cases} \end{cases}$$

Note that U and U' are andlike and orlike uninorms, respectively. Moreover, both formulas can be obtained from Lukasiewicz's t-norm and s-norm shown in Tables 2 and 3, respectively, with the neutral element $e = 1/2$. For more information on rough sets and uninorms, see [22,23].

3   Weighted Priority Uninorm Petri Nets

Let U_{min} and U_{max} denote the classes of uninorms of the form (1) and (2), respectively, defined in Sect. 2.2, with the neutral element $e \in (0,1)$.

A tuple $N_U = (P, T, I, O, Pr, M_0, S, \alpha, \beta, \gamma, Op, \delta)$ is called a *weighted priority uninorm Petri net* (WPUP-net for short) if: (1) $P = \{p_1, p_2, ..., p_n\}$ is a set of places, $T = \{t_1, t_2, ..., t_m\}$ is a set of transitions; (2) $I \colon P \times T \to [0,1]$ is the input function, $O \colon T \times P \to [0,1]$ is the output function; (3) $Pr \colon T \to \mathbb{N}$ is the priority function ($\mathbb{N}$ is the set of natural numbers), $M_0 \colon P \to [0,1]$ is the initial marking; (4) $S = \{s_1, s_2, \ldots, s_n\}$ is a set of statements (P, T, S are pairwise disjoint),

$\alpha \colon P \to S$ is the statement binding function; (5) $\beta \colon T \to [0,1]$ is the truth degree function, $\gamma \colon T \to [0,1]$ is the threshold function; (6) $Op = U_{min} \cup U_{max}$ is the set of operators, $\delta \colon T \to Op \times Op \times Op$ is the operator binding function.

Let N_U be a WPUP. For $t \in T$: ${}^{\bullet}t = \{p : I(p,t) > 0\}$ is a set of input places of t, and $t^{\bullet} = \{p' : O(t,p') > 0\}$ is a set of all output places of t.

By tokens we mean the values of the function M_0. We assume that the set O_p contains uninorms. If the arc (p,t) connects p and t, then $I(p,t) > 0$, otherwise 0. The value $I(p,t)$ is called the input weight t and is denoted iw. Similarly, if the arc (t,p) connects t and p, then $O(t,p) > 0$, otherwise 0. The value $O(t,p)$ is called the output weight t and is denoted ow. Places are graphically represented by circles and transitions by rectangles. If the weight of the directed arc is 1, then 1 is not shown in the net graph, if the weight of the directed arc is 0, this arc is skipped in the net graph. In other cases, the weight is placed next to the arc in question. Priorities are placed next to transitions. We only consider two priority values: 0 and 1. If the priority is 0, we do not show it in the net graph. Transitions representing decision rules have a priority of 1, all others have a priority of 0. Each place contains one token. The token is placed inside the place. If a token has a value of 0 in a given place, then that place is empty. Each statement is associated with only one place. The statement is placed next to the place. The values of $\beta(t)$ and $\gamma(t)$ are shown in the net graph under the transition t. The first value is interpreted as the accuracy factor of the rule represented by t. The second value affects the activation of the transition. If the value of the transition firing condition is not less than the threshold value, the transition can be fired. The value of the function δ is the triple of operators (In, Trs, Out), this triple is placed under the transition. In is the input operator, Trs is the transmission operator, and Out is the output operator. Operator In aggregates tokens placed in the input places of the transition to which it is assigned. The role of the Trs and Out operators is to aggregate the value received from In with the values of the remaining net parameters and send the generated value to all transition output places to which these three operators are assigned. We assume that the input operators belong to U_{min} or U_{max}, while the other two belong to U_{min} and U_{max}, respectively. These operators are used to describe the dynamics of the WPUP net. Arc weight values and β function values are calculated from the data table and interpreted using rough set theory concepts (see Sect. 4). However, values of the threshold function γ are set by the domain expert.

WPUP-net dynamics are defined by a *firing rule*, and the net evolution is represented by a *sequence of fired transitions*.

Let N_U be a WPUP-net. A marking of N_U is a function $M \colon P \to [0,1]$.

Firing Rule. Let $N_U = (P, T, S, I, O, \alpha, \beta, \gamma, Op, \delta, M_0)$ denotes a WPUP-net, M be a marking of N_U, $t \in T$, ${}^{\bullet}t = \{p_{i1}, p_{i2}, \ldots, p_{ik}\}$ be a set of input places for t, $\beta(t) \in (0,1]$, and $\delta(t) = (In, Trs, Out)$. A transition t is *enabled* (or *ready to fire*) for marking M if for all $p \in {}^{\bullet}t$: $In((iw_{i1} \cdot M(p_{i1}), iw_{i2} \cdot M(p_{i2})), \ldots, iw_{ik} \cdot M(p_{ik})) \geq \gamma(t) > 0$, where iw_{ij} is an input weight of an arc (p_{ij}, t) and $M(p_{ij})$ is a marking of a place p_{ij} for $j = 1, 2, \ldots, k$.

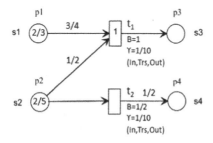

Fig. 1. A WPUP-net with the initial marking $M_0 = (2/3, 2/5, 0, 0)$.

If $\bullet t$ consists of only one place p, we assume that a transition t is enabled for M if the following condition is true: $iw \cdot M(p) \geq \gamma(t) > 0$, where iw is an input weight of arc (p, t). A net transition can be fired when it is enabled.

According to the definition of WPUP-net, transitions can be assigned priorities, which means that if two or more transitions are enabled simultaneously in a given marking, the transitions with the highest priority will be activated first [24].

Formula of the Next Marking. Firing an enabled transition t by a marking M results in a new marking M' defined by

$$M'(p) = \begin{cases} Out(ow \cdot Trs(In(iw_{i1} \cdot M(p_{i1}), iw_{i2} \cdot M(p_{i2}), ..., iw_{ik} \cdot M(p_{ik})), \beta(t)), \\ M(p)) \text{ if } p \in t^\bullet \\ M(p) \text{ otherwise} \end{cases}$$

where ow is an output weight of arc (t, p).

Example 3. For the WPUP-net of Fig. 1, we have: $P = \{p_1, p_2, p_3, p_4\}$; $T = \{t_1, t_2\}$; $I(p_1, t_1) = iw_1 = 3/4$, $I(p_2, t_1) = iw_2 = 1/2$, $I(p_3, t_1) = iw_3 = 0$, $I(p_4, t_1) = iw_4 = 0$, $I(p_1, t_2) = iw_5 = 0$, $I(p_2, t_2) = iw_6 = 1$, $I(p_3, t_2) = iw_7 = 0$, $I(p_4, t_2) = iw_8 = 0$; $O(t_1, p_1) = ow_1 = 0$, $O(t_1, p_2) = ow_2 = 0$, $O(t_1, p_3) = ow_3 = 1$, $O(t_1, p_4) = ow_4 = 0$, $O(t_2, p_1) = ow_5 = 0$, $O(t_2, p_2) = ow_6 = 0$, $O(t_2, p_3) = ow_7 = 0$, $O(t_2, p_4) = ow_8 = 1/2$; $Pr(t_1) = 1$, $Pr(t_2) = 0$; $M_0 = (2/3, 2/5, 0, 0)$; $S = \{s_1, s_2, s_3, s_4\}$; $\alpha(p_1) = s_1$, $\alpha(p_2) = s_2$, $\alpha(p_3) = s_3$, $\alpha(p_4) = s_4$; $\beta(t_1) = 1$, $\beta(t_2) = 1/2$; $\gamma(t_1) = \gamma(t_2) = 1/10$; $O_p = \{In, Trs, Out\}$, where In and Trs are interpreted as uninorm U, and Out as uninorm U' from Example 2; $\delta(t_1) = \delta(t_2) = (In, Trs, Out)$. Notice that t_1 and t_2 are ready to fire by the initial marking M_0. This is because: $In(iw_1 \cdot M(p_1), iw_2 \cdot M(p_2)) = In(3/4 \cdot 2/3, 1/2 \cdot 2/5) = In(1/2, 1/5) = max(0, 1/2 + 1/5 - 1/2) = 1/5 \geq \gamma(t_1) = 1/10$ and $iw_6 \cdot M(p_2) = 1 \cdot 2/5 \geq \gamma(t_2) = 1/10$. Only t_1 will be fired because it has priority higher than priority t_2. Firing transition t_1 with the initial marking M_0 leaves this marking unchanged. This is due to the fact that: $Trs(In(iw_1 \cdot M(p_1), iw_2 \cdot M(p_2)), \beta(t_1)) = Trs(1/5, 1) = min(1/5, 1) = 1/5$ and $Out(ow_3 \cdot Trs(1/5, 1), M_0(p_3)) = Out(1 \cdot 1/5, 0) = Out(1/5, 0) = max(0, 1/5 + 0 - 1/2) = max(0, -3/10) = 0$.

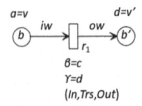

Fig. 2. A WPUP-net model of the type 1 rule after firing r_1.

Remark. Here and in the rest of the paper, instead of $\beta(t) = b, \gamma(t) = c$, where t is a transition and b, c values from the unit interval $[0,1]$, we will use the abbreviations $\beta = b, \gamma = c$.

4   WPUP-Net Representation of Rules

This section describes the three types of rules, including a list of parameters that characterize them (cf. [11, 25]).

Let $DT = (U, A \cup \{d\})$ denote a given decision table, and $\mathrm{DESC}(A, V_a)$ be the set of its conditional formulas.

Type 1. (A simple rule.) r_1: $(a = v) \to (d = v')$ $[b; \sigma(a), cov(r_1); acc(r_1)]$ with descriptors $a = v$ and $d = v'$ such that $a = v \in \mathrm{DESC}(A, V_a)$ and $v' \in V_d$, a truth degree value b of $a = v$, significance $\sigma(a)$ of attribute a given by the formula for $\sigma_2(a)$, a coverage factor $cov(r_1)$ and an accuracy factor $acc(r_1)$ of rule r_1. The method for calculating these parameters can be found in Subsect. 2.1.

The WPUP-net structure of rule r_1 is shown in Fig. 2, where $iw = \sigma(a)$ is an input weight of r_1, and $ow = cov(r_1)$ is an output weight of r_1. If iw is greater than ow, the first link is stronger than the second. However, value $c = \beta(r_1)$ is interpreted as $acc(r_1)$. As before, the higher the value of β, the more robust the rule. Value $d = \gamma(r_1)$ represents a threshold that requires degree of truth value $a = v$ to be not less than d. In, Trs, Out are operators assigned to r_1, which are the appropriate uninorms described in the WPUP-net definition (Sect. 3). According to Fig. 2 the value of a token at an output place of r_1 is calculated as $b' = ow \cdot Trs(b \cdot iw, c)$ if $b \cdot iw \geq d$.

If a rule's antecedent or successor contains an input operator In, it is called a *compound rule*.

Type 2. (A compound rule in the antecedent.) r_2 : $(a_1 = v_1)$ In $(a_2 = v_2)$ $\cdots$ In $(a_k = v_k) \to (d = v')$ $[b_1, b_2, \ldots, b_k; \sigma^1(a), \sigma^2(a), \ldots, \sigma^k(a), cov(r_2); acc(r_2)]$ with descriptors $a_1 = v_1$, $a_2 = v_2$, ..., $a_k = v_k$, $d = v'$ and truth degree values b_1, b_2, ..., b_k, b', respectively. A token value b' at an output place of r_2 is calculated as follows (Fig. 3): $b' = Trs(In(b_1 \cdot iw_1, b_2 \cdot iw_2, \ldots, b_k \cdot iw_k), c)) \cdot ow)$ if $In(b_1 \cdot iw_1, b_2 \cdot iw_2, \ldots, b_k \cdot iw_k) \geq d$. Note that in this case, the operator In can be interpreted as either andlike uninorm or orlike one.

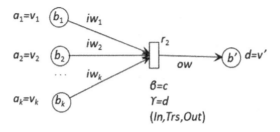

Fig. 3. A WPUP-net model of the type 2 rule after firing r_2.

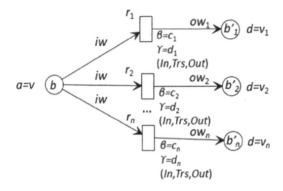

Fig. 4. A WPUP-net model of the type 3 rule after firing r_3.

Type 3. (A compound rule in the successor.) r_3: $(a = v) \rightarrow (d = v_1)$ *In* $(d = v_2) \cdots$ *In* $(d = v_n)$ $[b;\ \sigma^1(a), \sigma^2(a), \dots, \sigma^n(a), cov^1(r_3), cov^2(r_3), \dots, cov^n(r_3);\ acc^1(r_3), acc^2(r_3), \dots, acc^n(r_3)]$ with descriptors $a = v$, $d = v_1$, $d = v_2$, $\dots$, $d = v_n$, and a truth degree value b of $a = v$. A token value b'_j for each output place of r_3 is calculated in the following way (Fig. 4): $b'_j = ow_j \cdot Trs(b \cdot iw, c_j)$ if $b \cdot iw \geq d_j$, $j = 1, \dots, n$.

We assume that for the rules of type 3, *In* is interpreted as an andlike uninorm.

Remark. In each of the formulas presented above, in the case of nonzero markings of output places, the final value of the token b'' should be calculated according to the formula: $b'' = Out(b', M(p'))$, where b' is the token value calculated as described above for each of the considered rule types, and $M(p')$ denotes the nonzero marking of the output place p'.

5   An Algorithm

In this section, we introduce an algorithm that describes how a WPUP-net should work. It can be seen that in each computation of the net N_U built on the basis of a given decision table DT, two phases can be distinguished. In the first phase, tokens are set in input places of transitions representing conditional rules, not necessarily in all of them. In its second phase, the algorithm transfers tokens between places on the net as quickly as possible. This phase is implemented by the part of the net that represents all true and acceptable rules in the decision table DT. Net computation ends when the net model has proposed a decision or there are no transitions representing conditional rules ready to fire that have not been used before.

Algorithm 1: WPUP-net model for a given DT.

Input   : WPUP-net N_U.
Output: Final decision.
begin
Set the marking of N_U;
A: **if** *decision transitions are ready to fire* **then**
$\quad \llcorner$ fire them simultaneously and go to B;

if *condition transitions are ready to fire and they have not been fired yet* **then**
$\quad |$ fire them simultaneously and go to A
else
$\quad \llcorner$ Go to C;
B: Read the final decision
C: **end.**

---

6   An Illustrative Example

In this section, we will illustrate our methodology with the decision table in Example 1 representing a flu diagnosis system.

After transforming the rules with parameters from Example 1 into Petri nets (see Sect. 4) and combining them in common places, we obtain the resulting WPUP-net shown in Fig. 5. This net contains seven places and ten transitions. Among all places, there are five places (from p_1 to p_5) corresponding to conditional descriptors in the rules (these are starting places; selecting these places starts a diagnosis process), the remaining two refer to decision descriptors (these are treated as decision places whose nonzero tokens indicate the proposed decisions and their calculated degree of credibility). The set of transitions contains five transitions (from t_1 to t_5), which represent conditional rules (from r_1 to r_5), while the rest represent decision rules. Directed arcs connecting places with transitions and vice versa (elements of the sets $P \times T$ and $T \times P$ along with their weights on the arcs are illustrated in Fig. 5. In the initial net marking, p_2 and p_5 contain nonzero tokens, the rest are empty. Place p_2 includes token 3/4 which represents the truth degree of descriptor T = high, and place p_5 contains token 1/2 which represents the truth degree of descriptor H = yes. Statement set S contains all descriptors (conditional and decision) appearing in the set of rules from Example 1. The initial markings of p_2 and p_5 are nonzero, the rest are empty. The marking of p_2 is equal to 3/4 and represents the truth degree of

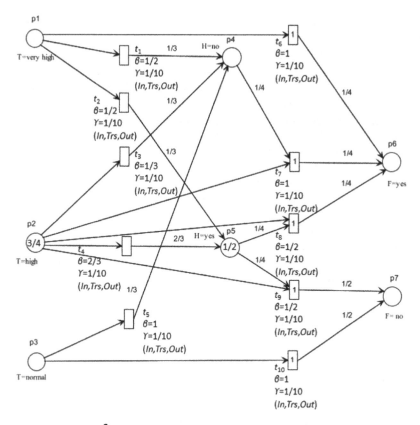

Fig. 5. A WPUP-net model for the rules in Example 1 with the initial marking $M_0 = (0, 3/4, 0, 0, 1/2, 0, 0)$.

the descriptor $T = \text{high}$, and the marking of p_5 is equal to $1/2$ and represents the truth degree of the descriptor $H = \text{yes}$. Statement set S consists of all the descriptors (conditional and decision) that appear in the rule set in Example 1. The elements of S correspond uniquely to their places in the net, as shown in the figure. The role of the degree of truth function β is to assign the value 1 to transitions t_5, t_6, t_7, t_{10}, the value $1/2$ to transitions t_1, t_2, t_8, t_9, the value $1/3$ to t_3 and finally the value $2/3$ to t_4. Threshold function γ assigns $1/10$ to each transition. Set of operators O_p contains three operators In, Trs and Out interpreted respectively as uninorms U, U and U' described in Example 2, but now we assume that the neutral element $e = 1/8$. Operator binding function δ assigns the triple form (In, Trs, Out) to each net transition. When evaluating statements attached to p_2 and p_5, we notice that t_3, t_4, t_8 and t_9 are ready to fire in the initial marking. Transitions t_8 and t_9 will run first because their priorities are higher than the priorities of the other two transitions. After firing t_8 and t_9 simultaneously or one at a time in any order, the net operation stops at $1/4$ and $1/2$ corresponding to decisions $F = \text{yes}$, $F = \text{no}$, respectively. Due to the fact that the degree of truth of the statement $F =$ no is greater than the degree of truth of the latter, the net model proposes the statement that there is no flu in the case under consideration.

Now consider the case where only p_2 is marked as before, and the other starting places are empty. It can be checked that in such a situation no transition representing a decision rule is ready to be fired, while t_3 and t_4 are ready. After firing these transitions, we will get a marking at which you can see that t_8 and t_9 are ready to fire. When these transitions are fired simultaneously or one at a time in any order, we get a marking where the net computation ends with a decision proposal indicating no flu and with the same degree of credibility as before. This example shows that the proposed net model can also work in the absence of tokens in some input places of decision transitions in the initial marking, which may result in the inability to make a decision immediately after starting its work. Sometimes missing information is obtained. This is the case when, after starting the model, conditional transitions are ready to fire, which can generate the necessary tokens. This was in our example. Detailed calculations related to the description of net operations have been omitted. They are similar to those described in Example 3 (Sect. 3).

Remark. All drawings of Petri net models in this paper were made in PNeS [8].

7   Conclusion and Further Work

In the paper, we presented a hybrid methodology that allows you to build concurrent models of decision-making systems based on knowledge extracted from empirical data stored in a given decision table. A new type of Petri net was used to represent the decision-making system for diagnosing flu cases. In this example, the operation of the model was analyzed in terms of the decisions it proposed, with particular emphasis on the situation when the input data of the model did not allow for immediate decision-making due to their incompleteness. The effect of such action is obtained by applying true and acceptable rules in the construction of the model along with the appropriate organization of its work. The expected functioning of the model became possible additionally due to the introduction of differentiated transition priorities and the appropriate interpretation of three transition operators in the uninorm class, responsible for the dynamics of the model's behavior.

Due to the fact that in many real-world situations it is difficult to determine the exact membership value or degree of truth, in further research we intend to focus on interval data rather than exact data [26]. For this purpose, in the WPUP-net model, we intend to replace the classical uninorms with interval uninorms and check experimentally what positive changes both in operation and in the effectiveness and usefulness of the proposed decisions can be obtained. Another problem of interest to us concerns the formulation of requirements under which net models of this type are deterministic (cf. [27]).

Acknowledgement. The authors thank the anonymous reviewers for their helpful comments.

Disclosure of Interests. The authors have no competing interests to declare that are relevant to the content of this article.

References

1. Munakata, T.: Fundamentals of the New Artificial Intelligence. Beyond Traditional Paradigms. Springer, Heidelberg (1998)
2. Pawlak, Z.: Rough sets. Int. J. Comput. Inf. Sci. **11**, 341–356 (1982)
3. Zadeh, L.A.: Fuzzy sets. Inf. Control **8**, 338–353 (1965)
4. Petri, C.A.: Komunikation mit Automaten. Schriften des IIM Nr. 2 Bonn (1962)
5. Pawlak, Z.: Rough Sets - Theoretical Aspects of Reasoning About Data. Kluwer, Dordrecht (1991)
6. Skowron, A.: Synthesis of adaptive decision systems from experimental data. Artif. Intell. Appl. **28**, 220–238 (1995)
7. Skowron, A., Suraj, Z.: A parallel algorithm for real-time decision making: a rough set approach. J. Intell. Inf. Syst. **7**, 5–28 (1996)
8. Suraj, Z., Grochowalski, P.: PNeS in modelling, control and analysis of concurrent systems. In: Ramanna, S., Cornelis, C., Ciucci, D. (eds.) IJCRS 2021. LNCS (LNAI), vol. 12872, pp. 279–293. Springer, Cham (2021). https://doi.org/10.1007/978-3-030-87334-9_24
9. Lipp, H.P.: Application of a fuzzy Petri net for controlling complex industrial processes. In: Proceedings of IFAC 1984, pp. 471–477 (1984)
10. Cardoso, J., Camargo, H. (eds.): Fuzziness in Petri Nets. Springer, Heidelberg (1999)
11. Liu, H.-C., You, J.-X., Li, Z.W., Tian, G.: Fuzzy Petri nets for knowledge representation and reasoning: a literature review. Eng. Appl. Artif. Intell. **60**, 45–56 (2017)
12. Jiang, W., Zhou, K.-Q., Sarkheyli-Hägele, A., Zain, A.M.: Modeling, reasoning, and application of fuzzy Petri net model: a survey. Artif. Intell. Rev. **55**, 6567–6605 (2022)
13. Yager, R.R., Rybalov, A.: Uninorm aggregation operators. Fuzzy Sets Syst. **80**, 111–120 (1996)
14. De Baets, B., Fodor, J.: Van Melle's combining function in MYCIN is a representable uninorms: an alternative proof. Fuzzy Sets Syst. **104**, 133–136 (1999)
15. Yager, R.: Uninorms in fuzzy systems modeling. Fuzzy Sets Syst. **122**, 167–175 (2001)
16. Mesiarová, A.: Multi-polar t-conorms and uninorms. Inform. Sci. **301**, 227–240 (2015)
17. Yager, R.R., Rybalov, A.: Bipolar aggregation using the uninorms. Fuzzy Optim. Decis. Making **10**, 59–70 (2011)
18. Suraj, Z.: A hybrid approach to approximate real-time decision making. In: Proceedings of FUZZ-IEEE 2021, Luxembourg, pp. 71–78. IEEE (2021)
19. Dubois, D., Prade, H.: A review of fuzzy sets aggregation connectives. Inform. Sci. **36**, 85–121 (1985)
20. Li, Y., Shi, Z.: Remarks on uninorms aggregation operators. Fuzzy Sets Syst. **114**, 377–380 (2000)

21. Klement, E.P., Mesiar, R., Pap, E.: Triangular Norms. Kluwer, Dordrecht (2000)
22. Pawlak, Z., Skowron, A.: Rudiments of rough sets. Inf. Sci. **177**, 3–27 (2007)
23. Fodor, J.C., Yager, R.R., Rybalov, A.: Structure of uninorms. Int. J. Uncertain Fuzziness Knowl.-Based Syst. **5**, 411–427 (1997)
24. Hack, M.: Decidability questions for petri nets. Ph.D. dissertation, MIT, Cambridge, Massachusetts (1975)
25. Ha, M.H., Li, Y., Wang, X.F.: Fuzzy knowledge representation and reasoning using a generalized fuzzy Petri net and a similarity measure. Soft. Comput. **11**(4), 323–327 (2007)
26. Moore, R.E., Kearfott, R.B., Cloud, M.J.: Introduction to Interval Analysis. SIAM, Philadelphia (2009)
27. Suraj, Z.: On selected properties of uninorm petri nets and their application in modeling knowledge-based systems. Procedia Comput. Sci. **225**, 155–164 (2023)

Simulation Model for Application of the SDN Concept in IMS/NGN Network Transport Stratum

Sylwester Kaczmarek[1] [ID], Maciej Sac[1]([✉]) [ID], and Jakub Adrych[2]

[1] Faculty of Electronics, Telecommunications and Informatics, Gdańsk University of
Technology, Narutowicza 11/12, 80-233 Gdańsk, Poland
{kasyl,Maciej.Sac}@eti.pg.edu.pl
[2] ASH-TECH, Wiosenna 2, 86-065 Łochowo, Poland

Abstract. The paper presents a simulation model allowing examination of cooperation between two currently used telecommunication networks concepts: IP Multimedia Subsystem/Next Generation Network (IMS/NGN) and Software-Defined Networking (SDN). Application of the SDN architecture elements in IMS/NGN networks will enable unified control and management of transport resources for various transport technologies and equipment manufacturers. However, such a cooperation is a new concept requiring verification, which is the aim of this paper. The structure of the modeled multidomain network and details about the simulator operation are described. Tests proving correctness of its operation are carried out. Selected research results regarding mean Call Set-up Delay and mean Call Disengagement Delay in the considered network are presented demonstrating that the cooperation between IMS/NGN and SDN is possible.

Keywords: SDN · NGN · IMS · simulation model · call processing performance

1 Introduction

Growing demand for higher bit rates, development of the Internet of Things (IoT) concept, multimedia services and continuous progress in network architecture solutions (including 5G mobile network) require advanced research to guarantee Quality of Service (QoS) and efficiency of the proposed solutions.

The currently used Next Generation Network (NGN) [1] concept consists of service stratum with applications and transport stratum performing transmission and switching functions. NGN service stratum is based on the servers defined in the IP Multimedia Subsystem (IMS) [2] concept. Hence the name "IMS/NGN" is commonly used. In NGN transport stratum any technology supporting transmission of IP packets is allowed.

The concept of Software-Defined Networking (SDN) [3, 4] was proposed primarily due to the diversity of telecommunications equipment related to transport stratum and the need to automate resource management and traffic control in this stratum. Recently, these features have become particularly important due to the Covid-19 pandemic, which

L. Franco et al. (Eds.): ICCS 2024, LNCS 14833, pp. 235–250, 2024.
https://doi.org/10.1007/978-3-031-63751-3_16

was associated with a huge amount of new and dynamically changing traffic (concerning distance learning, remote work, remote handling of official matters, etc.).

Due to these aspects, it is highly desirable to apply the SDN concept in existing telecommunications network architectures, such as the IMS/NGN network. However, the cooperation of the above-mentioned solutions has not been standardized or verified by the scientific community. This led the authors of this paper to address this problem. Due to complexity of the IMS/NGN network structure (especially for a multidomain network) and service scenarios as well as the desire to precisely describe the interaction of this network with the SDN concept, the authors decided to use the methods of computational science and develop a simulation model. The implemented simulator allows evaluation of mean Call Set-up Delay (mean CSD, $E(CSD)$) and mean Call Disengagement Delay (mean CDD, $E(CDD)$) for all types of successful call scenarios performed in a multidomain IMS/NGN network cooperating with the SDN concept. These parameters belong to a set of standardized Call Processing Performance (CPP) parameters [5, 6] describing control performance in a telecommunications network, which is important for both network operators and users.

The rest of the paper is organized as follows. A review of the related work is provided in Sect. 2. Section 3 describes the assumptions regarding the modeled network structure and service scenarios. It also contains details about the structure and operation of the implemented simulation model. Section 4 is devoted to functional tests of the simulator. It also provides selected obtained results. Section 5 concludes the paper.

2   Related Work

The aim of the performed review was to find other works related to the topic of this paper, i.e. regarding simulation models for the IMS/NGN network cooperating with the SDN concept. As already mentioned, such cooperation is not described in standardization documents. During the review, emphasis was placed on finding solutions that take into account both the NGN network service stratum based on the IMS concept and the transport stratum using the SDN architecture. Unfortunately, no such works were found, even when cooperation of pure IMS architecture with SDN was considered, without including NGN standards.

The review indicated that academic and industrial communities focus on practical verification of cooperation between IMS/NGN network and the SDN concept (testbeds). The available works can be divided into the following categories:

- proposals for network architecture and service scenarios, but without their implementation [7],
- practical implementations (testbeds) focusing on resource control using SDN (without taking into account the service stratum) [8],
- practical implementations (testbeds) of cooperation between pure IMS and SDN, without considering NGN standards [9–12]

A common feature of the above-mentioned works is the lack of consideration of ITU-T standards for NGN networks (e.g. related to the RACF resource control unit), as well as the lack of access to the source code, which prevents further work on these projects. Therefore, the authors of this paper decided to address these issues and create their own testbed for IMS/NGN network using the SDN concept [13], which does not have the disadvantages of the other mentioned solutions.

The advantage of testbeds is the possibility of practical verification of selected issues related to network operation in a laboratory environment. However, the problem is testing more complex network structures (e.g. a multidomain network) and service scenarios. For this purpose, the best solutions are simulation models that are flexible in configuration and reflect phenomena occurring in real networks. Therefore, the implemented testbed [13] was used to confirm the assumed message exchange procedures and provide the data (among others message lengths and processing times) necessary to develop a simulation model for a multi-domain IMS/NGN network based on the SDN concept. This model is described in the next section.

3 Simulation Model

In this paper, a multidomain IMS/NGN network belonging to two operators is assumed (Fig. 1). Each operator has its own IMS/NGN domain, therefore, the terms domain and operator will be used interchangeably with a similar meaning. Each domain includes elements of service stratum and transport stratum. Their names are appended with domain numbers (1 and 2).

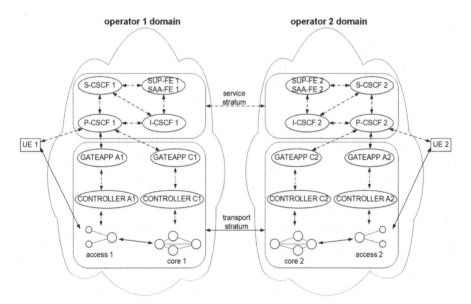

Fig. 1. Structure of the modeled multidomain IMS/NGN network utilizing the SDN concept. Physical connections are marked using solid lines with arrows, logical connections – dashed lines with arrows.

Service stratum is responsible for controlling the process of delivering services to users via their terminals called User Equipment (UE). This stratum uses mainly SIP [14] and Diameter [15] communication protocols. It is based on the elements taken from the IMS concept, such as:

- Call Session Control Function (CSCF) servers: Proxy-CSCF (P-CSCF, they exchange messages with UEs), Serving-CSCF (S-CSCF, the main servers handling all calls) and Interrogating-CSCF (I-CSCF, servers used in multidomain calls),
- Service User Profile Functional Entity/Service Authentication and Authorization Functional Entity (SUP-FE/SAA-FE): the database storing information about user location and subscription of services; used for, among others, authentication, authorization and accounting.

The role of IMS/NGN transport stratum is to provide resources necessary for the services requested by the users. It is assumed that transport stratum of each operator is based on the SDN architecture and includes an access and a core network. The resources of these networks (programmable switches) are managed by separate SDN controllers. "A" and "C" letters, which are used in transport stratum elements' names, indicate the network type, e.g. CONTROLLER A1 manages resources of access network in domain 1. The SDN concept unifies the protocol (the OpenFlow protocol [16]) used for resource control and management for different technologies and equipment. As a result, both classical packet networks and optical networks can be applied in IMS/NGN and managed in a unified way.

Consequently, the cooperation of IMS/NGN network and the SDN concept can bring many benefits. However, it has not been standardized. In the ITU-T standards for IMS/NGN networks [17], the P-CSCF server generates resource reservation and release requests to the transport resource control unit (called the Resource and Admission Control Function, RACF) using the Diameter protocol. In the SDN concept applications determine the required resource operations and use the API of a given controller, which adds, modifies or deletes entries in flow tables of programmable switches. To ensure IMS/NGN and SDN interoperability, it is therefore necessary to add an additional element (called the Gateway Application or simpler Gateapp) that translates messages generated by P-CSCF to these expected by the SDN controller and vice versa. It is important that SDN controller API is not standardized and depends on controller implementation. For the purpose of this work, one of the most popular API solutions was assumed (HTTP [18] REST API), which is used, e. g., in the ONOS controller [19].

Sixteen different service scenarios are assumed in the modeled network, including user registration as well as voice calls within one or two domains. Intra-domain calls may involve one or multiple access areas. The first case requires resource reservation only in the access network of a given domain, the second – also in the core network of a given domain. Inter-domain calls require resource reservation in the access and core networks of both domains. All the above-mentioned scenarios can be generated in domain 1 and 2. For voice calls, possibility of lack of resources in individual transport networks is taken into account, resulting in unsuccessful scenarios. Figure 2 presents a message exchange scenario for a successful intra-operator call performed in domain 1 with both UEs connected to the same access areas.

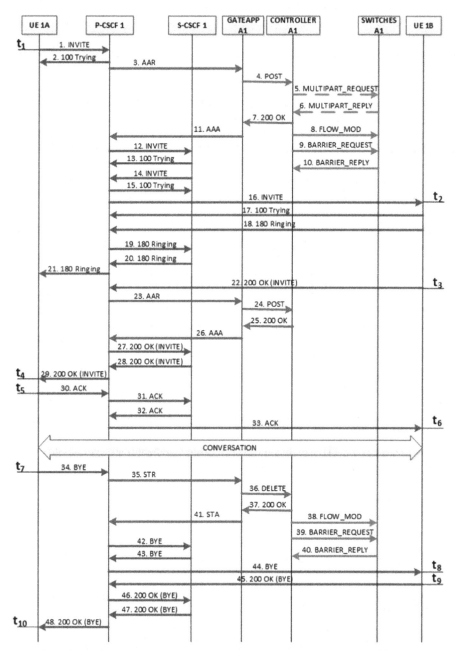

Fig. 2. Message exchange scenario for a successful intra-operator call performed in domain 1 with both UEs connected to the same access areas. Different colors are used for different signaling protocols: SIP – blue, Diameter – red, HTTP – pink, OpenFlow – green.

To increase readability of this scenario, the user terminal generating call set-up request (UE 1A) and the one receiving this request (UE 1B) are represented by separate blocks. In Fig. 1, one block (UE 1) represents all terminals connected to the network of operator 1.

The scenario depicted in Fig. 2 starts with sending a call set-up request (message 1) from UE 1A to P-CSCF 1, which is confirmed by P-CSCF 1 (message 2). P-CSCF 1 sends an AAR message (message 3) to GATEAPP A1 to reserve transport resources of access 1 network for the requested call. This message is translated to HTTP POST request (message 4) and sent to CONTROLLER A1. If necessary, optional messages 5–6 are send by the SDN controller to all programmable switches on path (SWITCHES A1), to determine availability of resources for the requested service. If the controller already has the knowledge about the necessary resources, the scenario goes to messages 7–10. Message 7 is the confirmation about transport resource availability send to GATEAPP A1 and messages 8–10 perform proper changes in flow tables of all programmable switches on path. GATEAPP A1 translates message 7 to message 11 and sends it to P-CSCF 1. After that, the call set-up request is sent over the network (to S-CSCF 1 and P-CSCF 1 again) until it reaches UE 1B (messages 12–17). UE 1B rings (messages 18–21) and accepts the call (message 22). At this time messages 23–26 are used to confirm that the previously reserved resources will be used and traffic regarding the requested call will be sent. This step does not require communication between the SDN controller and programmable switches. Subsequently, the 200 OK (INVITE) message is sent over the network to UE 1A (messages 27–29), which confirms its receipt to UE 1B (messages 30–33). This starts a conversation between both end users, which is ended by UE 1A by sending a BYE message to P-CSCF 1 (message 34). This results in communication between P-CSCF 1, GATEAP A1, CONTROLLER A1 and SWITCHES A1 for releasing the resources allocated to the disengaged call (messages 35–41). After that the BYE message is sent from P-CSCF 1 to UE 1B (through S-CSCF 1 and P-CSCF 1 again; messages 42–44), which confirms call disengagement (messages 45–48).

A very important aspect are the t_1–t_{10} times marked in Fig. 2. According to the ITU-T definitions [5, 6], they can be used to calculate Call Set-up Delay (*CSD*) and Call Disengagement Delay (*CDD*):

$$CSD = (t_2 - t_1) + (t_4 - t_3) + (t_6 - t_5) \tag{1}$$

$$CDD = (t_8 - t_7) + (t_{10} - t_9) \tag{2}$$

Values of *CSD* and *CDD* concern one voice call. In our simulation model they are gathered separately for all types successful call scenarios performed in the modeled network and then averaged. The considered scenarios include:

- successful intra-operator calls performed in domain 1 and 2 with both UEs connected to the same access areas (scenarios b1 and b2),
- successful intra-operator calls performed in domain 1 and 2 with both UEs connected to different access areas (scenarios d1 and d2),
- successful inter-operator calls originated in domain 1 and 2 (scenarios f1 and f2).

Message exchange for the b1 scenario is presented in Fig. 2. The b2 scenario is analogous but performed in domain 2. The d1 scenario involves multiple access areas. Thus, comparing to the b1 case, resource reservation and release in core network of operator 1 are necessary. These procedures are similar to those presented in Fig. 2 (messages 3–11 and 35–41). The most complex are successful inter-operator call scenarios (f1 and f2), requiring resource reservation in access and core transport networks in both domains. Due to lack of space they are not provided. Message flow for the f1 scenario for a multidomain IMS/NGN network without cooperation with SDN can be found in [20].

Structure of the implemented discrete-event simulation model software is presented in Fig. 3. The simulator was developed in the OMNeT++ simulation framework [21] by thoroughly extending our previous model regarding IMS/NGN service stratum [22]. This required adding new modules related to SDN-based transport stratum (GATEAPP, CONTROLLER, SWITCH) and including them in network operation logic.

To increase readability of Fig. 3, only elements of domain 1 and the "global" module common to the entire simulated network (providing global variables, collecting partial simulation results and calculating final results, providing diagnostic functions) are included. The structure of domain 2 is a mirror image of domain 1. Figure 3 includes two programmable switches for access network and core network, as this configuration will be further used in research (Sect. 4).

Blue, red, pink, and green lines with arrows indicate communication between different network elements. The meaning of individual colors is the same as in Fig. 2. Black lines with arrows are used to indicate internal communication between simulation modules forming particular network elements. Italic font in Fig. 3 is used to denote the names of simple modules that perform basic operations defined in the C++ language. Simple modules are grouped into compound modules (bold font), which act as elements of the simulated network (Figs. 1 and 2). For each compound module, the name of the implemented network element and the name of the compound module itself (in brackets) are given. Individual compound modules (e.g. SWITCH) are used many times to implement the operation of various network elements, which is determined by the parameters provided during their initialization.

The implemented compound modules have a common structure. They consist of a "*_main" simple module and a set of "l2tran" simple modules. "*_main" simple modules constitute the main logic of the simulator. They are responsible for handling messages received from other modules and so-called selfmessages (generated by the same module and used to manage simulation events). As a result of handling, received messages may be delayed, their fields may be changed, and new messages may be generated depending on the implemented service scenarios. Most "*_main" simple modules implement a queue and a message processor. The exceptions are the "ue_main" and "supfe_saafe_main" modules, which respond to each received message with a certain delay (without queuing). "l2tran" simple modules participate in communication between elements of the modeled network. Their number in each compound module is equal to the number of other compound modules with which it communicates. The role of each "l2tran" simple module is to buffer messages outgoing from a given compound module if communication link is busy. Messages incoming to a given compound module pass through the "l2tran" modules transparently.

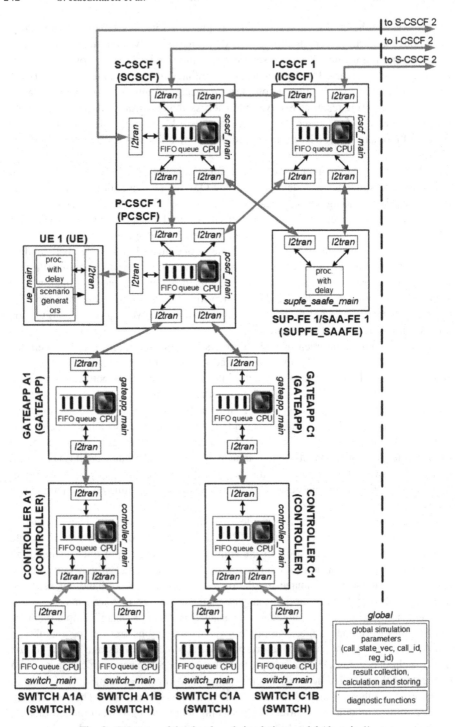

Fig. 3. Structure of the developed simulation model (domain 1).

In the developed simulation model, messages are sent between the elements of the modeled network in accordance with service scenarios for IMS/NGN and SDN networks. The sent messages contain fields defined in standardization documents. Modules process messages (they among others include delays) similarly to elements of a real network. Therefore, the implemented simulator reflects the phenomena occurring in real network.

UE modules in particular domains (UE 1 and UE 2) are responsible for generating call set-up and registration requests. All exchanged messages, apart from the fields defined by the standards, contain fields for carrying information about the t_1-t_{10} times necessary for calculation of *CSD* (1) and *CDD* (2) parameters. When a message passes through subsequent network elements in accordance with the given scenario, these fields are appropriately filled with the values of current simulation time. When a given call is set-up and disengaged correctly (scenarios b1, b2, d1, d2, f1, f2), all t_1–t_{10} times are filled, making it possible to calculate *CSD* and *CDD* values for this call.

The final simulation results are mean *CSD* ($E(CSD)$) and mean *CDD* ($E(CDD)$) values obtained separately for all successful call scenarios along with corresponding confidence intervals. Scenarios' names are added as indexes to these output variables, e.g. $E(CSD)_{b1}$. To obtain final results simulation time is divided into one warm-up period (related to the achievement of a steady state by the simulated network, it is not taken into account when analyzing the results) and a definable number of measurement periods. The Student's *t*-distribution is used to determine confidence intervals when processing measurement data obtained from the simulations. It is used in situations where the standard deviation of the population is unknown and estimated on the basis of measurements and the number of measurement periods is relatively small (below 10).

The simulator provides three possible conditions for ending simulations: exceeding a given simulation time, generating a given number of call set-up requests, reaching values of confidence intervals below a given threshold (this condition is checked periodically). All aspects of simulations are configurable using an *.ini file. The most important input variables of the implemented simulator are presented in Table 1.

Table 1. The most important input variables of the implemented simulator.

Variable name	Description	Default value
sim-time-limit	Maximum simulation duration time [s]	36000
warmup-period	Warp-up period duration time [s]	1250
call_num_max	Maximum number of generated calls	1000000000

(*continued*)

Table 1. (*continued*)

Variable name	Description	Default value
meas_per_num	Number of measurement periods	5
conf_level	Confidence level	0.95
conf_interv_max	Threshold value for confidence intervals, may be directly given in [s] or set in relation to the obtained mean values (in [%])	5%
delay	Base delay value in [s] defined separately for all modeled network elements. For SUPFE_SAAFE compound modules it applies to all messages. For other modules it concerns a base message (processing time of other messages is proportional according to the ak variable, which is described below). The base messages are as follows: SIP INVITE for CSCF servers and UEs, Diameter AAR for GATEAPP modules, HTTP POST for CONTROLLER modules, FLOW_MOD for SWITCH modules	0.001
ak	Vector describing how long other messages are processed in particular modules in relation to the base messages. It contains 32 values	Default values of ak and mess_length are not provided due space limitations
mess_length	Vector with message lengths in [B], containing 39 values	
link_datarate	Link bandwidth in [bps], can be defined separately for all links	50000000
link_length	Link length in [m], can be defined separately for all links	200000

(*continued*)

Table 1. (*continued*)

Variable name	Description	Default value
res_info_prob	Probability of controller having information about the resource state in programmable switches so that there is no need to send messages 5–6 from Fig. 2. This value can be defined separately for each controller	0.7
res_unav_prob	Probability of resource unavailability in particular programmable switch. This value can be defined separately for each switch	0
intrad_call_intensity	Intra-operator call set-up request intensity in [1/s], can be defined separately for each domain	50
interd_call_intensity	Inter-operator call set-up request intensity in [1/s], can be defined separately for each domain	50
registr_intensity	Registration request intensity in [1/s], can be defined separately for each domain	50
multiple_access_areas_ratio	Ratio of intra-operator calls concerning multiple access areas to all generated intra-operator calls, can be defined separately for each domain	0.5

4 Tests and Results

Verification of correctness of the simulation model was carried out in several stages. The first tests were performed while extending the previous version of the simulator [22] to the current version supporting the SDN architecture. Each time a coherent piece of software was developed (for example, a function within a module), the written code was carefully analyzed for errors and tested using different sets of input data. The results of the tested software fragments were displayed in the simulator console. If necessary, corrections were made to the code and tests were repeated until correct operation was achieved.

After implementing the target simulator software, final tests were performed by running simulations in graphical mode (Fig. 4) and analyzing simulations logs using the

Sequence Chart Tool built into OMNeT++ (Fig. 5). These steps made it possible to check correctness of implementation of all service scenarios. For this reason the simulator was configured in such a way that particular scenarios did not overlap and could be analyzed separately.

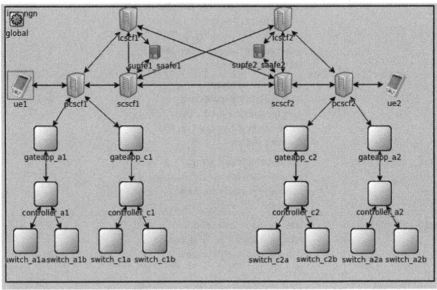

```
Initializing module ims_ngn.switch_c2b.switch_main, stage 0
INFO: RNG: 25
** Event #1  t=0  ims_ngn.ue1.ue_main (ue, id=91)  on selfmsg New_IntraD_Call (om
INFO:Wyslano wiadomosc: (SIP)INVITE len=930B; in transit from ims_ngn.ue1.ue_main
INFO: Call_Id: 1
INFO: From: 1
INFO: To: 1
INFO: Via: ue1
INFO: Multiple_Access_Areas: 0
INFO: Kind: 0
INFO: Length: 930
INFO:Stan polaczenia:
INFO: Call_Id: 1
INFO: From: 1
INFO: To: 1
INFO: Multiple_Access_Areas: 0
INFO: State: 1
INFO: Resource_State_A1: 0
INFO: Resource_State_C1: 0
INFO: Resource_State_A2: 0
INFO: Resource_State_C2: 0
```

Fig. 4. Graphical simulation mode (a fragment of window) – structure of the simulated network (up) and events occurring during the simulation (down).

Graphical simulation mode (Fig. 4) contains, among others, a window visualizing the structure of the simulated network and communication between the modules, as well as a console displaying information about events occurring during the simulation. In addition to the information provided in the console by default, displaying contents of each sent and received message was implemented. Using these functionalities and running simulations "step-by-step", correctness of message passing through network

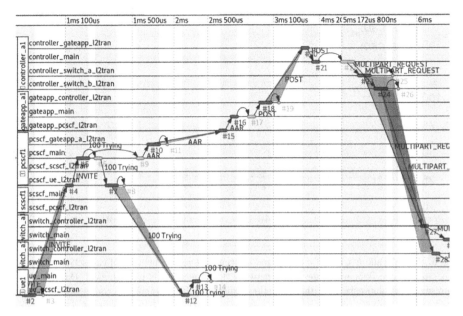

Fig. 5. Log from the performed simulation presented in graphical form using the Sequence Chart Tool (a fragment of window).

elements was checked for all service scenarios along with message handling procedures (processing times, changes in the content of message fields). Particular attention was paid to completion of t_1-t_{10} times, which are used to calculate the *CSD* (1) and *CDD* (2) parameters and, consequently, the final simulation results.

The above-mentioned aspects were additionally verified using simulation logs and the Sequence Chart Tool [21]. The optionally recorded simulation log allows generating a diagram of message transitions between the simulator modules (Fig. 5), as well as a detailed analysis of events in text form. This was very useful for final confirmation of correct operation of the simulation model. As a result, it was demonstrated that the functionality of all network elements and all assumed service scenarios were correctly implemented.

In addition to the described functional tests, the paper includes selected results obtained using the developed simulator (Fig. 6) running in dedicated text mode (without graphical interface). The presented results concern the values of CPP parameters for the b1 and f2 scenarios ($E(CSD)$ and $E(CDD)$) versus intensity of call set-up requests generated in one domain λ_{sum} = intrad_call_intensity + interd_call_intensity. For each measuring point, the component intensities (intrad_call_intensity and interd_call_intensity) were equal to each other. The following values of the base delay (the "delay" input variable from Table 1) were used in the research: 0.5 ms for all CSCF servers and GATEAPP elements, 10 ms for all SUP-FE/SAA-FE units, 1ms for other network elements. The remaining simulation input parameters were set to their default values (Table 1).

It can be noticed that $E(CSD)$ and $E(CDD)$ times raise with increasing λ_{sum} values and the modeled network can handle up to 90 call set-up requests per second generated

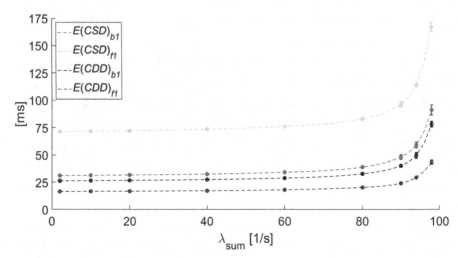

Fig. 6. $E(CSD)$ and $E(CDD)$ for the b1 and f1 scenarios versus intensity of call set-up requests generated in one domain (in the entire network, total intensity is twice larger). Dashed lines represent interpolation between the simulation results, which are points with confidence intervals.

in each domain (above this value the network begins to be overloaded and may be unstable). Additionally, the values of CPP parameters for more complicated multidomain call scenarios (f1) are higher than for scenarios within one domain (b1). Moreover, due to the more complex call set-up process, for particular call scenarios presented in Fig. 6 (b1 and f1) mean CSD times are greater than mean CDD times. The obtained results are as expected and confirm proper operation of the implemented simulator.

5 Conclusions

The paper presents a simulation model for a multidomain IMS/NGN telecommunications network based on the SDN concept in the transport stratum. No other simulation models for such a telecommunication network solution are provided by the scientific community. The proposed model enables a comprehensive analysis of the IMS/NGN/SDN architecture. It takes into account a wide set of service scenarios generated in both domains (registration, intra- and inter-operator calls), as well as parameters of network elements and traffic sources (including the probability of transport resource unavailability resulting in call set-up failure). The above mentioned scenarios and parameters were analyzed and verified in a laboratory testbed [13]. The output variables of the simulator are mean Call Set-up Delay ($E(CSD)$) and mean Call Disengagement Delay ($E(CDD)$) provided separately for all types of successful call scenarios. They are a subset of standardized Call Processing Performance parameters important for network users and operators.

The simulator was developed in the OMNeT++ environment and has a modular structure enabling its easy modification and extension. It was subjected to detailed functional tests at the stage of source code developing (partial tests) and after this process was completed (final tests). Using various tools, the correctness of message passing through network elements and message handling procedures in these elements (e.g. delays and changes in message fields) were checked for all service scenarios. The performed tests demonstrated that the functionality of all network elements and all assumed service scenarios are correctly implemented. Consequently, it can be stated that the described simulator reflects the phenomena taking place in real network.

Additionally, selected research results obtained using the simulator were presented. They confirmed the expected relationships between $E(CSD)$ and $E(CDD)$ values for intra- and inter-operator calls. They demonstrated that for the assumed parameter values, the modeled network is able to handle even 90 call set-up requests per second generated in each domain, without overload.

The presented research and test results allowed achieving the aim of this paper. It was demonstrated that the concept of integrating IMS/NGN and SDN works correctly and can be used in practice. However, it is necessary to comprehensively examine the properties of this concept, including the influence of SDN on quality in both IMS/NGN service and transport stratum [23], which will be carried out using the developed simulation model.

Additional research using the presented simulator is planned in order to collect experience and develop an analytical model for a multidomain IMS/NGN network based on the SDN concept in the transport stratum. This will make a step to move from analyzing the operation of this network to designing its resources in order to ensure appropriate quality parameters.

Disclosure of Interests.   The authors have no competing interests to declare that are relevant to the content of this article.

References

1. General overview of NGN. ITU-T Recommendation Y.2001 (2004)
2. IP Multimedia Subsystem (IMS); Stage 2 (Release 18), 3GPP TS 23.228 v18.4.0 (2023)
3. Framework of Software-Defined Networking, ITU-T Recommendation Y.3300 (2014)
4. Bolanowski, M., Gerka, A., Paszkiewicz, A., Ganzha, M., Paprzycki, M.: Application of genetic algorithm to load balancing in networks with a homogeneous traffic flow. In: Mikyška, J., de Mulatier, C., Paszynski, M., Krzhizhanovskaya, V.V., Dongarra, J.J., Sloot, P.M. (eds.) Computational Science – ICCS 2023. ICCS 2023. LNCS, vol. 14074. Springer, Cham (2023). https://doi.org/10.1007/978-3-031-36021-3_32
5. Call processing performance for voice service in hybrid IP networks, ITU-T Recommendation Y.1530 (2007)
6. SIP-based call processing performance, ITU-T Recommendation Y.1531 (2007)
7. Katov, A.N., Anggorojati, B., Kyriazakos, S., Mihovska, A.D., Prasad, N.R.: Towards Internet of services - SDN-enabled IMS architecture for IoT integration. In: Proceedings of 18th international symposium on wireless personal multimedia communications, WPMC 2015, Hyderabad, India (2015)
8. Kang, S., Yoon, W.: SDN-based resource allocation for heterogeneous LTE and WLAN multi-radio networks. J. Supercomput. **72**(4), 1342–1362 (2016)
9. Tranoris, C., Denazis, S., Mouratidis, N., Dowling, P., Tynan, J.: Integrating OpenFlow in IMS networks and enabling for future internet researchand experimentation. In: Galis, A., Gavras, A. (eds.) The Future Internet. FIA 2013. LNCS, vol. 7858. Springer, Berlin, Heidelberg (2013). https://doi.org/10.1007/978-3-642-38082-2_7

10. Khairi, S., Raouyane, B., Bellafkih, M.: Towards enhanced QoS management SDN-based for next generation networks with QoE evaluation: IMS use case. J. Mobile Multimedia **13**(3–4), 183–196 (2017)
11. Liu, Z., Wang, Q., Lee, J.-O.: A SDN controller enabled architecture for the IMS. In: Proceedings of 20th asia-pacific network operations and management symposium, APNOMS 2019, pp. 1–4, Matsue, Japan (2019)
12. Tang, C.-S., Twu, C.-Y., Ju, J.-H., Tsou, Y.-D.: Collaboration of IMS and SDN to enable new ICT service creation. In: Proceedings of the 16th asia-pacific network operations and management symposium, pp. 1–4, Hsinchu, Taiwan (2014)
13. Kaczmarek, S., Sac, M., Bachorski, K.: Implementation of IMS/NGN transport stratum based on the SDN concept. Sensors **23**(12), 5481 (2023)
14. Rosenberg, J., et al.: SIP: Session initiation protocol, IETF RFC 3261 (2002)
15. Fajardo, V., et al.: Diameter Base Protocol, IETF RFC 6733 (2012)
16. OpenFlow Switch Specification Version 1.5.1, ONF TS-025 (2015)
17. Resource Control Protocol No. 1, Version 3 — Protocol at the Rs Interface between Service Control Entities and the Policy Decision Physical Entity, ITU-T Recommendation Q.3301 (2013)
18. Fielding, R., Nottingham, M., Reschke, J.: HTTP/1.1, IETF RFC 9112 (2022)
19. Open Network Operating System (ONOS) SDN Controller. https://opennetworking.org/onos/. Accessed 1 March 2024
20. Kaczmarek, S., Sac, M.: Traffic model of a multidomain IMS/NGN. Telecommun. Rev. Telecommun. News **8–9**, 1030–1038 (2014)
21. OMNeT++ Documentation. https://omnetpp.org/documentation. Accessed 1 March 2024
22. Kaczmarek, S., Sac, M.: Verification of the analytical traffic model of a multidomain IMS/NGN using the simulation model. In: Grzech, A., Borzemski, L., Świątek, J., Wilimowska, Z. (eds.) Information systems architecture and technology: proceedings of 36th international conference on information systems architecture and technology – ISAT 2015 – Part II. AISC, vol. 430. Springer, Cham (2016). https://doi.org/10.1007/978-3-319-285 61-0_9
23. Kaczmarek, S., Litka, J.A.: Impact of SDN controller's performance on quality of service. IEEE Access **12**, 8262–8282 (2024)

Cascade Training as a Tree Search with Dijkstra's Algorithm

Dariusz Sychel$^{(\boxtimes)}$ [iD], Aneta Bera [iD], and Przemysław Klęsk [iD]

Faculty of Computer Science and Information Technology,
West Pomeranian University of Technology, ul. Żołnierska 49, 71-210 Szczecin, Poland
{dsychel,abera,pklesk}@zut.edu.pl

Abstract. We propose a general algorithm that treats cascade training as a tree search process working according to *Dijkstra's algorithm* in contrast to our previous solution based on the *branch-and-bound* technique. The reason behind the algorithm change is reduction of training time. This change does not affect in anyway the quality of the final classifier. We conduct experiments on cascades trained to become face or letter detectors with Haar-like features or Zernike moments being the input information, respectively. We experiment with different tree sizes and different branching factors. Results confirm that training times of obtained cascades, especially for large heavily branched trees, were reduced. For small trees, the previous technique can sometimes achieve better results but the difference is negligible in most cases.

Keywords: Cascade of classifiers · Dijkstra's algorithm · Training Time Reduction · Tree search

1 Introduction

Dijkstra's algorithm is a useful tool for finding the shortest paths between nodes in a graph. Many practical applications of this algorithm or its modifications can be found, e.g. airport automated guided vehicles (AGV) path optimization [16], evacuation route optimization under real-time toxic gas dispersion through computational fluid dynamics (CFD) simulation and Dijkstra's algorithm [15] or judgment of railway transportation path presented in [4].

Cascades of classifiers [13,14] were designed to work as classifying systems operating under two conditions: (1) very large number of incoming requests, (2) significant classe imbalance. A cascade should vary its computational effort depending on the contents of an object to be classified. Objects that are obvious negatives (non-targets) should be recognized fast, using only a few features extracted. Targets, or objects resembling them, are allowed to employ more features and time for computations. We remark that the optimization problem we try to solve in this research (and the previous one [9]) is to build such a cascade that minimize the *expected number of features* applied by the cascade (formal definition show in Sect. 2.3).

L. Franco et al. (Eds.): ICCS 2024, LNCS 14833, pp. 251–265, 2024.
https://doi.org/10.1007/978-3-031-63751-3_17

Despite the development of deep learning, recent literature shows that cascades of classifiers are still applied in detection systems or batch classification jobs e.g. GPGPU-based (General-Purpose Graphics Processing Unit) parallel version of eyes detecting cascade was used for driving an intelligent wheelchair [7]. Cascade with improved memory consumption was applied for remote sensing tasks [12]. Cascades of classifiers were also used for kitchen safety monitoring [2] or for object detection for robot cars [6]. Moreover authors of [1] show that classifier cascade can be significantly faster than YOLO (you only look once) classifier without substantial reduction of accuracy. Their comparison was conducted on the driver drowsiness detection problem.

In our previous work [10] we provided and proved a theoretical result demonstrating that the presence of slack between the constant per-stage requirements (on accuracy measures) used in the original cascade algorithm and actual rates observed while learning, allows to introduce new *relaxed* requirements for each successive stage and still complete the training procedure successfully. The relaxed requirements can be met more easily, using fewer features. This creates a potential possibility to reduce the *expected number of features* used by an operating cascade. Taking advantage of the relaxation, we proposed new stage-wise training algorithms that apply two approaches: uniform or greedy. They differ in the way the slack accumulated so far becomes "consumed" later on. Results obtained by the greedy algorithm (UGM-G) were better in most cases and this variant became the default one for further research.

This leads us to a new general algorithm that treats cascade training as a tree search process working according to the *branch-and-bound* technique [9]. Successive tree levels correspond to successive cascade stages. Sibling nodes represent variants of the same stage with different number of features applied. We provided suitable formulas for lower bounds on the expected value to be optimized. While searching, we observe suitable lower bounds on partial expectations and prune tree branches that cannot improve the best-so-far result. Once the search is finished, one of the paths from the root to some terminal node indicates the cascade with the smallest expected number of features. Both exact and approximate variants of the approach were formulated in [9]. Our results confirmed shorter operating times of cascades obtained owing to the reduction in the number of extracted features. The main contribution of this paper is a new training algorithm for cascade training. It is performed via a tree search procedure that uses *Dijkstra's algorithm* in order to reduce the training time compared to the *branch-and-bound* technique that was used in the previous work. For our purposes we consider the *single-source single-destination* variant of Dijkstra's algorithm [5] (rather than the single-source *all* shortest paths). Additionally, the reduction of training time does not have any negative impact on the *expected number of features* in obtained final cascades. Both approaches the new and the old one (exact branch-and-bound version) result in exactly the same cascades. The new technique is recommended for large heavily branched trees. In case of small trees, the previous technique can sometimes achieve better results in terms of training time but the difference is in fact negligible in most cases.

2 Preliminaries

2.1 Notation

Throughout this paper we use the following notation:

- K—number of cascade stages,
- $n = (n_1, n_2, \ldots, n_K)$—numbers of features used on successive stages,
- $(a_1, a_2, \ldots, a_K)$—FAR (false alarm rates) values on successive stages,
- $(d_1, d_2, \ldots, d_K)$—sensitivities (detection rates) on successive stages,
- A—required FAR for the whole cascade,
- D—required detection rate (sensitivity) for the whole cascade,
- $a_{\max} = A^{1/K}$—per-stage FAR requirement,
- $d_{\min} = D^{1/K}$—per-stage sensitivity requirement,
- $F = (F_1, F_2, \ldots, F_K)$—ensemble classifiers on successive stages (the cascade),
- A_k—FAR observed up to k-th stage of cascade ($A_k = \prod_{1 \leqslant i \leqslant k} a_i$),
- D_k—sensitivity observed up to k-th stage of cascade ($D_k = \prod_{1 \leqslant i \leqslant k} d_i$),
- θ_k—decision threshold for classifier F_k, it should be set to the minimal value that satisfies $d_k \geqslant d_{\min}$,
- $(p, 1 - p)$—true probability distribution of classes (unknown in practice),
- $\mathcal{D}, \mathcal{V}$—training and validation data sets,
- $\#$—set size operator (cardinality of a set),
- $\|$—concatenation operator (to concatenate cascade stages).

The probabilistic meaning of relevant quantities is as follows. The final requirements (A, D) demand that: $P(F(\mathbf{x}) = + \,|\, y = -) \leqslant A$ and $P(F(\mathbf{x}) = + \,|\, y = +) \geqslant D$, whereas false alarm and detection rates observed on particular stages are, respectively, equal to:

$$
\begin{aligned}
a_k &= P\left(F_k(\mathbf{x}) = + \,|\, y = -, F_1(\mathbf{x}) = \cdots = F_{k-1}(\mathbf{x}) = +\right), \\
d_k &= P\left(F_k(\mathbf{x}) = + \,|\, y = +, F_1(\mathbf{x}) = \cdots = F_{k-1}(\mathbf{x}) = +\right).
\end{aligned} \tag{1}
$$

2.2 Classical Cascade Training Algorithm (Viola-Jones Style)

The classical cascade training algorithm with constant per-stage requirements can be presented with the pseudo-code below (Algorithm 1).

2.3 Expected Number of Extracted Features

Cascade performance is directly dependent on the average number of features used per window regardless of the learning method, therefore there is a direct connection between the expected number of features and detection time. To support this claim Table 1 show impact of expected number of features on detection time for three example classifiers obtained in our previous work [10].

Algorithm 1. VJ-style training algorithm for cascade of classifiers

---

   procedure TRAINVJCASCADE($\mathcal{D}$, A, D, K, $\mathcal{V}$)

      From $\mathcal{D}$ take subsets $\mathcal{P}$, $\mathcal{N}$ with positive and negative examples, respectively.

      $F := ()$ ▷ initial cascade — empty sequence

      $a_{max} := A^{1/K}$, $d_{min} := D^{1/K}$, $A_0 := 1$, $D_0 := 1$, $k := 0$.

      while $A_k > A$ **do**

         $n_{k+1} := 0$, $F_{k+1} := 0$, $A_{k+1} := A_k$, $a_{k+1} := A_{k+1}/A_k$.

         while $a_{k+1} > a_{max}$ **do**

            $n_{k+1} := n_{k+1} + 1$.

            Train new weak classifier f using $\mathcal{P}$ and $\mathcal{N}$

            $F_{k+1} := F_{k+1} + f$.

            Adjust decision threshold θ_{k+1} for F_{k+1} to satisfy d_{min} requirement.

            Use cascade $F\|F_{k+1}$ on validation set $\mathcal{V}$ to measure A_{k+1} and D_{k+1}.

            $a_{k+1} := A_{k+1}/A_k$.

         $F := F\|F_{k+1}$.

         if $A_{k+1} > A$ **then**

            $\mathcal{N} := \emptyset$.

            Use cascade F to populate set $\mathcal{N}$ with false detections

                 sampled from non-target images.

         $k := k + 1$

      return $F = (F_1, F_2, \ldots, F_k)$.

Table 1. "Face detection"—impact of expected number of features on detection time ($A = 0.001, D = 0.95$).

Cascade		Expected value	FAR	Sensitivity	Detection time	
					image [ms]	window [μs]
n_k a_k	9 0.2468, 16 0.2548, 21 0.2234, 22 0.2635, 39 0.2606	14.7220	0.000964	0.9510	88	0.675
n_k a_k	9 0.2468, 18 0.2214, 26 0.2299, 30 0.2468, 38 0.2370	15.3571	0.000735	0.9520	89	0.680
n_k a_k	9 0.2468, 17 0.2516, 32 0.2450, 29 0.2303, 29 0.2798	15.7243	0.000980	0.9510	90	0.687

Definition-Based Formula. A cascade stops operating after a certain number of stages. Therefore the possible outcomes of the random variable of interest, describing the disjoint events, are: n_1, $n_1 + n_2$, ..., $n_1 + n_2 + \cdots + n_K$. Hence, by the definition of expected value, the expected number of features can be calculated as follows:

$$E(n) = \sum_{1 \leqslant k \leqslant K} \left(\sum_{1 \leqslant i \leqslant k} n_i \right) \left(p \Big(\prod_{1 \leqslant i < k} d_i \Big)(1 - d_k)^{[k < K]} + (1 - p) \Big(\prod_{1 \leqslant i < k} a_i \Big)(1 - a_k)^{[k < K]} \right), \tag{2}$$

where $[\cdot]$ is an indicator function.

Incremental Formula and Its Approximation. By grouping the terms in (2) with respect to n_k the following alternative formula can be derived:

$$E(n) = \sum_{1 \leqslant k \leqslant K} n_k \left(p \prod_{1 \leqslant i < k} d_i + (1-p) \prod_{1 \leqslant i < k} a_i \right). \tag{3}$$

In practical applications the true probability distribution underlying the data is unknown. Since the probability p of the positive class is very small, the expected value can be accurately approximated using only the summands related to the negative class as follows:

$$\widehat{E}(n) = \sum_{1 \leqslant k \leqslant K} n_k \prod_{1 \leqslant i < k} a_i \approx E(n). \tag{4}$$

In our previous works [9,10] we focused on creating an algorithm that tries to decrease this quantity compared to the original Viola and Jones algorithms. Because the solution proposed in this paper is a modification of [9], the expected value also plays a critical role in it, as is used to establish the priority of nodes in the priority queue used in Dijkstra's algorithm.

2.4 Relaxed Per-Stage Requirements

Instead of constant per-stage requirements proposed in the original approach we continue to use the greedy variant of relaxed per-stage requirements, proposed by us in [10], since applying them results in cascades with lower expected number of features. As a reminder:

Theorem 1. *The presence of slack between constant per-stage requirements* (a_{max}, d_{min}) *and actual rates* (a_k, d_k), $k = 1, \ldots, K$, *observed during cascade training —*

$$a_k = (1 - \epsilon_k) a_{max}, \qquad\qquad d_k = (1 + \delta_k) d_{min}, \tag{5}$$

where ϵ_k, δ_k *represent slack variables denoting small numbers—allows to introduce new relaxed requirements for each successive stage and carry out a training procedure that still satisfies the final requirements* (A, D) *for the whole cascade. In particular, when the k-th stage is done, the following two pairs of relaxed bounds (uniform and greedy) can be applied for the $(k+1)$-th stage:*

$$a_{k+1} \leqslant \frac{a_{max}}{\left(1 - \epsilon_{\leqslant k}\right)^{1/(K-k)}}, \qquad\qquad d_{k+1} \geqslant \frac{d_{min}}{\left(1 + \delta_{\leqslant k}\right)^{1/(K-k)}}, \tag{6}$$

or

$$a_{k+1} \leqslant \frac{a_{max}}{1 - \epsilon_{\leqslant k}}, \qquad\qquad d_{k+1} \geqslant \frac{d_{min}}{1 + \delta_{\leqslant k}}, \tag{7}$$

where $1 - \epsilon_{\leqslant k} = \prod_{1 \leqslant i \leqslant k} (1 - \epsilon_i)$ *and* $1 + \delta_{\leqslant k} = \prod_{1 \leqslant i \leqslant k} (1 + \delta_i)$.

For proof see [10].

The greedy per-stage requirements (UGM-G) can be expressed in terms of A, D constants and a_i, d_i rates observed so far, that is for $i \leqslant k$, as follows:

$$a_{\max,k+1} = \frac{A^{\frac{k+1}{k}}}{\prod_{1 \leqslant i \leqslant k} a_i}, \qquad d_{\min,k+1} = \frac{D^{\frac{k+1}{k}}}{\prod_{1 \leqslant i \leqslant k} d_i}. \tag{8}$$

2.5   Cascade Training as a Tree Search

When cascade training is treated as a tree search process, the root of the tree represents an empty cascade. Successive tree levels correspond to successive cascade stages. Each non-terminal tree node have an odd number of children nodes. They represent variants of a subsequent stage with slightly different number of features. The children are processed from left to right until the stop condition is met.

The size of the tree can be controlled by two integer parameters L and C, predefined by the user. To keep the tree fairly small, the branching of variants shall take place only at L top-most levels, e.g. $L = 2$. At those levels the branching factor is equal to C, an odd number, e.g. $C = 5$ (mandatory middle node, $\frac{C-1}{2}$ nodes created by removing features from it, $\frac{C-1}{2}$ nodes created by adding new features). At deeper levels the branching factor is one.

Pruning Search Tree Using Current Partial Expectations—Exact Branch-and-Bound
During an ongoing tree search (combined with cascade training) one can observe *partial* values for the expected value of interest—formula (4). Suppose a new $(k + 1)$-th stage has been completed, revealing n_{k+1} features. The formula (9)

$$\widehat{E}\big((n_1,\ldots,n_{k+1})\big) = \sum_{1 \leqslant j \leqslant k} n_j \prod_{1 \leqslant i < j} a_i + n_{k+1} \prod_{1 \leqslant i < k+1} a_i = \widehat{E}\big((n_1,\ldots,n_k)\big) + n_{k+1} \prod_{1 \leqslant i < k+1} a_i. \tag{9}$$

expresses the partial expectation for the extended cascade in an incremental manner. It should be clear that whenever a partial expectation for some tree branch is greater than (or equal to) the best-so-far exact expectation, say

$$\widehat{E}\big((n_1,\ldots,n_{k+1})\big) \geqslant \widehat{E}^*,$$

then there is no point in pursuing that branch further down the tree[1]. In other words, pruning can be applied because formula (9) provides a lower bound on the final unknown expectation. Figure 1 provides a symbolic illustration of a search tree with pruning.

---

[1] Initial value of $\widehat{E}^*$ is set to ∞, after first cascade satisfying (A, D) requirements finish its training, $\widehat{E}^*$ represents its expected number of features.

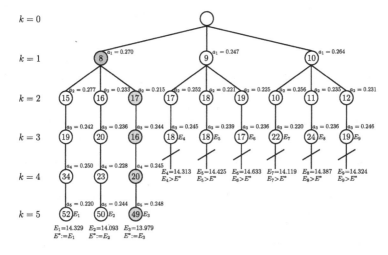

Fig. 1. Cascade training as a tree search with pruning—example illustration.

Algorithm 2. Training cascade of classifiers via tree search with exact pruning

---

procedure TRAINTREECASCADE($\mathcal{D}$, A, D, K, k, $\mathcal{V}$, F, C, L, F^*, $\widehat{E}^*$)

From $\mathcal{D}$ take subset $\mathcal{P}$ with all positive examples, and subset $\mathcal{N}$ with all negative examples.

Train stage for middle child: $F_{k+1,0}$:=TRAINSTAGE($\mathcal{P}$, $\mathcal{N}$, K, k, $\mathcal{V}$, F).

Use cascade $F\|F_{k+1,0}$ on validation set $\mathcal{V}$ to measure $A_{k+1,0}$ and $D_{k+1,0}$.

if $k > L$ **then**
  $C := 1$.

for $c := -1, -2, \ldots, -\lfloor C/2 \rfloor$ **do**   ▷ left children
  Create $F_{k+1,c}$ by cloning $F_{k+1,c+1}$.
  Remove most recent weak classifier from $F_{k+1,c}$.
  Adjust decision threshold $\theta_{k+1,c}$ for $F_{k+1,c}$ to satisfy $d_{\min,k+1}$ requirement.
  Use cascade $F\|F_{k+1,c}$ on validation set $\mathcal{V}$ to measure $A_{k+1,c}$ and $D_{k+1,c}$.

for $c := 1, 2, \ldots, \lfloor C/2 \rfloor$ **do**   ▷ right children
  Create $F_{k+1,c}$ by cloning $F_{k+1,c-1}$.
  Train new weak classifier f using $\mathcal{P}$ and $\mathcal{N}$.
  $F_{k+1,c} := F_{k+1,c} + f$.
  Adjust decision threshold $\theta_{k+1,c}$ for $F_{k+1,c}$ to satisfy $d_{\min,k+1}$ requirement.
  Use cascade $F\|F_{k+1,c}$ on validation set $\mathcal{V}$ to measure $A_{k+1,c}$ and $D_{k+1,c}$.

for $c := -\lfloor C/2 \rfloor, \ldots, 0, \ldots, \lfloor C/2 \rfloor$ **do**   ▷ all children
  Calculate expectation $\widehat{E}$ for cascade $F\|F_{k+1,c}$ using (9).
  if $A_{k+1,c} > A$ and $\widehat{E} < \widehat{E}^*$ **then**
    Prepare new training set $\mathcal{D}_{k+1,c}$ and new validation set $\mathcal{V}_{k+1,c}$.
    $(F^*, \widehat{E}^*)$:=TRAINTREECASCADE($\mathcal{D}_{k+1,c}$, A, D, K, $k+1$, $\mathcal{V}_{k+1,c}$, $F\|F_{k+1,c}$,
                                      L, C, E^*, F^*)
  else if $A_{k+1,c} \leqslant A$ and $\widehat{E} < \widehat{E}^*$ **then**
    $\widehat{E}^* := \widehat{E}$, $F^* := F\|F_{k+1,c}$.
    return $(F^*, \widehat{E}^*)$.
return (F^*, E^*).

The outermost recursion call (initial call) for Algorithm 2 is

$$\text{TrainTreeCascade}(\mathcal{D}, A, D, K, 0, \mathcal{V}, (), C, L, \text{null}, \infty).$$

The TRAINSTAGE function in Algorithm 2 correspond to a single execution of external while loop inside Algorithm 1. It results in a single ensemble trained using per-stage requirements. The requirements can be calculated as standard geometric means (classical VJ-style), leading to constant per-stage requirements for the whole training, or as updated geometric means (UGM).

Pruning Search Tree Using Expectation Predictions—Approximate Branch-and-Bound

As we have shown in [9] the training time can be reduced even more by the following approximate branch-and-bound approach. When the stage $k + 1$ is completed, we get to know two new pieces of information: n_{k+1} and a_{k+1}. That second piece is not needed to calculate formula (9) for stage $k+1$, but it is needed for stage $k + 2$. Therefore, the only unknown preventing us from calculating the exact partial expectation for stage $k + 2$ is n_{k+2}. We propose a formula that allows to approximate this value: $n_{k+2} \geqslant \alpha\, n_{k+1}$.

Dijkstra's Algorithm (Single-Source Single-Destination) for Cascade Tree Search

As can be noticed, the approach presented so far is based on a depth-first traversal accelerated by the branch-and-bound technique. One should realize that when the search process is conducted recursively from left to right, the training time can in some cases be significantly longer (e.g. if the solution is in the far right part of tree).

In order to overcome the above disadvantage, we postulate to replace the previous traversal order with the *Dijkstra's algorithm* in which the priority of each node is equal to its partial expected value on the number of features. This means that tree nodes with lower $\hat{E}$ values shall be visited sooner, and therefore the most promising nodes shall be always evaluated first regardless of their position the cascade search tree. The new training procedure is presented below as Algorithm 3.

Object *node* in Algorithm 3 represents a single tree node. Every *node* object contains the following fields: C—number of children created after the node's training is finished, k—tree level (equivalent to the cascade stage), A_k, D_k—FAR/sensitivity for the cascade ended at that node.

The way of creating new children nodes is the same as in Algorithm 2. Once the middle child is created, the left siblings are created by removing weak classifiers from the middle child, whereas the right siblings must undergo further training in order to add new weak classifiers (and features).

Algorithm 3. Training cascade of classifiers via tree search with Dijkstra's algorithm

---

procedure TRAINTREECASCADEDIJKSTRA($\mathcal{D}$, A, D, K, $\mathcal{V}$, C, L)
　From $\mathcal{D}$ take subset $\mathcal{P}$ with all positive examples, and subset $\mathcal{N}$ with all negative examples.
　Insert *root* (empty node with $\widehat{E} = 0$) into priority queue *open*.
　while *open* not empty **do**
　　Take *node* with the smallest $\hat{E}$ from *open*.
　　$node.C := 1$.
　　if $node.k < L$ **then**
　　　$node.C := C$.
　　if $node.A_k > A$ or $node.D_k < D$ **then**
　　　if $node.k < K$ **then**
　　　　Train C *children* of current *node* according to TRAINTREECASCADE
procedure.
　　　　for all $c \in$ *children* **do**
　　　　　if $node.k + 1 < K$ **then**
　　　　　　Prepare new training set $\mathcal{D}_{k+1,c}$ and new validation set $\mathcal{V}_{k+1,c}$.
　　　　　　Calculate expectation $\widehat{E}$ for cascade $F\|F_{k+1,c}$ using (9).
　　　　　　Insert child c into priority queue *open*.
　　　Insert *node* into list *closed*.
　　else if $node.A_k < A$ and $node.D_k > D$ **then**
　　　return *node*

---

3 Experiments

Similar to our previous work [9,10] research was conducted on machine with Intel Core i7-4790K 4/8 cores/threads, 8MB cache. In all experiments we apply *RealBoost+bins* [8] as the main learning algorithm, producing ensembles of weak classifiers as successive cascade stages. Each weak classifier is based on a single selected feature. In letter "A" detection task we used computer fonts prepared by T.E. de Campos et al. [3]. In face detection task for training purpose we used faces cropped from 3000 images, looked up using Google Images search engine. Test set contains faces from Essex facial images collection [11]. More details about experimental setup can be found in mentioned articles.

3.1 Average Time per Node

As our previous work has shown, the further a weak classifier is in a cascade the more time it takes to train it. This is mainly due to the time needed to perform the resampling of the training and validation sets. With each weak classifier added to a cascade its FAR value decreases. Because of this, it becomes more and more difficult to find an image window misclassified as a positive.

Table 2 shows how the node average training time and the average resampling times at given stages looked like in experiments. Average times were calculated based on cascades trained to satisfy final requirement equal to $A = 0.001$ (FAR) and $D = 0.95$ (sensitivity). The decrease in the node average training time for higher K values is associated with a smaller number of features per stage for more extensive cascades (that satisfy same final requirements), which can be observed in the Table 5. When it comes to resampling times, the average times shown in the Table 2 confirm a significant increase of time needed for successive stages.

Table 2. Average node training and resampling times per tree level (A = 0.001, D = 0.95)

Stages	Avg. Training Time [s]		Avg. Resampling Time per Stage [s]								
	per Stage	Total	1	2	3	4	5	6	7	8	9
HAAR-like features											
K = 5	280	1 400	36	332	2 011	7 286					
K = 10	164	1 639	6	34	106	271	537	913	1 916	4 553	8 087
Zernike moments											
K = 10	2	19	16	41	52	129	224	430	848	1 739	3 701

3.2 Training Times

The conducted research shows that the proposed approach works especially well for large cascades. In the case of experiments with face detection (Table 3) we can see that for a high value of FAR $A = 0.01$ even for relatively small cascades with only 5 stages Dijkstra's algorithm allowed to reduce the training time, but for more demanding settings $A = 0.001$ the obtained results are similar or worse to the ones achieved by pruning technique with high α factor. With the increase of the tree size the effectiveness of Dijkstra's algorithm (in terms of training time reduction) rises significantly. It is also worth recalling at this point that the pruning algorithm proposed in [9] approximates the partial expected value, therefore setting a high value of α increases the risk of omitting the cascade with the lowest expected value. On the other hand, Dijkstra's algorithm guarantees finding the optimal solution (i.e. the cascade with the minimal expected value contained in the current tree).

In the case of the second group of experiments (Table 4)—detection of letter 'A', we can notice that the results are similar. Because only few Zernike moments per stage were needed in order to detect letter 'A' with $A = 0.001$, we could not test trees with higher values of C, L and K parameters. Instead we decided to train more demanding classifiers with $A = 0.0001$ in order to show the impact of the proposed solution.

The approach proposed in this paper works particularly well for large multi-level trees with a high number of nodes. In practice, this allows a larger combination of cascades to be searched, without significantly increasing the training time compared to the classical cascade learning approach.

Table 3. Training time comparison (Face Detection—HAAR-like features)

Training algorithm	K	Splits (L)	Children per split (C)	Nodes	A	D	Trained Nodes DFS + pruning $\alpha = 0.0$	$\alpha = 1.2$	Dijkstra open	close	Training time DFS + pruning $\alpha = 0.0$	$\alpha = 1.2$	Dijkstra
							Face Detection (HAAR-like features)						
UGM-G	5	1	3	15	0.01	0.95	13	13	11	9	3 130s	2 925s	2 564s
UGM-G	5	1	3	15	0.001	0.95	12	11	12	11	13 375s	11 624s	11 960s
UGM-G	5	1	5	25	0.001	0.95	16	13	17	13	12 238s	10 525s	16 543s
UGM-G	5	1	7	35	0.001	0.95	19	14	20	14	12 039s	10 061s	15 957s
UGM-G	5	2	3	39	0.001	0.95	30	25	27	22	43 642s	30 232s	96 319s
UGM-G	10	1	3	30	0.001	0.95	24	23	17	15	27 916s	25 929s	14 076s
UGM-G	10	1	5	50	0.001	0.95	40	36	25	21	50 963s	49 467s	20 075s
UGM-G	10	1	7	70	0.001	0.95	46	41	31	25	51 679s	50 406s	25 427s
UGM-G	10	2	3	84	0.001	0.95	56	48	31	23	98 053s	90 897s	23 975s
UGM-G	15	1	3	45	0.001	0.95	44	43	33	31	65 381s	58 799s	19 421s
UGM-G	15	1	5	75	0.001	0.95	65	61	47	43	94 344s	90 863s	25 231s
UGM-G	15	1	7	105	0.001	0.95	85	79	58	52	130 437s	121 311s	28 468s
UGM-G	15	2	3	129	0.001	0.95	86	78	60	53	122 668s	106 741s	31 116s

Table 4. Training time comparison (Letter 'A' Detection—Zernike moments)

Training algorithm	K	Splits (L)	Children per split (C)	Nodes	A	D	Trained Nodes DFS + pruning $\alpha = 0.0$	$\alpha = 1.2$	Dijkstra open	close	Training time DFS + pruning $\alpha = 0.0$	$\alpha = 1.2$	Dijkstra
							Letter 'A' Detection (Zernike moments)						
UGM-G	10	1	3	30	0.001	0.95	21	21	14	12	14 575s	14 211s	5 349s
UGM-G	10	1	3	30	0.0001	0.95	18	12	18	17	50 054s	43 121s	47 961s
UGM-G	15	1	3	45	0.0001	0.95	32	31	20	18	196 150s	192 421s	52 912s

3.3 Prunning Efficiency

The results presented in Table 5 show the number of nodes visited by each method. This table is an extension of a table from [9] with additional results from Dijkstra's algorithm. In the case of Dijkstra's algorithm, two new values were reported: *open*—means how many nodes were generated and added to the priority queue (nodes in the open queue were trained, but their descendants were not yet created and hence data resampling was not yet needed), *closed*— represents the number of nodes that have been fully processed. The values in the *closed* column are always non-greater than the values of *open* column. It should be remarked that closed nodes counts correspond to number of trained nodes reported for the previous algorithms (open nodes are of lighter computational costs). The difference between closed and open counts tells us how many children were created but never visited.

In this experiment we decided to use three α values: $\alpha = 0.0$ is equivalent to the exact pruning method, $\alpha = 0.8$ to pruning with low risk of missing the optimal solution, $\alpha = 1.2$ corresponds to a more aggressive pruning. Higher values of α were not used because of high risk of missing the optimal solution.

As we can see, the proposed algorithm, thanks to the use of a priority queue, allowed for a significant reduction in the number of nodes compared to the exact pruning method. This difference is caused by the fact that in case of the branch-and-bound method the effectiveness of the approach depends on the order in which children nodes are visited (similar to other branch-and-bound based methods its computational complexity is $O(C^K)$, in optimistic case it is $\Omega(C^{\frac{1}{2}K})$, average complexity is $\Theta(C^{\frac{3}{4}K})$), while in case of Dijkstra's method we directly compare expected number of features of all reachable nodes. In most cases the number of nodes visited was also smaller than the one obtained from the greedy pruning with $\alpha = 1.2$.

Table 5. Face detection (Haar-like features)—comparison of search tree-based approaches for different values of parameter C and L.

Training algorithm	Cascade	E(n)	Validation FAR	Validation sensitivity	DFS + pruning exact	approximate α=0.0	approximate α=0.8	approximate α=1.2	Dijkstra open	Dijkstra close
Requirement:			10^{-3} = 0.001	0.95						
VJ	9, 18, 26, 30, 38; 0.2468, 0.2214, 0.2299, 0.2468, 0.2370	15.36	0.00074	0.9520						
TREE-C3-L1 VJ	8, 16, 22, 35, 55; 0.2703, 0.2329, 0.2188, 0.2419, 0.2487	14.37	0.00083	0.9520		12/15	11/15	10/15	12	11
TREE-C3-L1 UGM	8, 16, 22, 27, 37; 0.2703, 0.2329, 0.2188, 0.2601, 0.2439	14.21	0.00087	0.9510		12/15	11/15	10/15	12	11
TREE-C3-L1 UGM-G	8, 16, 22, 25, 40; 0.2703, 0.2329, 0.2188, 0.2713, 0.2646	14.20	0.00099	0.9510		12/15	11/15	11/15	12	11
TREE-C3-L2 VJ	8, 16, 22, 35, 55; 0.2703, 0.2329, 0.2188, 0.2419, 0.2487	14.37	0.00083	0.9520		30/39	26/39	23/39	27	23
TREE-C3-L2 UGM	8, 16, 22, 27, 37; 0.2703, 0.2329, 0.2188, 0.2601, 0.2439	14.21	0.00087	0.9510		29/39	25/39	24/39	28	21
TREE-C3-L2 UGM-G	8, 16, 22, 25, 40; 0.2703, 0.2329, 0.2188, 0.2713, 0.2646	14.20	0.00099	0.9510		30/39	26/39	25/39	27	22
TREE-C5-L1 VJ	7, 18, 23, 30, 40; 0.2770, 0.2134, 0.2500, 0.2426, 0.2408	13.93	0.00086	0.9520		17/25	15/25	15/25	18	14
TREE-C5-L1 UGM	7, 18, 21, 26, 43; 0.2770, 0.2134, 0.2535, 0.2577, 0.2456	13.78	0.00095	0.9510		17/25	15/25	14/25	17	14
TREE-C5-L1 UGM-G	7, 18, 17, 33, 30; 0.2770, 0.2134, 0.2648, 0.2514, 0.2452	13.62	0.00097	0.9510		16/25	15/25	13/25	17	13
VJ	4, 6, 16, 13, 11, 17, 14, 21, 24, 45; 0.45, 0.45, 0.48, 0.42, 0.50, 0.47, 0.45, 0.49, 0.50, 0.47	12.37	0.00051	0.9560						
TREE-C3-L1 VJ	4, 6, 16, 13, 11, 17, 14, 21, 24, 45; 0.45, 0.45, 0.48, 0.42, 0.50, 0.47, 0.45, 0.49, 0.50, 0.47	12.37	0.00051	0.9560		24/30	23/30	22/30	20	18
TREE-C3-L1 UGM	3, 6, 5, 10, 11, 14, 24, 15, 15, 18; 0.76, 0.34, 0.45, 0.50, 0.48, 0.45, 0.52, 0.50, 0.54, 0.51	11.51	0.00088	0.9510		19/30	17/30	17/30	20	18
TREE-C3-L1 UGM-G	4, 6, 5, 14, 8, 12, 17, 16, 14, 34; 0.45, 0.44, 0.59, 0.51, 0.49, 0.52, 0.50, 0.47, 0.54, 0.45	10.70	0.00090	0.9510		24/30	23/30	23/30	17	15
TREE-C3-L2 VJ	3, 6, 5, 10, 11, 15, 17, 30, 40, 19; 0.76, 0.34, 0.45, 0.50, 0.48, 0.45, 0.46, 0.48, 0.49, 0.50	11.59	0.00068	0.9550		49/84	44/84	42/84	42	34
TREE-C3-L2 UGM	3, 6, 5, 10, 11, 14, 24, 15, 15, 18; 0.76, 0.34, 0.45, 0.50, 0.48, 0.45, 0.52, 0.50, 0.54, 0.51	11.51	0.00088	0.9510		44/84	39/84	35/84	39	31
TREE-C3-L2 UGM-G	4, 6, 5, 14, 8, 12, 17, 16, 14, 34; 0.45, 0.44, 0.59, 0.51, 0.49, 0.52, 0.50, 0.47, 0.54, 0.45	10.70	0.00090	0.9510		56/84	55/84	48/84	31	23
TREE-C5-L1 VJ	4, 6, 16, 13, 11, 17, 14, 21, 24, 45; 0.45, 0.45, 0.48, 0.42, 0.50, 0.47, 0.45, 0.49, 0.50, 0.47	12.37	0.00051	0.9560		38/50	37/50	36/50	30	26
TREE-C5-L1 UGM	3, 6, 5, 10, 11, 14, 24, 15, 15, 18; 0.76, 0.34, 0.45, 0.50, 0.48, 0.45, 0.52, 0.50, 0.54, 0.51	11.51	0.00088	0.9510		33/50	30/50	29/50	29	25
TREE-C5-L1 UGM-G	4, 6, 5, 14, 8, 12, 17, 16, 14, 34; 0.45, 0.44, 0.59, 0.51, 0.49, 0.52, 0.50, 0.47, 0.54, 0.45	10.70	0.00090	0.9510		40/50	39/50	36/50	25	21

3.4 Detection Examples

Figure 2 presents examples of face detection obtained by classifiers with the lowest expected number of extracted features trained by tree search procedure with FAR values set to $A = 10^{-3}$ and $A = 10^{-4}$ respectively. Sensitivity for both classifiers was set to $D = 0.95$. The decision threshold for classifiers was set to 1.0.

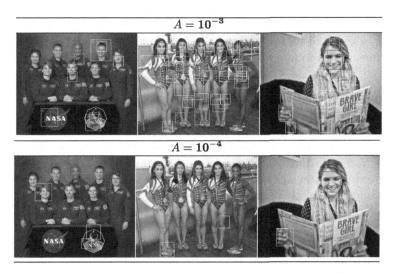

Fig. 2. "Face detection": detection examples. False alarms marked in yellow. (Color figure online)

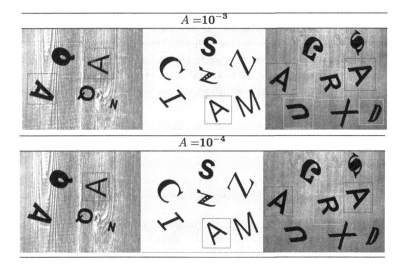

Fig. 3. "Synthetic A letters": detection examples.

Similarly, Fig. 3 shows examples of letter 'A' detection obtained by classifiers with the lowest expected number of extracted features trained by tree search procedure with FAR values set to $A = 10^{-3}$ and $A = 10^{-4}$ respectively. Sensitivity for both classifiers was also set to $D = 0.95$. The decision threshold for classifiers was set to 0.0.

It should be recalled that the cascades obtained by both approaches (Dijkstra's algorithm and branch-and-bound method) are identical, the difference is in the time needed to find solution.

4   Conclusions

Training a cascade of classifiers is a difficult optimization problem that, in our opinion, should be always carried out with a primary focus on the expected number of extracted features. This quantity reflects directly how fast an operating cascade is. In our previous research we propose to use a tree search-based training that allows to 'track' more than one variant of a cascade. This approach can be computationally expensive, but we have managed to reduce it with suitable branch-and-bound techniques.

The use of Dijkstra's algorithm for a tree search allows for further reduction of training times, especially in the case of complex trees with a large number of nodes. In addition, the aforementioned algorithm guarantees that the final cascade (returned as an outcome) always has the lowest expected value among all cascades contained in a given tree for the imposed settings of branching factor and tree depth (since priority queue in our approach sets nodes priority based on expected number of features in associated cascade and by taking into account that adding new stage to cascade can only increase expected value we can notice that after finding first cascade that satisfy requirements there is no possibility of finding other cascade that will improve the current expected value). This property is not necessarily satisfied in our previous method that use the approximate pruning.

References

1. Andrean, M.N., et al.: Comparing haar cascade and yoloface for region of interest classification in drowsiness detection. Jurnal Media Informatika Budidarma **8**(1), 272–281 (2024)
2. Bernabe, J.A., Dylan O. Catapang, J., Valiente, L.D.: Application of haar cascade classifier for kitchen safety monitoring. In: 2023 9th International Conference on Advanced Computing and Communication Systems (ICACCS), vol. 1, pp. 343–348 (2023)
3. de Campos, T.E., et al.: Character recognition in natural images. In: Proceedings of the International Conference on Computer Vision Theory and Applications, Lisbon, Portugal, pp. 273–280 (2009)
4. Di, J., Gao, R.: Research on railway transportation route based on dijkstra algorithm. In: Atiquzzaman, M., Yen, N., Xu, Z. (eds.) Big Data Analytics for Cyber-Physical System in Smart City, pp. 255–260. Springer, Singapore (2022). https://doi.org/10.1007/978-981-16-7469-3_28
5. Dijkstra, E.W.: A note on two problems in connexion with graphs. Numer. Math. **1**(1), 269–271 (1959)

6. Gharge, S., Patil, A., Patel, S., Shetty, V., Mundhada, N.: Real-time object detection using haar cascade classifier for robot cars. In: 2023 4th International Conference on Electronics and Sustainable Communication Systems (ICESC), pp. 64–70 (2023)
7. Ghorbel, A., Ben Amor, N., Abid, M.: GPGPU-based parallel computing of viola and jones eyes detection algorithm to drive an intelligent wheelchair. J. Signal Process. Syst. **94**(12), 1365–1379 (2022)
8. Rasolzadeh, B., et al.: Response binning: improved weak classifiers for boosting. In: IEEE Intelligent Vehicles Symposium, pp. 344–349 (2006)
9. Sychel, D., Klęsk, P., Bera, A.: Branch-and-bound search for training cascades of classifiers. In: Krzhizhanovskaya, V.V., et al. (eds.) ICCS 2020. LNCS, vol. 12140, pp. 18–34. Springer, Cham (2020). https://doi.org/10.1007/978-3-030-50423-6_2
10. Sychel, D., Klęsk, P., Bera, A.: Relaxed per-stage requirements for training cascades of classifiers. In: Frontiers in Artificial Intelligence and Applications – ECAI 2020, vol. 325, pp. 1523–1530. IOS Press (2020)
11. University of Essex: Face Recognition Data (1997). https://cswww.essex.ac.uk/mv/allfaces/faces96.html. Accessed 11 May 2019
12. Usilin, S.A., Slavin, O.A., Arlazarov, V.V.: Memory consumption and computation efficiency improvements of viola-jones object detection method for remote sensing applications. Pattern Recognit. Image Anal. **31**(3), 571–579 (2021)
13. Viola, P., Jones, M.: Rapid object detection using a boosted cascade of simple features. In: Conference on Computer Vision and Pattern Recognition (CVPR 2001), pp. 511–518. IEEE (2001)
14. Viola, P., Jones, M.: Robust real-time face detection. Int. J. Comput. Vision **57**(2), 137–154 (2004)
15. Wang, J., Yu, X., Zong, R., Lu, S.: Evacuation route optimization under real-time toxic gas dispersion through CFD simulation and Dijkstra algorithm. J. Loss Prev. Process Ind. **76**, 104733 (2022)
16. Zhou, Y., Huang, N.: Airport AGV path optimization model based on ant colony algorithm to optimize Dijkstra algorithm in urban systems. Sustain. Comput. Inform. Syst. **35**, 100716 (2022)

Optimizing Prescribed Burn Risk Management: A Computational and Economic Modeling Approach Using QUIC FIRE Simulations

Yeshvant Matey[1]($\boxtimes$), Raymond de Callafon[1], and Ilkay Altintas[2]

[1] Department of MAE, UCSD, La Jolla, CA 92093-0411, USA
[2] San Diego Supercomputer Center, La Jolla, CA 92093-0505, USA
{ymatey,callafon,ialtintas}@ucsd.edu

Abstract. This paper introduces a computational framework for optimizing vegetation removal, modelled via so-called blackline or fireline widths, to enhance efficiency and cost-effectiveness of a prescribed burn for planned reduction of vegetation density. The QUIC FIRE simulation tool is employed to conduct simulations across fireline widths ranging from 8 to 24 m in 2-m increments within a strategically chosen burn unit that covers the usecase of a wildland urban interface located around the region of Auburn, CA. Through visual analysis and quantitative cost function assessment, incorporating polynomial fit and the Broyden-Fletcher-Goldfarb-Shanno (BFGS) algorithm within a basin-hopping framework, an optimal fireline width is computed that minimizes costs, efforts and the risk of fire escapes. Findings indicate that strategic adjustments in fireline widths significantly influence the success of prescribed burns, underscoring the value of advanced simulation and optimization techniques. This work provides a foundational framework for subsequent studies, advocating for the development of dynamic, adaptive models that are scalable across varied ecological and geographical settings. Contributions extend to a computational and economic perspective on sustainable risk mitigation, underlining the pivotal influence of technology and advanced modeling in the evolution of prescribed burn strategies.

Keywords: Prescribed burns · wildfire management · computational optimization · QUIC FIRE simulation · blackline width

1 Introduction

In recent years, the United States has faced an alarming increase in wildfire incidents, marked by their growing intensity and frequency [5]. This increasing trend requires the urgent need for innovative fire management strategies to mitigate wildfire risks effectively. Among these strategies, prescribed burning has emerged as one of the prominent techniques [8]. By intentionally applying controlled fires, this planned treatment of vegetated land aims to manage vegetation

L. Franco et al. (Eds.): ICCS 2024, LNCS 14833, pp. 266–280, 2024.
https://doi.org/10.1007/978-3-031-63751-3_18

and reduce the accumulation of fuel, thereby protecting assets and lives from the devastating impact of uncontrolled wildfires [16].

The work in this paper utilizes QUIC FIRE, a simulation tool that is designed for modeling the spread and behavior of prescribed fire scenarios [13]. QUIC FIRE possesses the capability to accurately predict the behavior of fires across a spectrum of weather and terrain conditions. Understanding QUIC FIRE simulations requires knowledge of concepts such as the Burn Unit [11] and the Acceptable Fire Boundary. During Prescribed Burns, even in the most favorable weather conditions, there exists a possibility of fire breaching its Acceptable Fire Boundary [22]. This makes fuel management, particularly fuel removal, a key strategy to avoid fire escape [3]. A fundamental practice in fuel management during prescribed burns is the creation of a back line or fire line – a cleared strip of land intended to act as a barrier to stop the fire from spreading [7,24].

Various fuel removal strategies include ground-based mechanical whole tree, manual whole-tree, manual log, cut-to-length, and cable-operated systems. Following the comparative 2007-dollar analysis in [4], ground-based mechanical whole tree removal is the most cost-effective, averaging $0.1532 per m^2, while manual whole tree removal costs around $0.2526 per m^2. Cable systems, designed for challenging steep terrains, represent the higher end of the cost spectrum, with expenses starting at $0.6955 per m^2 for whole tree removal. Even with these capabilities, the risk of fire escape and consequent damage persists [23]. Also, excessive fuel removal, i.e. creating wide firelines to prevent the escape of fire can limit the use of prescribed burns [9,10,12]. This leads to a critical question: What is the most cost-effective fireline width that balances the costs of fuel removal and the potential damages from fires breaching prescribed boundaries?

This research explores the use of prescribed burns also called Interface burns [3,6] when practiced near residential areas to reduce wildfire risks, emphasizing the crucial role of fire lines, termed "black lines" in QUIC FIRE simulations. This paper provides a comprehensive cost analysis of fuel removal by analyzing various fuel removal methods and their associated costs and the characterization of costs associated with fire escape driven by fire simulation using QUIC-FIRE. It then formulates an optimization problem to define the optimal width of fuel removal, thereby guiding the creation of economically viable firelines.

By laying the groundwork for optimal trade-off concepts, this research establishes a foundational framework for advancing multi-dimensional control strategies in fireline creation and fuel management. This research enhances wildfire management discourse, providing insights and strategies to protect communities and ecosystems from wildfires.

2 QUIC-FIRE and Urban Interface Use Case

QUIC-Fire [11] is a simulation tool designed for the planning of prescribed burns. It offers a solution to understand and predict the complex interactions of fire with the atmosphere without the need for high-performance computing resources. This model integrates the 3-D rapid wind solver QUIC-URB [15,18] with a physics-based cellular automata fire spread model called FIRE-CA [1,2]. QUIC-Fire employs a probabilistic approach to simulate fire spread, where energy packets are dispersed based on local environmental conditions, yielding variable fire behavior across simulation iterations. This tool aids in prescribed fire management by enabling planners to design burn plans with complex ignition patterns and assess fire-atmosphere interactions.

To demonstrate the cost-benefit analysis of creating a fireline and taken into account the costs related to fire escape, this paper applies QUIC-Fire to a region around Auburn, CA. A visualization of this use case is summarized Fig. 1 and was chosen following an examination of Wildland Urban Interface (WUI) [17] zones where urban developments meet or intermingle with wildland vegetation. This analysis stands as the inaugural application of the proposed framework, constituting a novel initiative within the field. Therefore, it delves into uncharted areas, devoid of the comparative frameworks often available through established benchmarks or historical studies.

Fig. 1. Image describing real world zoning of use case scenario near Auburn, CA

The selected area is characterized by a sloped terrain interspersed with residential zones within the buffer area. These topographical and residential features underscore the necessity for rigorous fire containment strategies, making it a good case for investigation. To fully comprehend the analysis and the model itself, it is crucial to highlight and summarize the Quic-Fire simulation parameters used to describe the use case in Fig. 1:

- Wind Speed: Established at 32.18 kph to accurately simulate conditions that can affect the spread of fire.
- Wind Direction: Considered to be moving from left to right in Figure 1 or West to East i.e. 270°, with 15° perturbations.

- Topography and Vegetation: Incorporation of the terrain's features and existing vegetation information for prediction of fire dynamics.
- Ignition Pattern: The simulation adopts a head fire ignition pattern, chosen for its representation of how fire fronts typically advance with the wind. This approach is specifically relevant to the wind conditions and topographical layout of the study area as illustrated in Fig. 1.

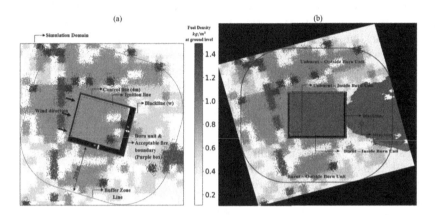

Fig. 2. (a) Image describing QUIC FIRE simulation domain, burn unit, acceptable fire boundary, buffer line, black line, and control line (b) Visual representation of the simulation domain depicting the blackline, the burnt and unburnt areas within and outside the burn unit, and the location of structures.

To further complete the terminology associated to QUIC-Fire simulations, terms such as Simulation Domain, Burn Unit, Buffer Zone, Acceptable Fire Boundary, Black Line, and Width are defined followed by a visual representation provided in Fig. 2(a). In particular, the following definitions are used:

- Simulation Domain: This term refers to the spatial extent under consideration for the simulation, capturing the prescribed burn area and its surroundings. It encompasses the entire landscape within which the fire dynamics are modeled.
- Burn Unit: This area denotes the specific parcel of land designated for the implementation of the prescribed burn within the simulation domain.
- Buffer Zone: A strategically established margin encircling the burn unit, designed to mitigate the risk of fire escape beyond the intended boundaries.
- Acceptable Fire Boundary: This indicates the maximum permissible boundary for fire spread, established to ensure the containment of the burn within predefined limits.
- Control Line and Black Line Width: This line serves as a narrow boundary along the wind inflow direction, established through the removal of fuel. In this specific scenario, its width is set to be 4 m. Additionally, the 'Black Line' is implemented as a preventive measure, placed strategically in alignment

with the wind direction and characterized by an absence (active removal) of fuel/vegetation. The 'Width' of the Black Line is subject to variation, serving as the key variable for prediction within this use case.

3   Cost Function Analysis

Spatial concepts of burned and none-burned areas within the simulation domain, and construction of the blackline, are graphically represented in Fig. 2(b). This visualization distinctly identifies regions inside and outside the burn unit, demarcates the blackline, and pinpoints structures, thereby offering clarity on the areas considered when calculating the associated costs of fuel removal. This section introduces a framework for systematically estimating blackline construction costs and penalties incurred when fires breach acceptable boundaries.

3.1   Fuel Removal Cost

The cost of fuel removal is comprehensively evaluated across the designated burn unit. This burn unit is depicted as a grid G with dimensions $x \times y$, which effectively represent the total area of the burn unit. For the purpose of this model, the following parameters are established:

– Base cost of fuel removal, $f(a, b)$, based on the slope at cell (a, b):

$$f(a, b) = \begin{cases} 0.3, & \text{if slope at } (a, b) < 40\% \\ 1.02, & \text{if slope at } (a, b) \geq 40\% \end{cases} \tag{1}$$

Ground-based mechanical and manual whole tree removal costs averaged \$0.1532 and \$0.2526 per m^2, respectively. For slopes above 40%, cable systems incurred costs starting at \$0.6955 per m^2 [4]. After adjusting for inflation from 2007–2023 [21], fuel removal costs for creating a fuel-free blackline zone are calculated. For slopes less than 40%, costs are averaged to \$0.3 per m^2. For steeper slopes over 40%, costs rise to \$1.02 per m^2.

– Fuel density factor at cell (a, b):

$$\text{factor}(a, b) = 1 + \frac{\text{fuel density at } (a, b) - \text{average fuel density}}{\text{average fuel density}} \tag{2}$$

The core of the model calculates the costs of constructing blacklines to a specified width, W, integrating both terrain-adjusted costs and fuel density factors. Formulas for blackline construction costs, such as $C^{(1)}_{\text{blackline}}(w)$, are aligned with prevailing wind directions. This method accounts for wind's impact on fire spread, facilitating strategic blackline placement and dimensioning for enhanced efficacy.

– Option 1 (North/South):

$$C_{\text{blackline}}^{(1)}(W) = \sum_{w=1}^{W} \left(\sum_{i=1}^{x} \mathbf{1}_{\{i+w \leq x\}} \cdot f(i+w, w) \cdot \text{factor}(i+w, w) \right.$$

$$\left. + \sum_{j=1}^{y} \mathbf{1}_{\{j+w<y, j+w \neq w\}} \cdot f(w, j+w) \cdot \text{factor}(w, j+w) \right) \quad (3)$$

where $\mathbf{1}_{\{.\}}$ is the indicator function, equaling 1 if the condition is true, and 0 otherwise.
– Options 2 to 4 are defined for East/West, South/North, West/East directions.

Control lines further enhance fire containment. The costs for these control lines, C_{control}, are established for a fixed width of $W_c = 4$, with adjustments based on their orthogonal orientation to the blackline. This method establishes a buffer zone along directions not influenced by wind.

$$C_{\text{control}} = (\text{Appropriate formula based on fixed width}$$
$$\text{and orthogonal orientation to blackline}) \quad (4)$$

The culmination of this model is the integration of the costs associated with both the creation of a blackline and control line constructions costs. The final cost, $C_{\text{fuelremoval}}(W)$, is found by combining blackline and control line costs up to width W and given by

$$C_{\text{fuelremoval}}(W) = C_{\text{control}} + C_{\text{blackline}}^{(o)}(W) \quad (5)$$

where o is the choice of option based on wind direction.

Following the formulation of the final cost $C_{\text{fuelremoval}}(W)$, Fig. 3 provides a visual representation of the fuel removal cost dynamics. This figure demonstrates an approximate logarithmic growth pattern between cumulative costs and width.

3.2 Penalty Costs for Fire Escape

Penalty costs are associated with both fire escape, bus also for vegetation not burned within the burn unit. For evaluation of penalty costs, the simulation domain S is defined within the bounds $(k_{\max}, l_{\max})$ of the pixels of the image. Each pixel within this domain is denoted by the coordinates (k, l), where k ranges from 1 to $k_{\max}$ and l from 1 to $l_{\max}$. The burn unit is represented by a grid G which is a subset of S, with G embedded within the simulation domain S.

To compute the penalty costs, several functions are introduced. The function $U(k, l)$ is defined to determine the membership of a pixel within the grid G:

$$U(k, l) = \begin{cases} 1 & \text{if } (k, l) \in G \\ 0 & \text{if } (k, l) \notin G \end{cases} \quad (6)$$

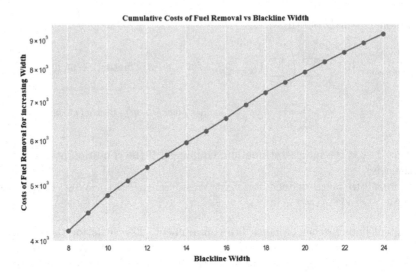

Fig. 3. Comparative analysis of fuel removal costs for blackline and control line constructions up to width W, y-axis is on log scale.

Furthermore, a fuel density evaluation function $B(k,l)$ is defined to ascertain whether a pixel is burnt or unburnt:

$$B(k,l) = \begin{cases} 1 & \text{if fuel density at } (k,l) = 0 \\ 0 & \text{if fuel density at } (k,l) \geq 0 \end{cases} \tag{7}$$

This function is applied over the entire simulation domain S. Additionally, the distance function $R(k,l)$ calculates the Euclidean distance from the center of grid G to each pixel in S:

$$R(k,l) = \sqrt{(x_c - k)^2 + (y_c - l)^2} \tag{8}$$

where (x_c, y_c) represent the coordinates of the center of G.

The penalty for land loss due to fire escape, denoted as $C_{\text{landloss}}(w)$, is calculated across the simulation domain S, encapsulated within the bounds $(k_{\max}, l_{\max})$, and is expressed as:

$$C_{\text{landloss}}(w) = \sum_{k=1}^{k_{\max}} \sum_{l=1}^{l_{\max}} f_w(k,l) \tag{9}$$

The function $f_w(k,l)$ is defined based on the conditions of pixel membership within grid G and the fuel density status of each pixel:

$$f_w(k,l) = \begin{cases} 0 & \text{if } U(k,l) = 1 \text{ and } B(k,l) = 1 \\ C & \text{if } U(k,l) = 1 \text{ and } B(k,l) = 0 \\ \alpha R(k,l) & \text{if } U(k,l) = 0 \text{ and } B(k,l) = 1 \\ 0 & \text{if } U(k,l) = 0 \text{ and } B(k,l) = 0 \end{cases} \tag{10}$$

In this context, α, representing the land value constant at \$1.09 per cell, adjusts the penalty for burnt pixels beyond the designated boundary based on distance. This constant is derived from averaging the 2022 values of United States farm real estate (\$0.9460 per m^2) and cropland (\$1.2572 per m^2) [20]. While this analysis utilizes average prices for concept demonstration, the framework allows for the adjustment of these prices to reflect specific regional values for more localized analyses.

To address the loss of important structures, the model incorporates a secondary penalty function. Each critical structure within the domain is indexed by t and located at pixel (k_t, l_t). The vicinity of each structure incurs an enhanced penalty, modeled by the function $d_t(k, l)$, which is a decreasing exponential function of the distance from the pixel to the structure:

$$d_t(k, l) = e^{-(\beta\sqrt{(k_t-k)^2+(l_t-l)^2})} \tag{11}$$

The decay constant β governs the influence range of the structural penalty in this case 30 m [19], decreasing with increasing distance from the structure. The cumulative penalty impact due to structures at pixel (k, l) is the sum of penalties from all structures t:

$$d(k, l) = \sum_t d_t(k, l) \tag{12}$$

The total penalty cost associated with structural loss, $C_{\text{structureloss}}(w)$, is then given by the sum of individual penalties across the domain:

$$C_{\text{structureloss}}(w) = \sum_{k=1}^{k_{\max}} \sum_{l=1}^{l_{\max}} p_w(k, l) \tag{13}$$

where $p_w(k, l)$ is defined as:

$$p_w(k, l) = \begin{cases} 0 & \text{if } U(k, l) = 1 \\ \gamma d(k, l) & \text{if } U(k, l) = 0 \text{ and } B(k, l) = 1 \\ 0 & \text{if } U(k, l) = 0 \text{ and } B(k, l) = 0 \end{cases} \tag{14}$$

The variable γ represents a constant associated with structural loss. Consequently, the total penalty cost is the sum of the land loss and structural loss costs:

$$C_{\text{penalty}}(w) = C_{\text{structureloss}}(w) + C_{\text{landloss}}(w) \tag{15}$$

Finally, the comprehensive cost for a prescribed burn at blackline width w encompasses both fuel removal and penalty costs:

$$C_{\text{burncost}}(w) = C_{\text{fuelremoval}}(w) + \theta * C_{\text{penalty}}(w) \tag{16}$$

where, θ is the dollar conversion constant that converts the penalty factor to dollar value. In this case, it has been assumed to be 1.

To illustrate the end result of the cosst analysis, Fig. 4(a) shows the spatial distribution of penalty cost factor, while Fig. 4(b) refers to costs associated with land and structure loss due to the fire escape illustrated earlier in Fig. 2(b).

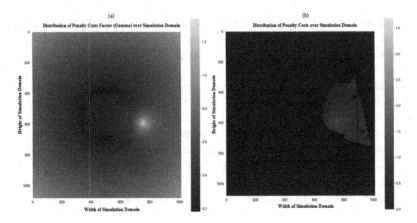

Fig. 4. Visual representation of the simulation domain depicting (a) cost factor and (b) costs calculated from fire escape and structure damage

4 Methodology for Optimization

In the optimization methodology, the cost function $C_{\text{burncost}}(w)$ in (16) can be calculated across a discrete set of potential widths (w) for blackline construction. In addition, variability in anticipated wind speed and wind direction can be used to find either an average or worst-case cost $C_{\text{burncost}}(w)$ and an optimal blackline width could then be chosen by the minimization of $C_{\text{burncost}}(w)$ over w. However, such optimization inevitably tends towards a trivial solution of maximal fuel removal (maximum w) when costs associate to fire escape are large. While such an approach effectively mitigates fire spread, it may not represent the most efficient allocation of resources.

To counter the propensity for trivial solutions, the optimization methodology integrates an additional effort cost component, analogous to regularization parameters in optimization problems. The additional effort cost is given $C_{\text{effort}}(w) = \delta \cdot (w - w_0)^2$, where w again represents the blackline width, w_0 is set to a value of 8 and δ is a coefficient adjusting the impact on the total cost and is set to a value of $\$135/m^2$. This incorporation of an effort cost is intended to ensure a more balanced approach to evaluating strategies advocating for a more judicious and effective deployment of fire management resources.

The cost evaluation has been summarized in Fig. 5. Specifically, Fig. 5(a) illustrates the penalty costs arising from prescribed burn escapes, with cost assessments per blackline width refined by introducing wind direction variations within a $270 \pm 15°$ range. Furthermore, Figure 5(b) reveals a non-linear correlation between total cost and blackline width, suggesting that optimal width aligns with a strategy of maximal fuel removal. Figure 5(c) displays the total costs.

$$C_{\text{fuelremoval}}(W) = C_{\text{control}} + C_{\text{blackline}}^{(o)}(W) + C_{\text{effort}}(W) \qquad (17)$$

including the additional effort cost, o is the choice of option based on wind direction.

Given the inherent variability in the simulation outcomes, a deterministic approach to determine the optimal blackline width is not advisable. Instead, a 6th order polynomial function is fitted to the data to capture the nuanced trends and provide a smooth approximation of costs over varying blackline widths. This is mathematically expressed as $C(W) = a_0 + a_1 W + a_2 W^2 + \cdots + a_6 W^6$, where $C(W)$ denotes the cost associated with a blackline of width W, and $a_0, a_1, \ldots, a_6$ are the coefficients determined by the fitting process. The rationale for selecting a polynomial function lies in its flexibility to model the non-linear relationships between the control measures and the resultant costs, thus ensuring a robust optimization process. This polynomial model offers a continuous, differentiable function, crucial for identifying cost minima with gradient-based optimization techniques.

The BFGS algorithm [14] emerges as a formidable quasi-Newton method for local minimization within the basin-hopping framework, crucial for the refined optimization of the cost function $C(w)$. It iteratively adjusts an estimation H of the inverse Hessian matrix alongside the position vector w, in accordance to the update rule $w_{\text{new}} = w_{\text{old}} - \alpha H \nabla f(w_{\text{old}})$, where α denotes the step size, ascertained via a line search adhering to Wolfe conditions. Simultaneously, the approximation H is updated by $H_{\text{new}} = (I - \rho s y^T) H_{\text{old}} (I - \rho y s^T) + \rho s s^T$, with $s = w_{\text{new}} - w_{\text{old}}$ and $y = \nabla f(w_{\text{new}}) - \nabla f(w_{\text{old}})$, encapsulating the changes in position and gradient. This method is integral to the basin-hopping strategy, which conducts a comprehensive exploration for the global minimum of $C(w)$, navigating beyond the barriers of local minima. This methodology offers a systematic way to determine the optimal blackline width, minimizing costs and efforts in prescribed burn operations.

5   Use Case Results

In the results section, findings from the simulation exercises conducted within the Auburn, CA, burn unit scenario are presented. Utilizing the QUIC FIRE simulation tool, a series of simulations were executed with specific parameters: a wind speed set at 32.18 kph, oriented from east to west, and a head fire ignition pattern. These parameters were chosen to simulate environmental conditions influencing wildfire spread in the designated area. An ensemble methodology, involving variations in the width of the black line across different simulations, evaluates the implications on cost and the efficacy of burn operations.

Figure 6 provides a visual sequence that conveys the outcomes of the ensemble simulations conducted within the context of the Auburn, CA, burn unit scenario, ranging from 8 m to 24 m in increments of 2 m, for 270° wind angle. Each depicted scenario offers a distinct visual representation, shedding light on the effect of blackline widths on prescribed burn control. This analysis necessitates minimum black line width to prevent fire escape, while also noting that widening the line beyond this minimum would incur excessive costs.

276   Y. Matey et al.

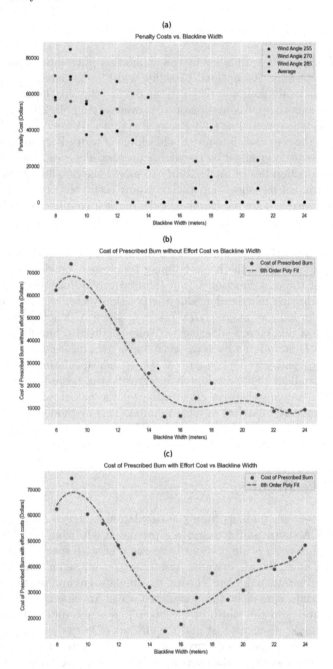

Fig. 5. Cost of Prescribed burn vs width of blackline

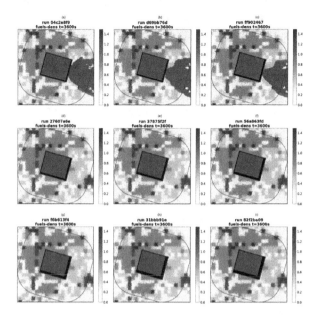

Fig. 6. Impact of Blackline widths on Fire Control in Auburn, CA. Panels a to i display the effects of blackline widths ranging from 8 m to 24 m in 2 m increments on prescribed burn spread for 270° wind angle

Building on the visual insights, Table 1 provides a quantitative assessment, detailing the financial impacts of different blackline widths. This methodical tabulation of burn costs linked to each width level offers a structured basis for informed decision-making in fire management strategies. It is important to consider that altering the blackline width may inadvertently increase local wind speeds, potentially escalating the risk of fire escape, even with wider blacklines.

Finally, Fig. 5 illustrates the crucial balance between the cost of prescribed burns and blackline width, pinpointing the optimal width for minimizing cost, efforts and fire damage. The comprehensive analysis and simulations detailed above culminate in identifying an optimal blackline width of 16 m as the most effective in balancing cost reduction, efforts and enhancing fire control measures for the prescribed burn scenario in Auburn, CA.

Table 1. Comprehensive Cost Analysis for Various Blackline widths Under Simulated Wind Conditions in Auburn, CA.

Blackline width (w)	Penalty Cost Average $(\theta \cdot C_{\text{penalty}})$	Cost of Fuel Removal $(C_{\text{fuelremoval}})$	Total Cost (C_{burncost})
8	58024.45	4285.95	62310.41
10	54271.44	6008.94	60280.38
12	39474.50	8737.10	48211.61
14	19334.57	12543.40	31877.97
16	0	17471.02	17471.02
18	13783.91	23588.35	37372.27
20	0.00	30693.04	30693.04
22	0.00	38905.80	38905.80
24	0.00	48205.51	48205.51

6   Conclusions

Embarking on an exploratory journey to unravel the complexities inherent in the planning of prescribed burns, this study has meticulously investigated the optimization of blackline widths, aiming to strike a delicate balance between operational efficacy and cost-efficiency. Through an exhaustive series of simulation exercises conducted within the Auburn, CA, burn unit scenario using the QUIC FIRE simulation tool, we have shed light on the intricate relationship between blackline width, fire control, and economic factors.

The ensemble methodology, exploring varying blackline widths has unequivocally demonstrated that strategic modifications in width can profoundly affect fire behavior and containment. This finding emphasizes the critical need for added precision in the planning and execution phases of prescribed burns. The simulations not only provided a qualitative insight into the impact of varying blackline widths on fire control efforts, but also paved the way for a quantitative analysis that meticulously outlined the financial ramifications.

Leveraging polynomial interpolation and the BFGS optimization algorithm within a basin-hopping framework facilitates the computation of optimal blackline width as a compromise between cost, effort and potential damage from fire escapes. This fusion of computational modeling and economic analysis marks a introduction of a framework in enhancing the effectiveness of prescribed burns as a wildfire management strategy, highlighting the potential of simulation-based optimization in achieving an optimal balance between operational efficiency and cost-effectiveness.

Reflecting on the objectives outlined in the introduction, this research has successfully:

– Established a foundational framework with an aim to advance multi-dimensional control strategies for fireline creation and fuel management.
– Developed a comprehensive cost-based analysis framework for prescribed burns, aiding in resource allocation and strategic planning.
– Introduced financial modeling within prescribed burn management domain by harnessing the capabilities of the QUIC FIRE simulation, thereby refining planning and cost estimation efforts.

This research highlights key opportunities for future work, including the use of computational and data driven technologies to aid prescribed burn strategies. Additionally, a more detailed economic model incorporating broader cost factors could improve prescribed burn planning. By integrating advanced simulations with economic analysis, this study contributes significantly to developing more effective and sustainable wildfire management strategies.

References

1. Achtemeier, G.L.: Field validation of a free-agent cellular automata model of fire spread with fire-atmosphere coupling. Int. J. Wildland Fire **22**(2), 148–156 (2012)
2. Achtemeier, G.L., Goodrick, S.A., Liu, Y.: Modeling multiple-core updraft plume rise for an aerial ignition prescribed burn by coupling daysmoke with a cellular automata fire model. Atmosphere **3**(3), 352–376 (2012)
3. Ager, A.A., Vaillant, N.M., Finney, M.A.: A comparison of landscape fuel treatment strategies to mitigate wildland fire risk in the urban interface and preserve old forest structure. Forest Ecol. Manag. **259** (2010). https://www.sciencedirect.com/science/article/pii/S0378112710000514
4. Arriagada, R.A., Cubbage, F.W., Abt, K.L., Huggett Jr, R.J.: Estimating harvest costs for fuel treatments in the west. Forest Prod. J. **58**(7), 24–30 (2008). https://www.proquest.com/scholarly-journals/estimating-harvest-costs-fuel-treatments-west/docview/214625926/se-2
5. Dennison, P.E., Brewer, S.C., Arnold, J.D., Moritz, M.A.: Large wildfire trends in the western united states, 1984-2011. Geophys. Res. Lett. **41**(8), 2928–2933 (2014). https://doi.org/10.1002/2014GL059576. https://agupubs.onlinelibrary.wiley.com/doi/abs/10.1002/2014GL059576
6. Gibbons, P., et al.: Land management practices associated with house loss in wildfires. PLoS ONE **7** (2012). https://api.semanticscholar.org/CorpusID:4677687
7. Graham, R.T., McCaffrey, S., Jain, T.B.: Science basis for changing forest structure to modify wildfire behavior and severity (2004). https://api.semanticscholar.org/CorpusID:140583948
8. Hesseln, H.: The economics of prescribed burning: a research review. Forest Sci. **46**(3), 322–334 (2000). https://doi.org/10.1093/forestscience/46.3.322
9. Ingalsbee, T.: Fuelbreaks for wildland fire management: a moat or a drawbridge for ecosystem fire restoration? Fire Ecol. **1**(1), 85–99 (2005). https://doi.org/10.4996/fireecology.0101085
10. Lemons, R.E., Prichard, S.J., Kerns, B.K.: Evaluating fireline effectiveness across large wildfire events in north-central washington state. Fire Ecol. **19**(1) (2023). https://doi.org/10.1186/s42408-023-00167-6

11. Linn, R., et al.: Quic-fire: a fast-running simulation tool for prescribed fire planning. Environ. Model. Softw. **125**, 104616 (2020). https://doi.org/10.1016/j.envsoft. 2019.104616

12. Macarena, O., y Silva Francisco, R., Ramón, M.J.: Fireline production rate of handcrews in wildfires of the spanish mediterranean region. Int. J. Wildland Fire **32**, 1503–1514 (2023). https://doi.org/10.1071/WF22087

13. Mercer-Smith, J.A.: Quic-fire: 3D fire-atmosphere feedback model for wildland fire management (2020). https://doi.org/10.2172/1650598. https://www.osti.gov/biblio/1650598

14. Nocedal, J.: Theory of algorithms for unconstrained optimization. Acta Numerica **1**, 199–242 (1992). https://doi.org/10.1017/S0962492900002270

15. Pardyjak, E.R., Brown, M.: QUIC-URB v. 1.1: Theory and user's guide. Los Alamos National Laboratory, Los Alamos, NM (2003)

16. Penman, T.D., et al.: Prescribed burning: how can it work to conserve the things we value? Int. J. Wildland Fire **20**, 721–733 (2011). https://api.semanticscholar.org/CorpusID:128823603

17. Radeloff, V.C., Helmers, D.P., Mockrin, M.H., Carlson, A.R., Hawbaker, T.J., Martinuzzi, S.: The 1990-2020 wildland-urban interface of the conterminous united states - geospatial data (2023). https://doi.org/10.2737/RDS-2015-0012-4.   https://usfs.maps.arcgis.com/home/item.html?id=454bddfa18784660a472685ac7965881. Financial support was provided by the USDA Forest Service under the National Fire Plan and by USDA Forest Service, Northern Research Station. Data obtained from the U.S. Census and USGS National Land Cover Data

18. Singh, B., Hansen, B.S., Brown, M.J., Pardyjak, E.R.: Evaluation of the QUIC-URB fast response urban wind model for a cubical building array and wide building street canyon. Environ. Fluid Mech. **8**, 281–312 (2008)

19. Syphard, A.D., Brennan, T.J., Keeley, J.E.: The role of defensible space for residential structure protection during wildfires. Int. J. Wildland Fire **23**, 1165–1175 (2014)

20. United States Department of Agriculture, National Agricultural Statistics Service: Land values 2022 summary. Technical Report ISSN: 1949-1867, United States Department of Agriculture, National Agricultural Statistics Service (2022)

21. U.S. Bureau of Labor Statistics: Consumer price index for all urban consumers (CPI-U): All items in U.S. city average, all urban consumers, not seasonally adjusted (2023). https://www.bls.gov/cpi/. Series Id: CUUR0000SA0. Data extracted on: February 22, 2024 (3:41:28 AM). Not Seasonally Adjusted

22. Weir, J.R., et al.: Prescribed fire: understanding liability, laws and risk. Technical report, Oklahoma Cooperative Extension Service (2020)

23. Weir, J.R., et al.: Liability and prescribed fire: perception and reality. Rangeland Ecol. Manag. **72**(3), 533–538 (2019). https://doi.org/10.1016/j.rama.2018.11.010. https://www.sciencedirect.com/science/article/pii/S1550742418301283

24. Wollstein, K., O'Connor, C., Gear, J., Hoagland, R.: Minimize the bad days: wildland fire response and suppression success. Rangelands **44**(3), 187–193 (2022). https://doi.org/10.1016/j.rala.2021.12.006. https://www.sciencedirect.com/science/article/pii/S019005282100122X. Changing with the range: Striving for ecosystem resilience in the age of invasive annual grasses

SESP-SPOTIS: Advancing Stochastic Approach for Re-identifying MCDA Models

Bartłomiej Kizielewicz[1] , Jakub Więckowski[2] , and Wojciech Sałabun[2(✉)]

[1] Research Team on Intelligent Decision Support Systems, Department of Artificial Intelligence and Applied Mathematics, Faculty of Computer Science and Information Technology, West Pomeranian University of Technology in Szczecin, ul. Żołnierska 49, 71-210 Szczecin, Poland
bartlomiej-kizielewicz@zut.edu.pl
[2] National Institute of Telecommunications, Szachowa 1, 04-894 Warsaw, Poland
{j.wieckowski,w.salabun}@il-pib.pl

Abstract. Multi-Criteria Decision Analysis (MCDA) is an interdisciplinary field that addresses decision-making problems that involve multiple conflicting criteria. MCDA methods are widely applied in various domains, including medicine, management, energy, and logistics. Despite their widespread use, MCDA techniques continuously evolve to address emerging challenges. This paper presents a new method called Stochastic Expected Solution Point SPOTIS (SESP-SPOTIS), for re-identifying MCDA models. SESP-SPOTIS conducts a stochastic search for the Expected Solution Point (ESP) which is then utilized within the Stable Preference Ordering Towards Ideal Solution (SPOTIS) framework. The study delves into comprehensive investigations of MCDA model re-identification and examines how the updated model influences the ranking of analyzed alternatives. Furthermore, the experiments were divided into training sets and tests to evaluate the similarity of the proposed approach, using two rank correlation coefficients, namely Weighted Spearman (r_w) and Weighted Similarity (WS). The results demonstrate that SESP-SPOTIS effectively re-identifies updated models and provides additional information from analysis as an ESP, thereby broadening knowledge and understanding in the decision-making process of the analyzed problem. By integrating machine learning models and stochastic optimization techniques, SESP-SPOTIS contributes to advancing the methodologies for MCDA model re-identification.

Keywords: MCDA · Re-identification · SPOTIS · PSO · Reference point

1 Introduction

Multi-Criteria Decision Analysis (MCDA) is an interdisciplinary field that solves decision-making problems with multiple criteria, often in conflict with each other. These methods are used in various areas of science and practice, such

© The Author(s), under exclusive license to Springer Nature Switzerland AG 2024
L. Franco et al. (Eds.): ICCS 2024, LNCS 14833, pp. 281–295, 2024.
https://doi.org/10.1007/978-3-031-63751-3_19

as medicine [10], energy [7], or logistics [24]. The relatively widespread use of MCDA techniques leads to the continuous development of new approaches that respond to emerging challenges. A Rank Reversal (RR) phenomenon occurs when one alternative is removed or added from a decision problem, leading to a change in the ranking of some other alternatives. This is one of the challenges in such classic methods as Technique for Order of Preference by Similarity to Ideal Solution (TOPSIS) [6] and VIseKriterijumska Optimizacija I Kompromisno Resenje (VIKOR) [5].

In 2014, a method known as the Characteristic Objects Method (COMET) was developed to deal with the rank reversal phenomenon by using characteristic objects and fuzzy logic to evaluate alternatives [19]. In 2016, the Reference Ideal Method (RIM) was created to evaluate various decision-making alternatives using a reference point (Reference Ideal) [2]. Like COMET, the RIM method was designed to be robust to the problem of reverse rankings. This was followed in 2020 by the Stable Preference Ordering Towards Ideal Solution (SPOTIS) method, characterized by its simplicity and resilience to reverse rankings [4]. However, these methods shared a common factor with the SPOTIS method because they were also based on reference points. In [4], Dezert et al. proposed, in addition to the classical approach used in many MCDA methods, that decision-makers determine a reference point rather than just inferring based on ideal point alternatives. This point was called the Expected Solution Point (ESP). However, due to such a point, another challenge arises in re-identifying such models.

Re-identification involves trying to re-map an existing model. It is based on a learning procedure to rank the considered decision variants [17]. For this purpose, machine learning models or optimization methods are used allowing to adjust the optimal parameters for MCDA methods to get the model as close to the original one as possible. The process of re-identifying decision models involves training the models. This process aims to find a new set of parameters that minimize prediction error or maximize the similarity of rankings. In this way, it is possible to effectively rebuild an unknown model to evaluate previously unconsidered alternatives in the context of a given decision problem. In addition, retrieving the lost parameters of the decision model, such as the weights of the criteria, will allow a more accurate analysis of the results obtained.

There are several previous works in which the focus was on the re-identification of MCDA models. In [13], the authors used an approach based on the re-identification of criteria weights using stochastic methods such as Genetic Algorithm (GA), Differential Evolution (DE), and Particle Swarm Optimization (PSO). On the other hand, the study [11] presented the possibility of re-identifying the MCDA model using the Stochastic IdenTifiCation Of Models (SITCOM) approach, which determined preference values for characteristic objects. The SITCOM method was further developed into the Dynamic SITCOM (D-SITCOM) approach [12], which additionally considered the search for characteristic values when creating characteristic objects. In addition, machine learning models such as MultiLayer Perceptron (MLP) [14] are also used for the re-identification process. However, no research has been conducted on the re-identification process using the expected solution point, which reflects the desired outcome gathered from decision-makers. Notably, the ESP concept has demonstrated high efficacy in addressing multi-criteria problems through its personalized assessment approach [21]. Consequently, the recognition of this information gap served as the primary motivation for undertaking the present study.

This paper combines a stochastic method named Particle Swarm Optimization (PSO) to search for a single reference point determined as ESP defined in the SPOTIS method. This approach becomes an alternative to the ISP-SPOTIS method, where difficulties are encountered in re-identifying the decision-makers preferences due to how the ISP is determined based on the boundaries of the decision problem. By combining the PSO and ESP-SPOTIS methods, it is possible to direct the re-identification process toward personalized decision-making, increasing the effectiveness of the decision models' determination. Moreover, the novelty of our study is the demonstration of the possibility of updating an already re-identified model, which does not apply to the previously introduced approaches. The work's main contribution is the possibility of finding an expectation point to create an analogous reference model. Moreover, the proposed approach provides additional information as an ESP that can help interpret the decision-maker's preferences.

The paper is organized as follows. The Sect. 2 presents a literature review on how decision-makers convey their preferences to MCDA methods. The Sect. 3 presents preliminaries of the SPOTIS approach used and the correlation coefficients of the rankings. The Sect. 4 presents a proposed approach for re-identifying MCDA models called SESP-SPOTIS and research on this approach. The Sect. 5 presents conclusions and future research directions.

2   Literature Review

Increasingly, research on Multi-Criteria Decision Analysis (MCDA) methods has focused on approaches related to processing decision-makers' preferences. In practice, the most commonly used technique is to assign weights to criteria that add up to unity. However, arbitrarily determining these values by decision-makers can be problematic. Consequently, criteria comparison methods are used to process their knowledge. The classic technique is the Analytic Hierarchy Process (AHP) method, which compares criteria using Saaty's scale. However, this approach is fraught with the paradox of reverse rankings, which can make its application unstable with constantly changing sets of alternatives.

However, the technique of comparing criteria is being developed in new approaches. The Best Worst Method (BWM) [18] focuses on comparing the best (best) and worst (worst) criteria in the context of the Multi-Criteria Decision Analysis problem under consideration. A similar approach is used in the Full Consistency Method (FUCOM) [15], which uses linear programming to determine criterion weights. However, these methods of determining weights only partially solve the problem of communicating the decision-maker's preferences.

Preference function modeling is an alternative method for conveying a decision-maker's preferences. One of the classic approaches that focuses on this methodology is the Preference Ranking Organization METHod for Enrichment of Evaluations (PROMETHEE) [1]. It allows the use of a variety of preference functions that can better reflect the specific preferences of the decision-maker. The PROMETHEE method evaluates alternatives based on the decision-maker's predefined preference functions, which allows for a more nuanced analysis than a simple assignment of weights. However, as in the case of transferring preferences through weights, the use of preference functions also does not guarantee that the MCDA technique is entirely immune to reverse rankings of alternatives.

Given the ability of decision makers to express preferences, the problem of re-identifying Multi-Criteria Decision Analysis (MCDA) models arises. There are many approaches to conveying preferences by the decision maker, such as using weights, pairwise comparisons, preference functions, or characteristic scores. This

Table 1. Overview of MCDA methods based on reference points.

Name	Acronym	Reference point	Ref.
Evaluation based on Distance from Average Solution	EDAS	Average solution for each criterion (AV)	[8]
Measurement of Alternatives and Ranking according to COmpromise Solution	MARCOS	Ideal solution (AI) Anti-ideal solutuion (AAI)	[23]
COmbinative Distance-based ASsessment	CODAS	Negative Ideal Solution (NS)	[9]
Stable Preference Ordering Towards Ideal Solution	SPOTIS	Ideal Solution Point (ISP) Expected Solution Point (ESP)	[4]
Technique for Order of Preferenceby Similarity to Ideal Solution	TOPSIS	Positive Ideal Solution (PIS) Neagative Ideal Solution (NIS)	[6]
VIekriterijumsko KOmpromisno Rangiranje	VIKOR	Maximum criteria values (f^*) Minimum criteria values (f^-)	[5]
Characteristic Objects METhod	COMET	Characteristic objects (CO)	[19]
Preference Ranking On the Basis of Ideal-average Distance	PROBID	Average solution ($\overline{A}$) Positive ideal solutions ($PISs$) Negative ideal solutions ($NISs$)	[26]
Compromise Ranking of Alternatives from Distance to Ideal Solution	CRADIS	Ideal Solution (TI) Anti-ideal Solution (TAI)	[16]
Election based on Relative Value Distances	ERVD	Reference points (μ) Positive ideal solution (PIS) Negative ideal solution (NIS)	[22]
Reference Ideal Method	RIM	Reference ideal (s_j)	[2]

paper will focus on one aspect of re-identifying MCDA models, specifically the search for a reference point.

More recent research has focused on exploring the possibilities offered by the decision maker's transfer of one or more reference points aimed at optimizing and better adjusting the decision model. Introducing a reference point enables the implementation of nonlinear preference modeling, which is a significant step forward in personalizing the decision-making process. Among the most prominent methods using reference points are the approaches already mentioned, such as Reference Ideal Method (RIM), Stable Preference Ordering Towards Ideal Solution (SPOTIS), and Characteristics Objects Method (COMET). In addition, it is worth noting that there is a wide range of methods based on reference points, as illustrated by the Table 1.

3   Preliminaries

3.1   SPOTIS

The SPOTIS method, which stands for Stable Preference Ordering Towards Ideal Solution, differs from other MCDA approaches by incorporating the notion of reference objects. While methods like TOPSIS and VIKOR establish these objects based on a decision matrix, SPOTIS requires explicitly defined data boundaries. By employing this strategy to outline the domain of the decision problem, it becomes feasible to stabilize the ranking of alternatives towards the Ideal Solution Point (ISP), thus mitigating the occurrence of the Rank Reversal (RR) paradox. Typically, the ISP is determined based on the values associated with each criterion type (e.g., cost or profit). Establishing data boundaries is a crucial step in the initial phase of applying this method. For each criterion C_j, it's essential to select the maximum S_j^{max} and minimum S_j^{min} bounds. The Ideal Solution Point S_j^* is defined as $S_j^* = S_j^{max}$ for profit criteria and $S^*j = Sj^{min}$ for cost criteria.

Additionally, as illustrated in [4], the SPOTIS approach allows for the utilization of any Expected Solution Point (ESP) in place of ISP. When employed, ESP generates a ranking specific to the subjectively chosen solution, proving beneficial when decision-makers seek a solution tailored precisely to a particular problem rather than a general ideal solution within the problem domain. The Expected Solution Point values S_j^* should be chosen within the defined bounds of the decision problem $[S_j^{min}, S_j^{max}]$. Subsequently, the ESP vector S^* should replace ISP in Eq. (2) during the SPOTIS calculation procedure [21]. Below, the subsequent steps of the SPOTIS method are outlined.

Step 1. Definition of decision matrix.

The decision matrix describes the characteristics of considered alternatives under selected criteria. The formal notation of the decision matrix could be defined as (1):

$$\mathbf{X} = \begin{bmatrix} x_{11} & x_{12} & \cdots & x_{1j} & \cdots & x_{1m} \\ x_{21} & x_{22} & \cdots & x_{2j} & \cdots & x_{2m} \\ \vdots & \vdots & \cdots & \vdots & \cdots & \vdots \\ x_{i1} & x_{i2} & \cdots & x_{ij} & \cdots & x_{im} \\ \vdots & \vdots & \cdots & \vdots & \cdots & \vdots \\ x_{n1} & x_{n2} & \cdots & x_{nj} & \cdots & x_{nm} \end{bmatrix} \tag{1}$$

where x_{ij} is the attribute value of the i-th alternative for j-th criterion.

Step 2. Calculation of the normalized distances from ISP (2):

$$d(A_i, S_j^*) = \frac{|S_{ij} - S_j^*|}{|S_j^{max} - S_j^{min}|} \tag{2}$$

Step 3. Calculation of weighted normalized distances $d(A_i, S^*) \in [0,1]$ as (3):

$$d(A_i, S^*) = \sum_{j=1}^{N} w_j d_{ij}(A_i, S_j^*) \tag{3}$$

Step 4. Ranking calculation.

The final ranking of alternatives should be determined based on the value of $d(A_i, S^*)$. Better evaluated decision variants have smaller values of $d(A_i, S^*)$, thus should be placed higher in the ranking.

3.2   Weighted Spearman's Correlation Coefficient

The Weighted Spearman's correlation coefficient (r_W), proposed by [3], extends the traditional Spearman coefficient by integrating weights. It computes the correlation between two rankings, both of size N, where x_i represents the position in the first ranking and y_i indicates the position in the second ranking (4).

$$r_W = 1 - \frac{6 \cdot \sum (x_i - y_i)^2 ((n - x_i + 1) + (n - y_i + 1))}{n \cdot (n^3 + n^2 - n - 1)} \tag{4}$$

3.3   WS Rank Similarity Coefficient

The Weighted Similarity (WS), proposed by [20], presents itself as an asymmetric measure of ranking similarity. In contrast to conventional methods, it places particular emphasis on alterations occurring at the top of rankings. Consequently, the correlation undergoes a significant decrease if, for example, there is an interchange between the first and last positions. The WS rank similarity coefficient is confined within the interval [0, 1], where zero indicates uncorrelated rankings, while a value of one signifies identical rankings. Computed for two rankings, x_i and y_i, both with a size of N, the similarity value is determined as (5):

$$WS = 1 - \sum \left(2^{-x_i} \frac{|x_i - y_i|}{\max |x_i - 1|, |x_i - N|} \right) \tag{5}$$

4   The Proposed Approach

This section will present a proposed approach called Stochastic Expected Solution Point SPOTIS (SESP-SPOTIS). This approach aims to identify the expected solution point using a stochastic method. It incorporates both the simplicity of the SPOTIS method, as discussed in the Sect. 3.1, and the re-identification capabilities of the decision model. In the context of possible nonlinearity associated with selecting the expected solution point, the re-identification approach becomes the answer to the problem of identifying nonlinear decision models. In the case of the present work, the stochastic technique used for this purpose is Particle Swarm Optimization (PSO). The main steps of this approach can be presented as follows:

Step 1. Select a dataset. The dataset should include the decision matrix of the given decision problem. Additionally, it should contain information such as weights criteria vectors (W), a criteria types vector (T), and a ranking vector (R).

Step 2. Select a stochastic optimization method. In this step, choose a stochastic method for solving the optimization problem and select its parameters. In this paper, Particle Swarm Optimization (PSO) was selected as the stochastic optimization method. PSO is a popular technique for stochastic optimization problems and allows for modeling flexible objective functions [27]. Its algorithm is presented in Algorithm 1.

Algorithm 1. Particle Swarm Optimization (PSO)

1: Initialize $X, V, P, P_{\text{value}}, G, G_{\text{value}}$
2: **for** $iteration \leftarrow 1$ to $max\_iterations$ **do**
3:     **for** each particle i **do**
4:         Update velocity and position
5:         Clip position to within bounds
6:         Evaluate objective function
7:         **if** $f_i > P_{\text{value}}[i]$ **then**
8:             Update personal best
9:         **end if**
10:     **end for**
11:     Update global best
12:     **if** convergence_criteria_met() **then**
13:         **break**
14:     **end if**
15: **end for**
16: **Output:** G, G_{value}

Step 3. Model training. Training the model is done using the stochastic optimization algorithm and the fitness function, which can be determined as presented in Algorithm 2.

Algorithm 2. Fitness Function

1: **procedure** FITNESS(*solutions*):
2:     base.esp ← *solutions*
3:     *preference* ← base(C, *solutions*, T)
4:     **return** rw(base.rank(*preference*), R)
5: **end procedure**

4.1   Re-identification - Exemplary Study Case

This paper presents a simple example of expert model re-identification in the context of multi-criteria decision analysis. Suppose a particular set of evaluated samples is evaluated by a multi-criteria decision analysis model. However, the multi-criteria decision analysis model itself and its parameters are unknown. In this case, when wanting to re-identify such a model, the SPOTIS method can be used. The SPOTIS method has two possible modeling routes. The first is to use an ISP point created based on the values derived from the model boundaries for each criterion. In such a model, the only possible representation of the decision-maker's preferences is expressed in terms of weights. The second way of modeling is to use the ESP point, which the decision-maker chooses. This point determines the most preferred alternative that we would like to obtain. Therefore, by taking the second route, it is possible to find an ESP point that can produce a similar model to the reference one.

Assume that there is a multi-criteria evaluation problem, where a decision matrix containing attribute values for 10 alternatives against two criteria is available. Suppose the evaluation model, which has not yet been applied, runs from 0 (smallest cutoff value) to 1 (largest cutoff value) for each criterion. In order to find an Expected Solution Point that will help us create a similar evaluation model, a PSO method will be used. This method is widely used for optimization problems and involves simulating the behavior of a swarm of particles in the solution search space to find the optimal point. To implement the PSO algorithm, an implementation of the MealPy library will be used, which provides ready-made tools for solving optimization problems. The version of the MealPy library used in this implementation of the library is 2.5.4 [25]. For this particular problem at hand, it is necessary to define a fitness function that will evaluate the quality of solutions in the context of ESP search. A detailed description of the PSO method and the fitness function can be found above. It is also worth mentioning the selected parameters for the PSO algorithm, such as the number of particles ($pop_{size} = 20$), weight coefficients ($c_1 = c_2 = 2.05$), maximum number of iterations ($epoch = 1000$), which have been adjusted for our specific problem in order to achieve optimal results.

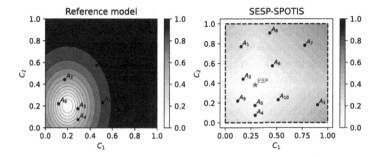

Fig. 1. Reference model and model re-identified using the SESP-SPOTIS approach.

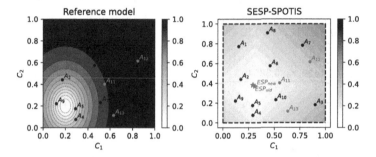

Fig. 2. Reference model and updated SEPS-SPOTIS model with 3 additional alternatives.

Using Fig. 1, the unidentified reference model and the SEPS-SPOTIS derived model are shown. It can be seen that in the case of the re-identification of such a model, the distribution of preferences was similarly mapped. In the case of the present re-identification, the rankings of the alternatives are very close to each other, where a Weighted Spearman correlation coefficient value of 0.96915 and a WS coefficient value of 0.94152 were obtained. On the other hand, referring to the distance of the found ESP from the extremum of the present unknown model, it is 0.20341. Such re-identification makes it possible to obtain similar evaluations to the reference evaluations, by which it is also possible to evaluate a new set of alternatives.

In the case of re-identification, it is also possible to update the SESP using the newly evaluated samples. Suppose that in addition to the set on which working also obtained additional evaluated alternatives. In this case, another re-identification of the reference model is possible. The SEPS retrieved in the previous re-identification is set as the search's starting point. Then the whole process looks identical, however, in the case of this training, the training set is a set of 13 alternatives, not 10 alternatives. The Fig. 2 shows the result of the re-identification of the model. The shift of the retrieved ESP relative to the old ESP can be seen. The distance between the two is 0.02232. In addition, the distance between the extremum of the unknown reference model and the retrieved new

ESP has been reduced and is 0.20159. This means that re-identifying the model using the SESP-SPOTIS approach makes it possible to increase the accuracy.

Using Fig. 3, a comparison of the re-identified models on a test set of 10 alternatives is presented. Figure 3a shows the comparison of the re-identified model on the 10 test alternatives with the reference model, while Fig. 3b shows the comparison of the updated re-identified model by 3 test alternatives with the reference model. It can be observed that for the updated re-identified model with 3 alternatives, the correlation is higher with the ranking obtained with the reference model than for the re-identified model with 10 alternatives. For the model updated with 3 alternatives, the correlation of the ranking with the reference model was based on the index r_w, a value of 0.93829, while using the coefficient WS, a value of 0.95179. Referring to the re-identified model on 10 alternatives without updating, its ranking correlation with the reference model is $r_w = 0.88540$ and $WS = 0.91429$.

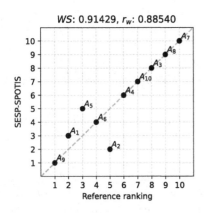 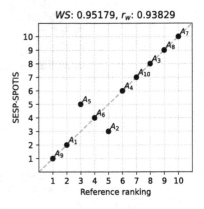

(a) Re-identification on 10 alternatives.   (b) Re-identification update with 3 alternatives.

Fig. 3. Comparison on a test set of rankings from the unknown and re-identified model (SESP-SPOTIS).

4.2 Effectiveness

This section will focus on a simulation study related to testing the proposed SESP-SPOTIS approach. The unknown MCDA model shown in the previous section is the reference model in this section. Figure 4 shows a two-dimensional histogram through which the distribution of ESP values for the two considered criteria can seen. This distribution was obtained from simulations for 1000 random training sets of size 10 alternatives. However, it can observed that the values of searched ESPs concentrated around the extremum of the unknown MCDA model. Extremes are often searched for due to the distribution of alternatives existing in the training set. The mean values of ESPs that were searched are for the criterion $ESP_{C_1}^{Mean} = 0.19164$ and the criterion $ESP_{C_2}^{Mean} = 0.30834$. The standard deviation among values for criterion $ESP_{C_1}^{STD} = 0.11450$ and criterion $ESP_{C_2}^{STD} = 0.13275$.

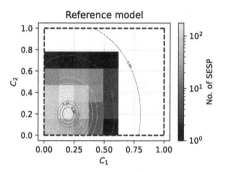

Fig. 4. Two-dimensional histogram of ESP values for the two considered criteria.

The SESP-SPOTIS approach was also compared with the ISP-SPOTIS approach, which uses an ideal point formed from the model's outliers. The Fig. 5 shows the distributions of the values of correlation coefficients for comparisons of rankings derived from the reference model and the SESP-SPOTIS and ISP-SPOTIS models. A significant difference in the accuracy of the approaches can be observed. The SESP-SPOTIS model searches for the expected point based on the evaluated alternatives, and its accuracy is much higher than that of the ISP-SPOTIS model, which is based only on the model boundaries. The average values of WS and r_w obtained by the SESP-SPOTIS model are 0.95492 and 0.94101, while the average values of WS and r_w obtained by the ISP-SPOTIS model are 0.46773 and -0.24206, respectively.

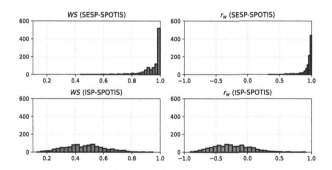

Fig. 5. Distributions of obtained values of correlation coefficients from comparisons of rankings with the reference model and SPOTIS models.

Algorithm 3. Demonstration of the experimental procedure.

1: **Input:** $N \leftarrow 1000$
2: **Input:** $criteria \leftarrow 2$
3: **for** num_alt in $[5, 10, 25, 50, 100]$ **do**
4:     coefficients_train,coefficients_test $\leftarrow$ Coefficients()
5:     **for** $i = 1$ to N **do**
6:         **1) Train results:**
7:         $alternatives\_train \leftarrow$ generate_alternatives(num_alt, $criteria$)
8:         $weights \leftarrow$ equal_weights($alternatives\_train$)
9:         $reference\_train \leftarrow$ ref($alternatives\_train$)
10:        $model \leftarrow$ sesp_spotis($alternatives\_train$, $reference\_train$, $weights$)
11:        $result \leftarrow$ model.pred($alternatives\_train$)
12:        coefficients_train.add(WS($result$, $reference\_train$))
13:        coefficients_train.add(rw($result$, $reference\_train$))
14:        **2) Test results:**
15:        $alternatives\_test \leftarrow$ generate_alternatives(10, $criteria$)
16:        $result \leftarrow$ model.pred($alternatives\_test$)
17:        $reference\_test \leftarrow$ ref($alternatives\_test$)
18:        coefficients_test.add(WS($result$, $reference\_test$))
19:        coefficients_test.add(rw($result$, $reference\_test$))
20:    **end for**
21:    save_coefficients(coefficients_train)
22:    save_coefficients(coefficients_test)
23: **end for**

Simulation studies were also performed for 1000 random sets of alternatives, where the number of alternatives $5, 10, 25, 50, 100$ was taken as the training set, while the test set of alternatives had a fixed size of 10 alternatives. The possibility of re-identifying the reference model was tested using the SESP-SPOTIS approach, where the comparison of the accuracy of the re-identified model with the reference model was verified using the correlation coefficients of WS and rw rankings. The Algorithm 3 shows the procedure of the conducted study.

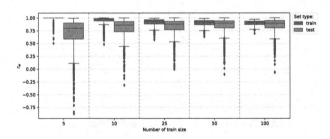

Fig. 6. Distribution of weighted Spearman rank correlation coefficient values (r_w) for reference model comparisons and SESP-SPOTIS for training and test sets.

Figure 6 displays results from the reference and re-identified models using the SESP-SPOTIS approach, represented by the weighted Spearman coefficient (r_w). Across test datasets, the average r_w was 0.689 to 0.842, and for training datasets, it was 0.899 to 0.978. Standard deviation varied from 0.161 to 0.309 for tests and 0.044 to 0.076 for training, with smaller datasets showing higher variability. The r_w range was -0.877 to -0.069 (test) and 0.479 to 1.000 (training). The analysis indicates strong correlation between models and reference, but smaller datasets had increased variability and reduced correlation.

Figure 7 presents results from the reference and re-identified models using the SESP-SPOTIS approach, shown by the WS correlation coefficient. For test sets, the mean WS ranged from 0.789 to 0.880, and for training sets, from 0.905 to 0.982. Standard deviation was between 0.084 to 0.163 for tests and 0.048 to 0.054 for training, with higher variability in smaller sets. Minimum WS values were 0.137 to 0.461 (test) and 0.643 to 0.672 (training), and maximum values were 0.884 to 1.000 (test) and 0.986 to 1.000 (training). The re-identified models showed high correlation with the reference, but smaller datasets had increased variability and decreased correlation.

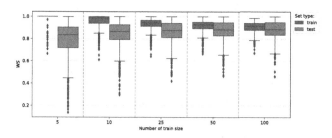

Fig. 7. Distribution of weighted rank correlation coefficient values (WS) for reference model comparisons and SESP-SPOTIS for training and test sets.

5   Conclusions

This paper introduces a novel re-identification approach for Multi-Criteria Decision Analysis (MCDA) models using the stochastic Expected Solution Point search method in SPOTIS. It explores updating re-identified models with new evaluations and evaluates accuracy through training and test sets using rank correlation coefficients r_w and WS. Results show the SESP-SPOTIS method identifies models close to the reference and provides valuable expected solution points for interpreting decision maker preferences.

Future research should consider developing this approach, taking into account uncertainty based on fuzzy sets and their generalizations. Other stochastic search methods should also be explored in the context of re-identifying MCDA models. In addition, it is worth considering the possibility of searching for multiple points of the expected solution under ESP.

Acknowledgments. The work was supported by the National Science Centre 2021/41/B/HS4/01296.

References

1. Brans, J.P., Nadeau, R., Landry, M.: L'ingénierie de la décision. Elaboration d'instruments d'aide à la décision. La méthode PROMETHEE. In l'Aide à la Décision: Nature, Instruments et Perspectives d'Avenir, pp. 183–213 (1982)
2. Cables, E., Lamata, M.T., Verdegay, J.L.: RIM-reference ideal method in multi-criteria decision making. Inf. Sci. **337**, 1–10 (2016). https://doi.org/10.1016/j.ins.2015.12.011
3. Dancelli, L., Manisera, M., Vezzoli, M.: On two classes of weighted rank correlation measures deriving from the Spearman's ρ. In: Giudici, P., Ingrassia, S., Vichi, M. (eds.) Statistical Models for Data Analysis, pp. 107–114. Springer, Heidelberg (2013). https://doi.org/10.1007/978-3-319-00032-9_13
4. Dezert, J., Tchamova, A., Han, D., Tacnet, J.M.: The SPOTIS rank reversal free method for multi-criteria decision-making support. In: 2020 IEEE 23rd International Conference on Information Fusion (FUSION), pp. 1–8. IEEE (2020). https://doi.org/10.23919/FUSION45008.2020.9190347
5. Duckstein, L., Opricovic, S.: Multiobjective optimization in river basin development. Water Resour. Res. **16**(1), 14–20 (1980). https://doi.org/10.1029/WR016i001p00014
6. Hwang, C.L., Yoon, K.: Methods for multiple attribute decision making. In: Hwang, C.L., Yoon, K. (eds.) Multiple Attribute Decision Making, pp. 58–191. Springer, Heidelberg (1981). https://doi.org/10.1007/978-3-642-48318-9_3
7. Kaya, İ, Çolak, M., Terzi, F.: Use of MCDM techniques for energy policy and decision-making problems: a review. Int. J. Energy Res. **42**(7), 2344–2372 (2018). https://doi.org/10.1002/er.4016
8. Keshavarz Ghorabaee, M., Zavadskas, E.K., Olfat, L., Turskis, Z.: Multi-criteria inventory classification using a new method of evaluation based on distance from average solution (EDAS). Informatica **26**(3), 435–451 (2015). https://doi.org/10.15388/Informatica.2015.57
9. Keshavarz Ghorabaee, M., Zavadskas, E.K., Turskis, Z., Antucheviciene, J.: A new combinative distance-based assessment (CODAS) method for multi-criteria decision-making. Econ. Comput. Econ. Cybern. Stud. Res. **50**(3) (2016)
10. Khan, I., Pintelon, L., Martin, H.: The application of multicriteria decision analysis methods in health care: a literature review. Med. Decis. Making **42**(2), 262–274 (2022). https://doi.org/10.1177/0272989X211019040
11. Kizielewicz, B.: Towards the identification of continuous decisional model: the accuracy testing in the SITCOM approach. Procedia Comput. Sci. **207**, 4390–4400 (2022). https://doi.org/10.1016/j.procs.2022.09.502
12. Kizielewicz, B., Jankowski, J.: Dynamic SITCOM: an innovative approach to re-identify social network evaluation models. In: 2023 18th Conference on Computer Science and Intelligence Systems (FedCSIS), pp. 1023–1027. IEEE (2023). https://doi.org/10.15439/2023F539
13. Kizielewicz, B., Paradowski, B., Więckowski, J., Sałabun, W.: Identification of weights in multi-cteria decision problems based on stochastic optimization (2022)

14. Kizielewicz, B., Wieckowski, J., Jankowski, J.: MLP-COMET-based decision model re-identification for continuous decision-making in the complex network environment. In: 2023 18th Conference on Computer Science and Intelligence Systems (FedCSIS), pp. 591–602. IEEE (2023). https://doi.org/10.15439/2023F5438

15. Pamučar, D., Stević, Ž, Sremac, S.: A new model for determining weight coefficients of criteria in MCDM models: full consistency method (FUCOM). Symmetry **10**(9), 393 (2018). https://doi.org/10.3390/sym10090393

16. Puška, A., Stević, Ž., Pamučar, D.: Evaluation and selection of healthcare waste incinerators using extended sustainability criteria and multi-criteria analysis methods. Environ. Dev. Sustain. 1–31 (2022). https://doi.org/10.1007/s10668-021-01902-2

17. Qin, Z., et al.: Are neural rankers still outperformed by gradient boosted decision trees? In: International Conference on Learning Representations (2020)

18. Rezaei, J.: Best-worst multi-criteria decision-making method. Omega **53**, 49–57 (2015). https://doi.org/10.1016/j.omega.2014.11.009

19. Sałabun, W.: The characteristic objects method: a new distance-based approach to multicriteria decision-making problems. J. Multi-Criteria Decis. Anal. **22**(1–2), 37–50 (2015). https://doi.org/10.1002/mcda.1525

20. Sałabun, W., Urbaniak, K.: A new coefficient of rankings similarity in decision-making problems. In: Krzhizhanovskaya, V.V., et al. (eds.) ICCS 2020. LNCS, vol. 12138, pp. 632–645. Springer, Cham (2020). https://doi.org/10.1007/978-3-030-50417-5_47

21. Shekhovtsov, A.: Decision-making process customization by using expected solution point. Procedia Comput. Sci. **207**, 4556–4564 (2022). https://doi.org/10.1016/j.procs.2022.09.519

22. Shyur, H.J., Yin, L., Shih, H.S., Cheng, C.B.: A multiple criteria decision making method based on relative value distances. Found. Comput. Decis. Sci. **40**(4), 299–315 (2015). https://doi.org/10.1515/fcds-2015-0017

23. Stević, Ž, Pamučar, D., Puška, A., Chatterjee, P.: Sustainable supplier selection in healthcare industries using a new MCDM method: measurement of alternatives and ranking according to compromise solution (MARCOS). Comput. Ind. Eng. **140**, 106231 (2020). https://doi.org/10.1016/j.cie.2019.106231

24. Ulutaş, A., Karaköy, Ç.: An analysis of the logistics performance index of EU countries with an integrated MCDM model. Econ. Bus. Rev. **5**(4), 49–69 (2019). https://doi.org/10.18559/ebr.2019.4.3

25. Van Thieu, N., Mirjalili, S.: MEALPY: an open-source library for latest metaheuristic algorithms in python. J. Syst. Architect. **139**, 102871 (2023). https://doi.org/10.1016/j.sysarc.2023.102871

26. Wang, Z., Rangaiah, G.P., Wang, X.: Preference ranking on the basis of ideal-average distance method for multi-criteria decision-making. Ind. Eng. Chem. Res. **60**(30), 11216–11230 (2021). https://doi.org/10.1021/acs.iecr.1c01413

27. Weiel, M., Götz, M., Klein, A., Coquelin, D., Floca, R., Schug, A.: Dynamic particle swarm optimization of biomolecular simulation parameters with flexible objective functions. Nat. Mach. Intell. **3**(8), 727–734 (2021). https://doi.org/10.1038/s42256-021-00366-3

BiWeighted Regular Grid Graphs—A New Class of Graphs for Which Graph Spectral Clustering is Applicable in Analytical Form

Mieczysław A. Kłopotek[✉], Sławomir T. Wierzchoń,
Bartłomiej Starosta, Dariusz Czerski, and Piotr Borkowski

Institute of Computer Science, Polish Academy of Sciences, ul. Jana Kazimierza 5,
01-248 Warsaw, Poland
{klopotek,stw,barstar,dcz,pbr}@ipipan.waw.pl
http://www.ipipan.waw.pl

Abstract. This paper presents a closed form solution to the eigen-problem of combinatorial graph Laplacian for a new type of regular grid graphs - biweighted grid graphs. Biweighted grid graphs differ from ordinary ones in that the weights along a single dimension are altering which adds complexity to the eigen-solutions and makes the graphs better testbed for potential applications.

Keywords: Artificial Intelligence · Machine Learning · Graph spectral clustering · Analytical solution of eigenvalue problem · Combinatorial Laplacian

1 Introduction

Present day artificial intelligence (AI) is linked tightly to machine learning (ML) solutions, enabling machines to learn from data and subsequently make predictions based on uncovered patterns in data. The ML tools used encompass unsupervised learning (or clustering) methods. One of the intensively developing clustering techniques is Graph Spectral Analysis, encompassing Graph Spectral Clustering (GSC). It works best for objects whose mutual relationships are described by a graph that connects them based on a similarity measure [16,20,21]. The concept of Graph Laplacians has been in use for a long time now. An extensive overview of early research can be found in the paper [13] by Merris from the year 1994. In this paper, the author uses *combinatorial Laplacian* $L = D - S$ of a graph G, where S is the adjacency matrix of G, and D is the (diagonal) degree matrix of G. A recent survey can be found in

Supported by Polish Ministry of Science.

the booklet [8] by Gallier, with a particular orientation towards applications in graph clustering[1].

While new GSA algorithms are developed, it is an important issue to have a sufficient amount of test data with clustering properties known in advance. One pathway to this goal is to identify graphs with analytical solutions. One such possibility, although quite simple, is provided by regular grid graphs. Currently, analytical solutions of the eigen-problem of grid Laplacians are known for unweighted grids, which can simulate structures where there are no intrinsic clusters, as well as for weighted grids with different weights in different directions that can simulate structures with known clusters (grid layers separated by links with the lowest weights).

In this paper, we present an analytical solution to the biweighted grid graph, that is one where along one direction the weights are alternating. This structure can be viewed as a better candidate for investigating clustering problems as a cluster can consist not of one but of two layers in a given direction.

The paper is structured as follows. In Sect. 2, a brief overview of related research is given. In Sect. 3, our solution to the biweighted grid graph problem is presented. In Sect. 4, some conclusions are presented.

2   Previous Work

Regular graph structures and their properties are of interest for several reasons, mostly for derivation of analytical graph properties [14]. In particular Ramachandran and Berman [15] exploit a priori knowledge of Laplacians of rectangular grid in investigations of properties of robotic swarms. Stankiewicz [18] discusses relation between the orientable genus of a graph (the minimum number of handles to be added to the plane in order to embed this graph without crossings) and the spectrum of its Laplacian. Cornelissen et al. [3] investigate gonality of curves using grid Laplacians. Merris [13] reviews numerous properties of grid graph Laplacians from the point of view of chemical applications. Cvetkovic et al. [4] write about application in mechanics (membrane vibration). They present explicit solutions to the combinatorial Laplacian eigen-problem (eigenvalues and eigenvectors) of the path-graph and as a consequence by the virtue of the construction of the two-dimensional grid graph as a product of path graphs also a solution to the rectangular grid graph combinatorial Laplacian. Cheung et al. [2] elaborate applications in image processing, with a particular interest in grid structures. Burden and Hedstrom [1] were interested in the eigenvalue spectrum of combinatorial Laplacians of grid graphs and derived them from the continuous Laplacian equations. Fiedler [7] established bounds for the second lowest eigenvalue of the combinatorial Laplacian (currently called Fiedler eigenvalue), while mentioning the formula of the Fiedler eigenvalue for the path graph. He also provided a theorem allowing to combine product graph eigenvalues from component graphs. Based on that paper, Anderson and Morey [9] derived explicit formulas

---

[1] For another overview of spectral clustering methods, see e.g. Chapter 5 of the book [21].

for combinatorial Laplacian eigenvalues of grid graphs, without referring to the continuous analogue.

Merris [13] recalls a number of previous results relevant to grid graphs, and also for other special graphs, like tree graphs. Spielman [17] proves explicit formulas for eigenvalues and eigenvectors for path graphs and grid graphs, without, however, caring about eigenvalues with multiplicity. Fan et al. [6] tackle the issue of signless Laplacians for bicyclic graphs. Edwards [5] found an explicit analytical solution to two-dimensional grid graph Laplacian eigenproblem showing the solution validity in case of eigenvalue ties. Kouachi [10] investigated eigenproperties of tridiagonal matrices, recalling multiple special cases, the topic relevant to path graphs.

The paper [11] presents analytical solutions to normalized and combinatorial Laplacians of grid graphs. The paper [12] investigates the relationship between various types of spectral clustering methods and their kinship to relaxed versions of graph cut methods, based on the closed (or nearly closed) form of eigenvalues and eigenvectors of unnormalized (combinatorial), normalized, and random walk Laplacian of multidimensional weighted and unweighted grids. It is demonstrated the GSA methods can be compared to (normalized) graph cut clustering only if the cut is performed to minimize the sum of weight square roots of removed edges, and not the sum of weights, as generally claimed. In the limit behaviour of combinatorial and normalized Laplacians was investigated showing that the eigenvalues of both converge to one another with increase of the number of nodes while their eigenvectors do not. It is also shown that the distribution of eigenvalues is not uniform in the limit, violating a fundamental assumption of compressive spectral clustering CSC [19].

3 Biweighted Grid Graphs

Let us define a one-dimensional biweighted grid graph as $G_{(n_1)(\mathfrak{w}_1)(\mathfrak{v}_1)}$ being a biweighted path graph of n_1 vertices with weight $\mathfrak{w}_1$ for any link in this graph from an odd node i to the even node $i+1$ and with weight $\mathfrak{v}_1$ for any link in this graph from an odd node i to the even node $i-1$. Further, let us define a d-dimensional biweighted grid graph as the weighted graph Cartesian product $G_{(n_1,...,n_d)(\mathfrak{w}_1,...,\mathfrak{w}_d)(\mathfrak{v}_1,...,\mathfrak{v}_d)}$ as $G_{(n_1,...,n_{d-1})(\mathfrak{w}_1,...,\mathfrak{w}_{d-1})(\mathfrak{v}_1,...,\mathfrak{v}_{d-1})} \times G_{(n_d)(\mathfrak{w}_d)(\mathfrak{v}_d)}$ where n_j is the number of layers in the jth dimension and $\mathfrak{w}_j, \mathfrak{v}_j$ are the alternating weights of links between layers in the jth dimension. Integer identities to nodes are assigned as in weighted grid graph.

3.1 Eigensolutions of Combinatorial Laplacians of Bi-weighted Grid Graphs - Path Graph Case

First, let us consider biweighted grid path, which is a one-dimensional graph. The biweighted grid graph treatment, like in the case of weighted grid graph is the product of biweighted grid paths. This means: For an even node e, its entries in the similarity matrix S are of the form $S_{e,e-1} = \mathfrak{w}$, $S_{e,e+1} = \mathfrak{v}$, and all other entries in the row are zeros (the column is accordingly filled). For an odd node o, its entries in the similarity matrix S are of the form $S_{o,o-1} = \mathfrak{v}$, $S_{o,o+1} = \mathfrak{w}$, and all other entries in the row are zeros (the column is accordingly filled). Therefore the Laplacian entries for an even non-border node e are of the form: $L_{e,e-1} = -\mathfrak{w}$, $L_{e,e} = \mathfrak{w} + \mathfrak{v}$, $L_{e,e+1} = -\mathfrak{v}$. The Laplacian entries for an odd non-border node o are of the form: $L_{o,o-1} = -\mathfrak{v}$, $L_{o,o} = \mathfrak{w} + \mathfrak{v}$, $L_{o,o+1} = -\mathfrak{w}$. The Laplacian entries for the first (hence border odd) node 1 are of the form: $L_{1,1} = \mathfrak{w}$, $L_{e,e+1} = -\mathfrak{w}$. The last node can be either even or odd. The Laplacian entries for the last even border node l_e are of the form: $L_{le,le-1} = -\mathfrak{w}$, $L_{le,le} = \mathfrak{w}$. The Laplacian entries for the last odd border node l_o are of the form: $L_{o,o-1} = -\mathfrak{v}$, $L_{o,o} = \mathfrak{v}$.

Other entries in the respective rows are zeros. Column entries follow the symmetry principle of L.

Let us illustrate the biweighted graph path with a small example, where $n = 5$, $\mathfrak{w} = 2$, $\mathfrak{v} = 3$. See Fig. 1.

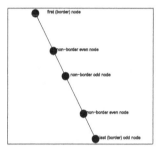

Fig. 1. An example of a biweighted path graph. Shorter edges illustrate higher edge weight ($\mathfrak{v} = 3$) and longer edges lower weight ($\mathfrak{w} = 2$)

The similarity matrix S and its combinatorial Laplacian L have the form

$$S = \begin{pmatrix} 0 & 2 & 0 & 0 & 0 \\ 2 & 0 & 3 & 0 & 0 \\ 0 & 3 & 0 & 2 & 0 \\ 0 & 0 & 2 & 0 & 3 \\ 0 & 0 & 0 & 3 & 0 \end{pmatrix}, \quad L = \begin{pmatrix} 2 & -2 & 0 & 0 & 0 \\ -2 & 5 & -3 & 0 & 0 \\ 0 & -3 & 5 & -2 & 0 \\ 0 & 0 & -2 & 5 & -3 \\ 0 & 0 & 0 & -3 & 3 \end{pmatrix}$$

Let us introduce some notation: Let n be the number of nodes on the path. If $\mathbf{v}$ is an eigenvector of L, then $\mathbf{v}' = L\mathbf{v}$. Let $\lambda_{s[z]}, \lambda_{c[z]}$ be the eigenvalues that we seek, where $z = 1, \ldots n$. The lower indexes s, c indicate which type of eigenvectors will be used, based on the sine (in case of s, see Eqs. (8, 9)) or cosine function (in case of c, see Eqs. (5, 6)). Let $\boldsymbol{\nu}_{s[z]}, \boldsymbol{\nu}_{c[z]}$ be the corresponding eigenvectors. Let $\nu_{s[z],[x]}, \nu_{c[z],[x]}$ with $x = 1, \ldots, n$ be the n components of the zth eigenvector $\boldsymbol{\nu}_{[z]}$.

Let

$$\delta_{\mathfrak{w}\mathfrak{v},[z]} = \frac{z2\pi}{n} \tag{1}$$

Let, if $\mathfrak{w} \geq \mathfrak{v}$ then

$$\delta_{\mathfrak{w},[z]} = \arctan \frac{\mathfrak{v}\sin(\delta_{\mathfrak{w}\mathfrak{v},[z]})}{\mathfrak{v}\cos(\delta_{\mathfrak{w}\mathfrak{v},[z]}) + \mathfrak{w}} \tag{2}$$

else

$$\delta_{\mathfrak{v},[z]} = \arctan \frac{\mathfrak{w}\sin(\delta_{\mathfrak{w}\mathfrak{v},[z]})}{\mathfrak{w}\cos(\delta_{\mathfrak{w}\mathfrak{v},[z]}) + \mathfrak{v}} \tag{3}$$

whereby the result of arctan is taken from the interval from the range $(-\frac{\pi}{2}, \frac{\pi}{2})$ so that the sine of that δ has the same sign as the sine of the right-hand side, whereby the other δ is computed from the relationship $\delta_{\mathfrak{w}\mathfrak{v},[z]} = \delta_{\mathfrak{w},[z]} + \delta_{\mathfrak{v},[z]}$.

We will prove in this section the following theorem.

Theorem 1. *The analytical solution of the eigenproblem for biweighted path graphs is of the following form.*

$$\lambda_{c[z]} = \mathfrak{w}(1 - \cos(\delta_\mathfrak{w})) + \mathfrak{v}(1 - \cos(\delta_\mathfrak{v})) \tag{4}$$

$$\nu_{c[z],[2x+1]} = \cos(\frac{1}{2}\delta_\mathfrak{v} + x\delta_{\mathfrak{w}\mathfrak{v}}) \tag{5}$$

$$\nu_{c[z],[2x]} = \cos(\frac{1}{2}\delta_\mathfrak{v} + (x-1)\delta_{\mathfrak{w}\mathfrak{v}} + \delta_\mathfrak{w}) \tag{6}$$

$$\lambda_{s[z]} = \mathfrak{w}(1 + \cos(\delta_\mathfrak{w})) + \mathfrak{v}(1 + \cos(\delta_\mathfrak{v})) \tag{7}$$

$$\nu_{s[z],[2x+1]} = -(-1)^{nodeid} \sin(\frac{1}{2}\delta_\mathfrak{v} + x\delta_{\mathfrak{w}\mathfrak{v}}) \tag{8}$$

$$\nu_{s[z],[2x]} = -(-1)^{nodeid} \sin(\frac{1}{2}\delta_\mathfrak{v} + x\delta_{\mathfrak{w}\mathfrak{v}} + \delta_\mathfrak{w}) \tag{9}$$

for $x = 0, 1, 2, \ldots$.

If n is even, then $z = 1 : n/2$ first $\lambda_{s[z]}$ eigenvalues and corresponding eigenvectors and first $z = 1 : n/2 - 1$ and $z = n$ $\lambda_{c[z]}$ eigenvalues and corresponding eigenvectors are taken. If n is odd, then $z = 1 : (n-1)/2$ first $\lambda_{s[z]}$ elements and $z = 1 : (n-1)/2$ and $z = n$ $\lambda_{c[z]}$ elements are taken. If $z = n$, then $\lambda_{c[z]}$ is equal zero and $\nu_{c[z]}$ is a constant vector.[2]

Subsequently, we will generally omit the z index unless it turns out to be necessary.

In the above example, the eigenvalues implied by Theorem 1 are:
$0.000, 0.912, 3.186, 6.814, 9.088$.
For eigenvalue 3.186 we have eigenvector $[0.939\ {-}0.556\ {-}0.962\ {-}0.038\ 0.618]$.
For eigenvalue 9.088 we have eigenvector $[0.240\ {-}0.849\ 0.997\ {-}0.765\ 0.377]$. For eigenvalue 6.814 we have eigenvector $[-0.345\ 0.831\ {-}0.272\ {-}0.999\ 0.786]$. For eigenvalue 0.912 we have eigenvector $[-0.971\ {-}0.528\ {-}0.072\ 0.645\ 0.926]$. For eigenvalue 0 we have eigenvector $[1\ 1\ 1\ 1\ 1]$.

---

[2] If n is even and $z = n/2$, then $\nu_{c[z]}$ is a zero vector and therefore respective $\lambda_{c[z]}$ is not used as a solution.

3.2   Eigensolutions of Combinatorial Laplacians of Bi-weighted Grid Graphs - Path Graph Case with Cosine Shaped Functions

As visible from Eqs. (5) and (6), our working hypothesis is that the eigenvector **v** elements are of the form

$$v_{nodeid} = \cos(\alpha_{nodeid}) \tag{10}$$

and the angles α_{nodeid} differ between neighbouring nodes by alternating either $\delta_\mathfrak{w}$ or $\delta_\mathfrak{v}$, when we have to do with a path graph.

Non-border Even Nodes. For an even node e, its eigenvector component amounts to $v_e = \cos(\alpha_e)$, its preceding (odd) node component is $v_{e-1} = \cos(\alpha_e - \delta_\mathfrak{w})$, its succeeding (odd) node component is $v_{e+1} = \cos(\alpha_e + \delta_\mathfrak{v})$.

Upon multiplication of the Laplacian matrix with the eigenvector the result for a non-border even e node would be

$$
\begin{aligned}
v'_e = & - \mathfrak{w}\cos(\alpha_e - \delta_\mathfrak{w}) + (\mathfrak{w} + \mathfrak{v})\cos(\alpha_e) - \mathfrak{v}\cos(\alpha_e + \delta_\mathfrak{v}) \\
= & \mathfrak{w}(\cos(\alpha_e) - \cos(\alpha_e - \delta_\mathfrak{w})) + \mathfrak{v}(\cos(\alpha_e) - \cos(\alpha_e + \delta_\mathfrak{v})) \\
= & \mathfrak{w}(\cos(\alpha_e) - \cos(\alpha_e)\cos(\delta_\mathfrak{w}) - \sin(\alpha_e)\sin(\delta_\mathfrak{w})) \\
& + \mathfrak{v}(\cos(\alpha_e) - \cos(\alpha_e)cos(\delta_\mathfrak{v}) + \sin(\alpha_e)\sin(\delta_\mathfrak{v})) \\
= & \cos(\alpha_e)(\mathfrak{w}(1 - \cos(\delta_\mathfrak{w})) + \mathfrak{v}(1 - \cos(\delta_\mathfrak{v}))) - \sin(\alpha_e)(\mathfrak{w}\sin(\delta_\mathfrak{w}) - \mathfrak{v}\sin(\delta_\mathfrak{v}))
\end{aligned}
$$

If we assume that

$$- \mathfrak{v}\sin(\delta_\mathfrak{v}) + \mathfrak{w}\sin(\delta_\mathfrak{w}) = 0 \tag{11}$$

then

$$v'_e = \cos(\alpha_e)(\mathfrak{w}(1 - \cos(\delta_\mathfrak{w})) + \mathfrak{v}(1 - \cos(\delta_\mathfrak{v}))) \tag{12}$$

Recall that the hypothesised eigenvector component v_e is of the form $\cos(\alpha_e)$. So the eigenvalue requirement is fulfilled here with the constant factor $\lambda = \mathfrak{w}(1 - \cos(\delta_\mathfrak{w})) + \mathfrak{v}(1 - \cos(\delta_\mathfrak{v}))$, as $\mathfrak{v}, \mathfrak{w}, \delta_\mathfrak{v}, \delta_\mathfrak{w}$ are constants. This fits Eq. (4).

Non-border Odd Nodes. For an odd node o, its eigenvector component amounts to $v_o = \cos(\alpha_o)$, its preceding (even) node component is $v_{o-1} = \cos(\alpha_o - \delta_\mathfrak{v})$, its succeeding (odd) node component is $v_{o+1} = \cos(\alpha_o + \delta_\mathfrak{w})$.

Hence upon multiplication of the Laplacian matrix with the eigenvector the result for a non-border odd o node would be

$$v'_o = -\mathfrak{v}\cos(\alpha_o - \delta_\mathfrak{v}) + (\mathfrak{w} + \mathfrak{v})\cos(\alpha_o) - \mathfrak{w}\cos(\alpha_o + \delta_\mathfrak{w})$$

That is

$$
\begin{aligned}
v'_o = & \mathfrak{v}(\cos(\alpha_o) - \cos(\alpha_o - \delta_\mathfrak{v})) + \mathfrak{w}(\cos(\alpha_o) - \cos(\alpha_o + \delta_\mathfrak{w})) \\
= & \mathfrak{v}(\cos(\alpha_o) - \cos(\alpha_o)\cos(\delta_\mathfrak{v}) - \sin(\alpha_o)\sin(\delta_\mathfrak{v})) \\
& + \mathfrak{w}(\cos(\alpha_o) - \cos(\alpha_o)\cos(\delta_\mathfrak{w}) + \sin(\alpha_o)\sin(\delta_\mathfrak{w})) \\
= & \cos(\alpha_o)(\mathfrak{v}(1 - \cos(\delta_\mathfrak{v})) + \mathfrak{w}(1 - \cos(\delta_\mathfrak{w}))) - \sin(\alpha_o)(\mathfrak{v}\sin(\delta_\mathfrak{v}) - \mathfrak{w}\sin(\delta_\mathfrak{w}))
\end{aligned}
$$

Recall that the hypothesised eigenvector component v_o is of the form $\cos(\alpha_o)$. As $\mathfrak{w}, \mathfrak{v}, \delta_\mathfrak{w}, \delta_\mathfrak{v}$ are constants, the condition of $\mathbf{v}$ being an eigenvector requires that $-\mathfrak{v}\sin(\delta_\mathfrak{v}) + \mathfrak{w}\sin(\delta_\mathfrak{w}) = 0$, as previously in Eq. (11). Then

$$v'_o = \cos(\alpha_o)(\mathfrak{v}(1 - \cos(\delta_\mathfrak{v})) + \mathfrak{w}(1 - \cos(\delta_\mathfrak{w}))) \tag{13}$$

and the λ factor is the same as above.

First Border Node. We need now to discuss the behaviour of the border nodes. It is an odd node, but without a preceding node.

$$v'_1 = \mathfrak{w}\cos(\alpha_1) - \mathfrak{w}\cos(\alpha_1 + \delta_\mathfrak{w}) = \mathfrak{w}(\cos(\alpha_1) - \cos(\alpha_1 + \delta_\mathfrak{w}))$$

Let us guess that $\alpha_1 = \frac{1}{2}\delta_\mathfrak{v}$ (Eq. (5)) . Under this assumption: $\cos(\alpha_1 - \delta_\mathfrak{v}) = \cos(\frac{1}{2}\delta_\mathfrak{v} - \delta_\mathfrak{v}) = \cos(-\frac{1}{2}\delta_\mathfrak{v}) = \cos(\frac{1}{2}\delta_\mathfrak{v}) = \cos(\alpha_1)$. Hence

$$
\begin{aligned}
v'_1 &= \mathfrak{v}(\cos(\alpha_1) - \cos(\alpha_1 - \delta_\mathfrak{v})) + \mathfrak{w}(\cos(\alpha_1) - \cos(\alpha_1 + \delta_\mathfrak{w}))\\
&= \mathfrak{v}(\cos(\alpha_1) - \cos(\alpha_1)\cos(\delta_\mathfrak{v}) - \sin(\alpha_1)\sin(\delta_\mathfrak{v}))\\
&\quad + \mathfrak{w}(\cos(\alpha_1) - \cos(\alpha_1)\cos(\delta_\mathfrak{w}) + \sin(\alpha_1)\sin(\delta_\mathfrak{w}))\\
&= \cos(\alpha_1)(\mathfrak{v}(1 - \cos(\delta_\mathfrak{v})) + \mathfrak{w}(1 - \cos(\delta_\mathfrak{w}))) - \sin(\alpha_1)(\mathfrak{v}\sin(\delta_\mathfrak{v}) - \mathfrak{w}\sin(\delta_\mathfrak{w}))
\end{aligned}
$$

Assuming again that $-\mathfrak{v}\sin(\delta_\mathfrak{v}) + \mathfrak{w}\sin(\delta_\mathfrak{w}) = 0$, then the eigenvalue requirement is fulfilled here with the same constant (λ) factor $(\mathfrak{w}(1 - \cos(\delta_\mathfrak{w})) + \mathfrak{v}(1 - \cos(\delta_\mathfrak{v})))$.

Last Even Border Node. So consider now the last node when the number of nodes is even. It can be either even or odd. Upon multiplication of the Laplacian matrix with the eigenvector the result for the last border even l_e node would be

$$v'_{le} = -\mathfrak{w}\cos(\alpha_{le} - \delta_\mathfrak{w}) + \mathfrak{w}\cos(\alpha_{le}) = \mathfrak{w}(\cos(\alpha_{le}) - \cos(\alpha_{le} - \delta_\mathfrak{w}))$$

Let us assume $\alpha_{le} = -\frac{1}{2}\delta_\mathfrak{v}$ (Eq. (6) in which case $v_{le} = \cos(\frac{1}{2}\delta_\mathfrak{v})$. Then clearly $\cos(\alpha_{le} + \delta_\mathfrak{v}) = \cos(-\frac{1}{2}\delta_\mathfrak{v} + \delta_\mathfrak{v}) = \cos(\frac{1}{2}\delta_\mathfrak{v}) = \cos(\alpha_{le})$. Therefore

$$
\begin{aligned}
v'_{le} &= \mathfrak{v}(\cos(\alpha_{le}) - \cos(\alpha_{le} + \delta_\mathfrak{v})) + \mathfrak{w}(\cos(\alpha_{le}) - \cos(\alpha_{le} - \delta_\mathfrak{w}))\\
&= \mathfrak{v}(\cos(\alpha_{le}) - \cos(\alpha_{le})\cos(\delta_\mathfrak{v}) + \sin(\alpha_{le})\sin(\delta_\mathfrak{v}))\\
&\quad + \mathfrak{w}(\cos(\alpha_{le}) - \cos(\alpha_{le})\cos(\delta_\mathfrak{w}) - \sin(\alpha_{le})\sin(\delta_\mathfrak{w}))\\
&= \cos(\alpha_{le})(\mathfrak{v}(1 - \cos(\delta_\mathfrak{v})) + \mathfrak{w}(1 - \cos(\delta_\mathfrak{w}))) + \sin(\alpha_{le})(\mathfrak{v}\sin(\delta_\mathfrak{v}) - \mathfrak{w}\sin(\delta_\mathfrak{w}))
\end{aligned}
$$

Assuming again that $-\mathfrak{v}\sin(\delta_\mathfrak{v}) + \mathfrak{w}\sin(\delta_\mathfrak{w}) = 0$, then the eigenvalue requirement is fulfilled here with the same constant (λ) factor $(\mathfrak{w}(1 - \cos(\delta_\mathfrak{w})) + \mathfrak{v}(1 - \cos(\delta_\mathfrak{v})))$. Let us assume now that $\alpha_{le} = -\frac{1}{2}\delta_\mathfrak{v} + \pi$ in which case $v_{le} = -\cos(\frac{1}{2}\delta_\mathfrak{v})$. Then clearly $\cos(\alpha_{le} + \delta_\mathfrak{v}) = \cos(-\frac{1}{2}\delta_\mathfrak{v} + \pi + \delta_\mathfrak{v}) = \cos(\frac{1}{2}\delta_\mathfrak{v} + \pi) = -\cos(\frac{1}{2}\delta_\mathfrak{v}) = \cos(\alpha_{le})$. By the same derivation as above, we find that the eigenvalue requirement is satisfied.

Last Odd Border Node. So consider now the last node, when the number of nodes is odd. It can be either even or odd. Upon multiplication of the Laplacian matrix with the eigenvector the result for the last border even l_o node would be

$$v'_{lo} = -\mathfrak{v}\cos(\alpha_{lo} - \delta_{\mathfrak{v}}) + \mathfrak{v}\cos(\alpha_{lo}) = \mathfrak{v}(\cos(\alpha_{lo}) - \cos(\alpha_{lo} - \delta_{\mathfrak{v}}))$$

Let us assume $\alpha_{lo} = -\frac{1}{2}\delta_{\mathfrak{w}}$ (Eq. (5)) in which case $v_{lo} = \cos(\frac{1}{2}\delta_{\mathfrak{w}})$. Then clearly $\cos(\alpha_{lo} + \delta_{\mathfrak{w}}) = \cos(-\frac{1}{2}\delta_{\mathfrak{w}} + \delta_{\mathfrak{w}}) = \cos(\frac{1}{2}\delta_{\mathfrak{w}}) = \cos(\alpha_{lo})$. Therefore

$$\begin{aligned}
v'_{lo} &= \mathfrak{w}(\cos(\alpha_{lo}) - \cos(\alpha_{lo} + \delta_{\mathfrak{w}})) + \mathfrak{v}(\cos(\alpha_{lo}) - \cos(\alpha_{lo} - \delta_{\mathfrak{v}})) \\
&= \mathfrak{w}(\cos(\alpha_{lo}) - \cos(\alpha_{lo})\cos(\delta_{\mathfrak{w}}) + \sin(\alpha_{lo})\sin(\delta_{\mathfrak{w}})) \\
&\quad + \mathfrak{v}(\cos(\alpha_{lo}) - \cos(\alpha_{lo})\cos(\delta_{\mathfrak{v}}) - \sin(\alpha_{lo})\sin(\delta_{\mathfrak{v}})) \\
&= \cos(\alpha_{lo})(\mathfrak{w}(1 - \cos(\delta_{\mathfrak{w}})) + \mathfrak{v}(1 - \cos(\delta_{\mathfrak{v}}))) + \sin(\alpha_{lo})(\mathfrak{w}\sin(\delta_{\mathfrak{w}}) - \mathfrak{v}\sin(\delta_{\mathfrak{v}}))
\end{aligned}$$

Assuming again that $-\mathfrak{w}\sin(\delta_{\mathfrak{w}}) + \mathfrak{v}\sin(\delta_{\mathfrak{v}}) = 0$, then the eigenvalue requirement is fulfilled here with the same constant (λ) factor ($\mathfrak{v}(1-\cos(\delta_{\mathfrak{v}})) + \mathfrak{w}(1-\cos(\delta_{\mathfrak{w}}))$). Let us assume now that $\alpha_{le} = -\frac{1}{2}\delta_{\mathfrak{w}} + \pi$ in which case $v_{le} = -\cos(\frac{1}{2}\delta_{\mathfrak{w}})$. Then clearly $\cos(\alpha_{le} + \delta_{\mathfrak{w}}) = \cos(-\frac{1}{2}\delta_{\mathfrak{w}} + \pi + \delta_{\mathfrak{w}}) = \cos(\frac{1}{2}\delta_{\mathfrak{w}} + \pi) = -\cos(\frac{1}{2}\delta_{\mathfrak{w}}) = \cos(\alpha_{le})$. By the same derivation as previously, we find that the eigenvalue requirement is satisfied.

3.3 Eigensolutions of Combinatorial Laplacians of Bi-weighted Grid Graphs - Path Graph Case with Sine Shaped Function

Our working hypothesis is that the eigenvector **v** elements are of the form

$$v_{nodeid} = -(-1)^{nodeid}\sin(\alpha_{nodeid}) \tag{14}$$

and the angles α_{nodeid} differ between neighbouring nodes by either $\delta_{\mathfrak{w}}$ or $\delta_{\mathfrak{v}}$, when we have to do with a path graph (Eq. (8)).

Non-border Even Nodes. For an even node e, its eigenvector component amounts to $v_e = -(-1)^e\sin(\alpha_e)$, its preceding (odd) node component is $v_{e-1} = -(-1)^{e-1}\sin(\alpha_e - \delta_{\mathfrak{w}})$ its succeeding (odd) node component is $v_{e+1} = -(-1)^{e+1}\sin(\alpha_e + \delta_{\mathfrak{v}})$. Upon multiplication of the Laplacian matrix with the eigenvector the result for a non-border even e node would be

$$\begin{aligned}
v'_e &= \mathfrak{w}(-1)^{e-1}\sin(\alpha_e - \delta_{\mathfrak{w}}) - (\mathfrak{w} + \mathfrak{v})(-1)^e\sin(\alpha_e) + \mathfrak{v}(-1)^{e+1}\sin(\alpha_e + \delta_{\mathfrak{v}}) \\
&= -\mathfrak{w}\sin(\alpha_e - \delta_{\mathfrak{w}}) - (\mathfrak{w} + \mathfrak{v})\sin(\alpha_e) - \mathfrak{v}\sin(\alpha_e + \delta_{\mathfrak{v}}) \\
&= -\mathfrak{w}(\sin(\alpha_e) + \sin(\alpha_e - \delta_{\mathfrak{w}})) - \mathfrak{v}(\sin(\alpha_e) + \sin(\alpha_e + \delta_{\mathfrak{v}})) \\
&= -\mathfrak{w}(\sin(\alpha_e) + \sin(\alpha_e)\cos(\delta_{\mathfrak{w}}) - \cos(\alpha_e)sin(\delta_{\mathfrak{w}})) \\
&\quad - \mathfrak{v}(\sin(\alpha_e) + \sin(\alpha_e)cos(\delta_{\mathfrak{v}}) + \cos(\alpha_e)\sin(\delta_{\mathfrak{v}})) \\
&= -\sin(\alpha_e)(\mathfrak{w}(1 + \cos(\delta_{\mathfrak{w}})) + \mathfrak{v}(1 + cos(\delta_{\mathfrak{v}}))) + cos(\alpha_e)(\mathfrak{w}sin(\delta_{\mathfrak{w}}) - \mathfrak{v}\sin(\delta_{\mathfrak{v}}))
\end{aligned}$$

If we assume that $-\mathfrak{v}\sin(\delta_{\mathfrak{v}}) + \mathfrak{w}\sin(\delta_{\mathfrak{w}}) = 0$, then

$$v'_e = -\sin(\alpha_e)(\mathfrak{w}(1 + \cos(\delta_{\mathfrak{w}})) + \mathfrak{v}(1 + cos(\delta_{\mathfrak{v}})))$$

The hypothesised eigenvector component v_e is of the form $-(-1)^e \sin(\alpha_e) = -\sin(\alpha_e)$. So the eigenvalue requirement is fulfilled here with the constant factor $\lambda = \mathfrak{w}(1 + \cos(\delta_\mathfrak{w})) + \mathfrak{v}(1 + \cos(\delta_\mathfrak{v}))$, as $\mathfrak{v}, \mathfrak{w}, \delta_\mathfrak{v}, \delta_\mathfrak{w}$ are constants.

Non-border Odd Nodes. For an odd node o, its eigenvector component amounts to $v_o = -(-1)^o \sin(\alpha_o)$, its preceding (even) node component is $v_{o-1} = -(-1)^{o-1} \sin(\alpha_o - \delta_\mathfrak{v})$ its succeeding (odd) node component is $v_{o+1} = -(-1)^{o+1} \sin(\alpha_o + \delta_\mathfrak{w})$. Hence upon multiplication of the Laplacian matrix with the eigenvector the result for a non-border odd o node would be

$$v'_o = \mathfrak{v}(-1)^{o-1} \sin(\alpha_o - \delta_\mathfrak{v}) - (\mathfrak{w} + \mathfrak{v})(-1)^o \sin(\alpha_o) + \mathfrak{w}(-1)^{o+1} \sin(\alpha_o + \delta_\mathfrak{w})$$
$$= \sin(\alpha_o)(\mathfrak{v}(1 + cos(\delta_\mathfrak{v})) + \mathfrak{w}(1 + \cos(\delta_\mathfrak{w}))) + \cos(\alpha_o)(-\mathfrak{v} \sin(\delta_\mathfrak{v}) + \mathfrak{w} \sin(\delta_\mathfrak{w}))$$

The hypothesised eigenvector component v_o is of the form $-(-1)^o \sin(\alpha_o) = \sin(\alpha_o)$. As $\mathfrak{w}, \mathfrak{v}, \delta_\mathfrak{w}, \delta_\mathfrak{v}$ are constants, the condition of **v** being an eigenvector requires that $-\mathfrak{v} \sin(\delta_\mathfrak{v}) + \mathfrak{w} \sin(\delta_\mathfrak{w}) = 0$, as previously. Then

$$v'_o = \sin(\alpha_o)(\mathfrak{v}(1 + cos(\delta_\mathfrak{v})) + \mathfrak{w}(1 + cos(\delta_\mathfrak{w})))$$

and the λ factor is the same as above.

First Border Node. We need now to discuss the behaviour of the border nodes. In order to ensure that also the eigen-property holds at the end points of the path, we have to take $\alpha_1 = \frac{1}{2}\delta_\mathfrak{v}$ for the first node. because in this case the result of the product with the Laplacian matrix ($L_{1,1} = \mathfrak{w}$, $L_{1,2} = -\mathfrak{w}$) with the eigenvector **v** at v_1 will amount to:

$$v'_1 = -(-1)^1 \sin(\alpha_1)\mathfrak{w} - (-1)^1 \sin(\alpha_1 + \delta_\mathfrak{w})(-\mathfrak{w})$$
$$= \sin(\frac{1}{2}\delta_\mathfrak{v})(\mathfrak{w}(1 + \cos(\delta_\mathfrak{w})) + \mathfrak{v}(1 + \cos(\delta_\mathfrak{v}))) + \cos(\frac{1}{2}\delta_\mathfrak{v})(\mathfrak{w} \sin(\delta_\mathfrak{w}) - \mathfrak{v} \sin(\delta_\mathfrak{v}))$$

Recall that $v_1 = -\sin(\frac{1}{2}\delta_\mathfrak{v})$ Assuming again that $-\mathfrak{v} \sin(\delta_\mathfrak{v}) + \mathfrak{w} \sin(\delta_\mathfrak{w}) = 0$, then the eigenvalue requirement is fulfilled here with the same constant (λ) factor ($\mathfrak{w}(1 + \cos(\delta_\mathfrak{w}) + \mathfrak{v}(1 + cos(\delta_\mathfrak{v})))$.

Last Odd Border Node. For an odd border node l_o, let us take $\alpha_{lo} = -\frac{1}{2}\delta_\mathfrak{w}$. Its eigenvector component amounts to $v_{lo} = -(-1)^{l_o} \sin(\alpha_{lo}) = \sin(\alpha_{lo})$, its preceding (even) node component is $v_{l_o-1} = -(-1)^{l_o-1} \sin(\alpha_{lo} - \delta_\mathfrak{v})$ Hence upon multiplication of the Laplacian matrix with the eigenvector the result for a last odd l_o node would be

$$v'_{lo} = \mathfrak{v}(-1)^{l_o-1} \sin(\alpha_{l_o} - \delta_\mathfrak{v}) - \mathfrak{v}(-1)^{l_o} \sin(\alpha_{l_o})$$
$$= \mathfrak{v} \sin(\alpha_{l_o} - \delta_\mathfrak{v}) + \mathfrak{v} \sin(\alpha_{l_o}) = \mathfrak{v}(\sin(\alpha_{l_o} - \delta_\mathfrak{v}) + \sin(\alpha_{l_o}))$$

Let us assume $\alpha_{lo} = -\frac{1}{2}\delta_{\mathfrak{w}}$ in which case $v_{lo} = \sin(\frac{1}{2}\delta_{\mathfrak{w}})$. Then clearly $\sin(\alpha_{lo} + \delta_{\mathfrak{w}}) = \sin(-\frac{1}{2}\delta_{\mathfrak{w}} + \delta_{\mathfrak{w}}) = \sin(\frac{1}{2}\delta_{\mathfrak{w}}) = -\sin(\alpha_{lo})$. Hence

$$
\begin{aligned}
v'_{lo} =&(\mathfrak{v}(\sin(\alpha_{l_o}) + \sin(\alpha_{l_o} - \delta_{\mathfrak{v}})) + \mathfrak{w}(\sin(\alpha_{l_o}) + \sin(\alpha_{l_o} + \delta_{\mathfrak{w}}))) \\
=&(\mathfrak{v}(\sin(\alpha_{l_o}) + \sin(\alpha_{l_o})\cos(\delta_{\mathfrak{v}}) - \cos(\alpha_{l_o})\sin(\delta_{\mathfrak{v}})) \\
&+ \mathfrak{w}(\sin(\alpha_{l_o}) + \sin(\alpha_{l_o})\cos(\delta_{\mathfrak{w}}) + \cos(\alpha_{l_o})\sin(\delta_{\mathfrak{w}}))) \\
=&\mathfrak{v}\sin(\alpha_{l_o}) + \mathfrak{v}\sin(\alpha_{l_o})\cos(\delta_{\mathfrak{v}}) - \mathfrak{v}\cos(\alpha_{l_o})\sin(\delta_{\mathfrak{v}}) \\
&+ \mathfrak{w}\sin(\alpha_{l_o}) + \mathfrak{w}\sin(\alpha_{l_o})\cos(\delta_{\mathfrak{w}}) + \mathfrak{w}\cos(\alpha_{l_o})\sin(\delta_{\mathfrak{w}}) \\
=&\mathfrak{v}\sin(\alpha_{l_o}) + \mathfrak{v}\sin(\alpha_{l_o})\cos(\delta_{\mathfrak{v}}) + \mathfrak{w}\sin(\alpha_{l_o}) + \mathfrak{w}\sin(\alpha_{l_o})\cos(\delta_{\mathfrak{w}}) \\
&- \mathfrak{v}\cos(\alpha_{l_o})\sin(\delta_{\mathfrak{v}}) + \mathfrak{w}\cos(\alpha_{l_o})\sin(\delta_{\mathfrak{w}}) \\
=&\sin(\alpha_{l_o})(\mathfrak{v}(1 + \cos(\delta_{\mathfrak{v}})) + \mathfrak{w}(1 + \cos(\delta_{\mathfrak{w}}))) + \cos(\alpha_{l_o})(-\mathfrak{v}\sin(\delta_{\mathfrak{v}}) + \mathfrak{w}\sin(\delta_{\mathfrak{w}}))
\end{aligned}
$$

Assuming that $-\mathfrak{w}\sin(\delta_{\mathfrak{w}}) + \mathfrak{v}\sin(\delta_{\mathfrak{v}}) = 0$, then the eigenvalue requirement is fulfilled here with the same constant (λ) factor ($\mathfrak{v}(1+\cos(\delta_{\mathfrak{v}})) + \mathfrak{w}(1+\cos(\delta_{\mathfrak{w}}))$). Let us assume $\alpha_{lo} = -\frac{1}{2}\delta_{\mathfrak{w}} + \pi$ in which case $v_{lo} = \sin(\frac{1}{2}\delta_{\mathfrak{w}} + \pi) = -\sin(\frac{1}{2}\delta_{\mathfrak{v}})$. Then clearly $\sin(\alpha_{lo} + \delta_{\mathfrak{w}}) = \sin(-\frac{1}{2}\delta_{\mathfrak{w}} + \pi + \delta_{\mathfrak{w}}) = \sin(\frac{1}{2}\delta_{\mathfrak{w}} + \pi) = -sin(-\frac{1}{2}\delta_{\mathfrak{w}} + \pi) = -\sin(\alpha_{lo})$. By the same derivation as above, we can show that the eigenvalue requirement is satisfied.

Last Even Border Node. If it is even, let us take $\alpha_{le} = -\frac{1}{2}\delta_{\mathfrak{v}}$.

For an even node l_e, its eigenvector component amounts to $v_{le} = -(-1)^{l_e}\sin(\alpha_{le}) = -\sin(\alpha_{le})$, its preceding (odd) node component is $v_{l_e-1} = -(-1)^{l_e-1}\sin(\alpha_{le} - \delta_{\mathfrak{w}}) = \sin(\alpha_{le} - \delta_{\mathfrak{w}})$

Hence upon multiplication of the Laplacian matrix with the eigenvector the result for a last even l_e node would be

$$
\begin{aligned}
v'_{le} =&\mathfrak{w}(-1)^{l_e-1}\sin(\alpha_{l_e} - \delta_{\mathfrak{w}}) - \mathfrak{w}(-1)^{l_e}\sin(\alpha_{l_e}) \\
=&-\mathfrak{w}\sin(\alpha_{l_o} - \delta_{\mathfrak{w}}) - \mathfrak{w}\sin(\alpha_{l_o}) = -\mathfrak{w}(\sin(\alpha_{l_o} - \delta_{\mathfrak{w}}) + \sin(\alpha_{l_o}))
\end{aligned}
$$

Let us assume $\alpha_{le} = -\frac{1}{2}\delta_{\mathfrak{v}}$ in which case $v_{le} = -\sin(\frac{1}{2}\delta_{\mathfrak{v}})$. Then clearly $\sin(\alpha_{le} + \delta_{\mathfrak{v}}) = \sin(-\frac{1}{2}\delta_{\mathfrak{v}} + \delta_{\mathfrak{v}}) = \sin(\frac{1}{2}\delta_{\mathfrak{v}}) = -\sin(\alpha_{le})$. Hence

$$
\begin{aligned}
v'_{le} =&-\mathfrak{w}(\sin(\alpha_{l_e}) + \sin(\alpha_{l_e} - \delta_{\mathfrak{w}})) - \mathfrak{v}(\sin(\alpha_{l_e}) + \sin(\alpha_{l_e} + \delta_{\mathfrak{v}}))) \\
=&-\sin(\alpha_{l_e})(\mathfrak{w}(1 + \cos(\delta_{\mathfrak{w}})) + \mathfrak{v}(1 + \cos(\delta_{\mathfrak{v}}))) - \cos(\alpha_{l_o})(-\mathfrak{w}\sin(\delta_{\mathfrak{w}}) + \mathfrak{v}\sin(\delta_{\mathfrak{v}}))
\end{aligned}
$$

Assuming that $-\mathfrak{w}\sin(\delta_{\mathfrak{w}}) + \mathfrak{v}\sin(\delta_{\mathfrak{v}}) = 0$, then the eigenvalue requirement is fulfilled here with the same constant (λ) factor ($\mathfrak{v}(1+\cos(\delta_{\mathfrak{v}})) + \mathfrak{w}(1+\cos(\delta_{\mathfrak{w}}))$). Let us assume $\alpha_{le} = -\frac{1}{2}\delta_{\mathfrak{v}} + \pi$ in which case $v_{le} = -\sin(\frac{1}{2}\delta_{\mathfrak{v}} + \pi) = \sin(\frac{1}{2}\delta_{\mathfrak{v}})$. Then clearly $\sin(\alpha_{le} + \delta_{\mathfrak{v}}) = \sin(-\frac{1}{2}\delta_{\mathfrak{v}} + \pi + \delta_{\mathfrak{v}}) = \sin(\frac{1}{2}\delta_{\mathfrak{v}} + \pi) = -sin(-\frac{1}{2}\delta_{\mathfrak{v}} + \pi) = -\sin(\alpha_{le})$. By the same derivation as in preceding section, we find that the eigenvalue requirement is satisfied.

3.4   Materialization of Assumptions

We have made the following assumptions so far:

- $-\mathfrak{v}\sin(\delta_\mathfrak{v}) + \mathfrak{w}\sin(\delta_\mathfrak{w}) = 0$ (Eq. (11)).
- Eigenvector elements for combinatorial Laplacian are of the form $v_{nodeid} = -(-1)^{nodeid}\sin(\alpha_{nodeid})$ (Sect. 3.3) or of the form $v_{nodeid} = \cos(\alpha_{nodeid})$ (Sect. 3.2)
- The first node angle is of the form $\alpha_1 = \frac{1}{2}\delta_\mathfrak{v}$ (pages 9, 7), this fits Eqs. (9, 8, 6, 5)
- The above angles in the analytical form are related as follows: $\alpha_o = \alpha_{o-1} + \delta_\mathfrak{v}$ for odd nodes o and $\alpha_e = \alpha_{e-1} + \delta_\mathfrak{v}$ for even nodes e. This fits Eqs. (9, 8, 6, 5)
- If the last node is an odd node l_o, the angle of which is of the form $\alpha_{l_o} = -\frac{1}{2}\delta_\mathfrak{w}$ or $\alpha_{l_o} = -\frac{1}{2}\delta_\mathfrak{w} + \pi$ (page 8).
- If the last node is an even node l_e, its angle is of the form $\alpha_{l_e} = -\frac{1}{2}\delta_\mathfrak{v}$ or $\alpha_{l_e} = -\frac{1}{2}\delta_\mathfrak{v} + \pi$ (page 7)
- Eigenvalues for combinatorial Laplacian are of the form $\mathfrak{w}(1+\cos(\delta_\mathfrak{w})) + \mathfrak{v}(1+\cos(\delta_\mathfrak{v}))$ (page 8) or $\mathfrak{w}(1-\cos(\delta_\mathfrak{w})) + \mathfrak{v}(1-\cos(\delta_\mathfrak{v}))$, correspondingly. (page 6)

We need to check whether and when these assumptions, for which we do not know whether they fit the biweighted graph theorem, are true. Additionally, to ensure that we have an analytical form for the eigenvalues and eigenvectors, we need to ensure that we have as many orthogonal eigenvectors as there are nodes. So let us check what the condition $-\mathfrak{v}\sin(\delta_\mathfrak{v}) + \mathfrak{w}\sin(\delta_\mathfrak{w}) = 0$ implies for $\delta_\mathfrak{w}$ and $\delta_\mathfrak{v}$. Denote $\delta_\mathfrak{wv} = \delta_\mathfrak{w} + \delta_\mathfrak{v}$. Then

$$-\mathfrak{v}\sin(\delta_\mathfrak{wv} - \delta_\mathfrak{w}) + \mathfrak{w}\sin(\delta_\mathfrak{w}) = 0$$

$$\tan(\delta_\mathfrak{w}) = \frac{\mathfrak{v}\sin(\delta_\mathfrak{wv})}{\mathfrak{v}\cos(\delta_\mathfrak{wv}) + \mathfrak{w}}$$

$$\delta_\mathfrak{w} = \arctan\frac{\mathfrak{v}\sin(\delta_\mathfrak{wv})}{\mathfrak{v}\cos(\delta_\mathfrak{wv}) + \mathfrak{w}}.$$

Similarly

$$\delta_\mathfrak{v} = \arctan\frac{\mathfrak{w}\sin(\delta_\mathfrak{wv})}{\mathfrak{w}\cos(\delta_\mathfrak{wv}) + \mathfrak{v}}$$

This fits Eqs. (3) and (2) However, $\delta_\mathfrak{v}, \delta_\mathfrak{w}$ shall not be computed simultaneously from the above formulas, but rather only one of them and the other from the sum $\delta_\mathfrak{wv} = \delta_\mathfrak{w} + \delta_\mathfrak{v}$ because of ambiguity of arctan in the range $[-\pi, \pi]$. If $\mathfrak{v}/\mathfrak{w} \geq 1$, then $\delta_\mathfrak{v}$ should be computed from the formula above, and otherwise $\delta_\mathfrak{w}$. The value of the computed δ should be taken from the range $(-\frac{\pi}{2}, \frac{\pi}{2})$ so that the sine of that δ has he same sign as the sine of the right-hand side.

We have the requirement:

$$\nu_{[z],[2x]} = \cos(\frac{1}{2}\delta_\mathfrak{v} + x\delta_{\mathfrak{w}\mathfrak{v}} + \delta_\mathfrak{w})$$

$$\nu_{[z],[2x+1]} = -(-1)^{nodeid}\sin(\frac{1}{2}\delta_\mathfrak{v} + x\delta_{\mathfrak{w}\mathfrak{v}})$$

$$\nu_{[z],[2x]} = -(-1)^{nodeid}\sin(\frac{1}{2}\delta_\mathfrak{v} + (x-1)\delta_{\mathfrak{w}\mathfrak{v}} + \delta_\mathfrak{w})$$

which needs to be aligned with the aforementioned restriction on the values of last noder angles. Note that we have introduced the restrictions for the last node: If it is even, then $\nu_{[z],[n]} = \cos(-\frac{1}{2}\delta_\mathfrak{v})$, and $\nu_{[z],[n]} = \sin(-\frac{1}{2}\delta_\mathfrak{v})$, respectively. Both may be true if $\frac{1}{2}\delta_\mathfrak{v} + (\frac{n}{2} - 1)\delta_{\mathfrak{w}\mathfrak{v}} + \delta_\mathfrak{w} = -\frac{1}{2}\delta_\mathfrak{v} + k2\pi$ for some k. Equivalently $\delta_\mathfrak{v} + (\frac{n}{2} - 1)\delta_{\mathfrak{w}\mathfrak{v}} + \delta_\mathfrak{w} = k2\pi$ or $\frac{n}{2}\delta_{\mathfrak{w}\mathfrak{v}} = k2\pi$ Also we have that if it is even, then $\nu_{[z],[n]} = \cos(-\frac{1}{2}\delta_\mathfrak{v} + \pi)$ or $\nu_{[z],[n]} = \sin(-\frac{1}{2}\delta_\mathfrak{v} + \pi)$ respectively. Both may be true if $\frac{1}{2}\delta_\mathfrak{v} + (\frac{n}{2} - 1)\delta_{\mathfrak{w}\mathfrak{v}} + \delta_\mathfrak{w} = -\frac{1}{2}\delta_\mathfrak{v} + \pi + k2\pi$ for some k. That is $\delta_\mathfrak{v} + (\frac{n}{2} - 1)\delta_{\mathfrak{w}\mathfrak{v}} + \delta_\mathfrak{w} = k2\pi + \pi$, so $\frac{n}{2}\delta_{\mathfrak{w}\mathfrak{v}} = k2\pi + \pi$. Summarizing both cases we get the criterion:

$$\frac{n}{2}\delta_{\mathfrak{w}\mathfrak{v}} = k'\pi$$

for some k'. Hence

$$\delta_{\mathfrak{w}\mathfrak{v}} = \frac{k'\pi}{0.5n}$$

as required by Eq. (1).

If n is odd, then $\nu_{[z],[n]} = \cos(-\frac{1}{2}\delta_\mathfrak{w})$ or $\nu_{[z],[n]} = \sin(-\frac{1}{2}\delta_\mathfrak{w})$ or $\nu_{[z],[n]} = \cos(-\frac{1}{2}\delta_\mathfrak{w} + \pi)$ or $\nu_{[z],[n]} = \sin(-\frac{1}{2}\delta_\mathfrak{w} + \pi)$. The first two may be true if $\frac{1}{2}\delta_\mathfrak{v} + \frac{n-1}{2}\delta_{\mathfrak{w}\mathfrak{v}} = -\frac{1}{2}\delta_\mathfrak{w} + k2\pi$, so $\frac{1}{2}\delta_{\mathfrak{w}\mathfrak{v}} + \frac{n-1}{2}\delta_{\mathfrak{w}\mathfrak{v}} = k2\pi$, so $\frac{n}{2}\delta_{\mathfrak{w}\mathfrak{v}} = k2\pi$. The last two if $\frac{1}{2}\delta_{\mathfrak{w}\mathfrak{v}} + \frac{n-1}{2}\delta_{\mathfrak{w}\mathfrak{v}} = -\frac{1}{2}\delta_\mathfrak{w} + \pi + k2\pi$. That is $\frac{n}{2}\delta_{\mathfrak{w}\mathfrak{v}} = k2\pi + \pi$ So, as previously $\frac{n}{2}\delta_{\mathfrak{w}\mathfrak{v}} = k'\pi$ for some k'. Hence $\delta_{\mathfrak{w}\mathfrak{v}} = \frac{k'\pi}{0.5n}$ as required by Eq. (1).

Note that in this way we get $2n$ eigenvectors and eigenvalues while only n eigenvalues/eigenvectors are possible. This means that there will be repetitions, which are easily detectable. In practice, the leading half of sine ad cosine related eigenvectors form the orthogonal basis plus the eigenvector related to the eigenvalue zero. More precisely: if n is even, then $1 : n/2$ first sine elements and $1 : n/2 - 1$ cosine elements plus nth element. If n is odd, then $1 : (n-1)/2$ first sine elements and $1 : (n-1)/2$ cosine elements plus nth element. If $z = n$, then cosine λ is equal zero and ν is a constant vector. If n is even and $z = n/2$, then cosine ν is a zero vector and hence respective λ should be ignored. The collapsing of eigenvalues can be explained as follows.

$$\lambda_{[z]} = \mathfrak{w}(1 - \cos(\delta_\mathfrak{w})) + \mathfrak{v}(1 - \cos(\delta_\mathfrak{v}))$$

$$= \mathfrak{w} + \mathfrak{v} - \sin(\delta_\mathfrak{w})(\mathfrak{w}\frac{\cos(\delta_\mathfrak{w})}{\sin(\delta_\mathfrak{w})} + \mathfrak{v}\frac{\cos(\delta_\mathfrak{v})}{\sin(\delta_\mathfrak{w})})$$

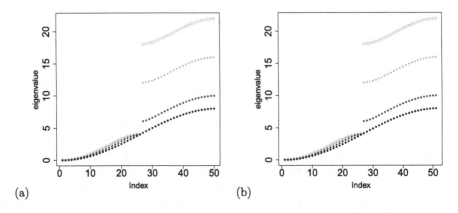

Fig. 2. Eigenvalue spectrum for (a) $n = 50$ and (b) $n = 51$. $\mathfrak{w} = 2$, whereby $\mathfrak{v} = 9$ in yellow, $\mathfrak{v} = 6$ in green, $\mathfrak{v} = 3$ in blue, $\mathfrak{v} = 2$ in black. (Color figure online)

Condition $-\mathfrak{v}\sin(\delta_\mathfrak{v}) + \mathfrak{w}\sin(\delta_\mathfrak{w}) = 0$ implies $\frac{\mathfrak{v}}{\sin(\delta_\mathfrak{w})} = \frac{\mathfrak{w}}{\sin(\delta_\mathfrak{v})}$. Therefore

$$\lambda_{[z]} = \mathfrak{w} + \mathfrak{v} - \mathfrak{w}\sin(\delta_\mathfrak{w})\left(\frac{\cos(\delta_\mathfrak{w})}{\sin(\delta_\mathfrak{w})} + \frac{\cos(\delta_\mathfrak{v})}{\sin(\delta_\mathfrak{v})}\right)$$

Now recall that $\tan(\delta_\mathfrak{w}) = \frac{\mathfrak{v}\sin(\delta_{\mathfrak{w}\mathfrak{v}})}{\mathfrak{v}\cos(\delta_{\mathfrak{w}\mathfrak{v}})+\mathfrak{w}}$ and $\tan(\delta_\mathfrak{v}) = \frac{\mathfrak{w}\sin(\delta_{\mathfrak{w}\mathfrak{v}})}{\mathfrak{w}\cos(\delta_{\mathfrak{w}\mathfrak{v}})+\mathfrak{v}}$. Therefore we get

$$\lambda_{[z]} = \mathfrak{w} + \mathfrak{v} - \mathfrak{w}\sin(\delta_\mathfrak{w})\left(\frac{\mathfrak{v}\cos(\delta_{\mathfrak{w}\mathfrak{v}})+\mathfrak{w}}{\mathfrak{v}\sin(\delta_{\mathfrak{w}\mathfrak{v}})} + \frac{\mathfrak{w}\cos(\delta_{\mathfrak{w}\mathfrak{v}})+\mathfrak{v}}{\mathfrak{w}\sin(\delta_{\mathfrak{w}\mathfrak{v}})}\right)$$

$$\lambda_{[z]} = \mathfrak{w} + \mathfrak{v} - \text{sign}(\sin(\delta_{\mathfrak{w}\mathfrak{v}}))\mathfrak{w}\sqrt{\frac{1}{1+\left(\frac{\cos(\delta_{\mathfrak{w}\mathfrak{v}})+\frac{\mathfrak{w}}{\mathfrak{v}}}{\sin(\delta_{\mathfrak{w}\mathfrak{v}})}\right)^2}} \frac{2\cos(\delta_{\mathfrak{w}\mathfrak{v}})+\frac{\mathfrak{v}}{\mathfrak{w}}\frac{\mathfrak{w}}{\mathfrak{v}}}{\sin(\delta_{\mathfrak{w}\mathfrak{v}})}$$

$$\lambda_{[z]} = \mathfrak{w} + \mathfrak{v} - \mathfrak{w}\sqrt{\frac{1}{1+(\frac{\cos(\delta_{\mathfrak{w}\mathfrak{v}})+\frac{\mathfrak{w}}{\mathfrak{v}}}{\sin(\delta_{\mathfrak{w}\mathfrak{v}})})^2}} \frac{2\cos(\delta_{\mathfrak{w}\mathfrak{v}})+\frac{\mathfrak{v}}{\mathfrak{w}}\frac{\mathfrak{w}}{\mathfrak{v}}}{|\sin(\delta_{\mathfrak{w}\mathfrak{v}})|}$$

Now consider $z' = n - z$. Then clearly the corresponding $\delta'_{\mathfrak{w}\mathfrak{v}} = 2\pi - \delta_{\mathfrak{w}\mathfrak{v}}$. As $\cos(2\pi - \alpha) = \cos(\alpha)$ and $|\sin(2\pi - \alpha)| = |\sin(\alpha)|$, we get that $\lambda_{[z']} = \lambda_{[z]}$. Similarly foir the sine based λs. Therefore we need to reject the eigenvalues and eigenvectors as described above. This completes the proof.

At the end, let us have a look at Fig. 2, presenting eigenvalue spectrum for an even number of nodes $n = 50$ and for an odd number of nodes $n = 51$. It illustrates the changes to the spectrum when the proportion of alternating weights changes. $\mathfrak{w}$ was fixed at weights 2, while $\mathfrak{v}$ takes on values 9,6,3 and 2. One sees that if both weights are equal, the spectrum id somehow "continuous", while increasing disproportions move one part of the spectrum upwards.

3.5   Multidimensional Case

For multidimensional grids the eigenvalues are sums of component eigenvalues and the eigenvector components are products of component eigenvector components, like in case of weighted eigenvectors and eigenvalues.

Note that contrary to unweighted and weighted grids, the eigenvector components for biweighted grids depend on the weights.

3.6   Eigensolutions of Unoriented Laplacians of Biweighted Grid Graphs

These are easily derived from the combinatorial Laplacian eigenvalues (identical) and eigenvectors (with alternating signs of components), just like in the unweighted and singly weighted case.

4   Conclusions

We have presented a closed-form method of computation of all eigenvalues and eigenvectors of a biweighted path grid graph for combinatorial Laplacians. Their properties may be of interest as generalisations of results of [5,11,12]. The closed-form formulas for eigenvalues and eigenvectors of bi-weighted grid graphs may be of high interest to researchers dealing with cluster analysis of graphs [8], especially with spectral cluster analysis, and compressive spectral clustering (CSC) [19]. While unweighted grid graphs can be considered as types of graphs that have no intrinsic cluster structure, the bi-weighted grid graphs can be considered as types of graphs that have either no intrinsic cluster structure (when the weights are equal) or the structure of which can be twisted in various ways. The weights permit to simulate node clusters not perfectly separated from each other, with various shades of this imperfection. This fact opens new possibilities for exploitation of closed-form solutions eigenvectors and eigenvalues of graphs while testing and/or developing such algorithms and exploring their theoretical properties. This is particularly true for tests on grids with billions of nodes where typical numerical procedures suffer from space and time problems.

As increasing interest in weighted graph Laplacians exists, it would be an interesting research topic to find also closed form solutions to Laplacians of weighted graphs with other weighting schemas than those assumed in this work. Also the results presented here may be a starting point for finding solutions for normalized, random walk and other Laplacians in the spirit of the paper [12].

As mentioned in Sect. 2, the biweighted grid graph problem may be viewed as a special case of tridiagonal matrices, investigated by [10], where in our case the sub- and superdiagonals are identical (though not constant). His theorems 3.2. and 3.4 are relevant here. Our results were derived independently of [10]. As one would expect, the derivation and the formulas are simpler, in particular for eigenvalues (e.g. no square rooting) and eigenvectors (e.g. either sine or cosine is computed once for each vector element). Our solution seems also be better suited for generalisations to normalised and random walk Laplacians because it follow the spirit of [12]. Normalised and random walk Laplacians cannot be handled by [10] as they are not tridiagonal matrices as defined in [10].

References

1. Burden, R.L., Hedstrom, G.W.: The distribution of the eigenvalues of the discrete laplacian. BIT Numer. Math. **12**(4), 475–488 (1972). https://doi.org/10.1007/BF01932957
2. Cheung, G., Magli, E., Tanaka, Y., Ng, M.K.: Graph spectral image processing. Proc. IEEE **106**(5), 907–930 (2018)
3. Cornelissen, G., Kato, F., Kool, J.: A combinatorial Li-Yau inequality and rational points on curves. Math. Ann. **361**(1), 211–258 (2015)
4. Cvetković, D.M., Doob, M., Sachs, H.: Spectra of Graphs: Theory and Application. Academic Press, Cambridge (1980)
5. Edwards, T.: The Discrete Laplacian of a Rectangular Grid (2013). https://sites.math.washington.edu/~reu/papers/2013/tom/DiscreteLaplacianofaRectangularGrid.pdf
6. Fan, Y.Z., Tam, B.S., Zhou, J.: Maximizing spectral radius of unoriented laplacian matrix over bicyclic graphs of a given order. Linear Multilinear Algebra **56**, 381–397 (2008)
7. Fiedler, M.: Algebraic connectivity of graphs. Czech. Math. J. **23**(98), 298–305 (1973)
8. Gallier, J.: Spectral Theory of Unsigned and Signed Graphs. Applications to Graph Clustering: a Survey. arXiv preprint arXiv:1601.04692 (2017)
9. Anderson Jr, W.N., Morley, T.D.: Eigenvalues of the laplacian of a graph. Linear Multilinear Algebra **18**(2), 141–145 (1985). https://doi.org/10.1080/03081088508817681
10. Kouachi, S.: Eigenvalues and eigenvectors of tridiagonal matrices. ELA. Electron. J. Linear Algebra **15**, 115–133 (2006)
11. Kłopotek, M.A., Wierzchoń, S.T., Kłopotek, R.A.: Analytical forms of normalized and combimnatorial laplacians of grid graphs. In: Proceedings of PP-RAI 2019, pp. 281–284 (2019)
12. Kłopotek, M.A., Wierzchoń, S.T., Kłopotek, R.A.: Weighted laplacians of grids and their application for inspection of spectral graph clustering methods. TASK Q. **25**(3), 329–353 (2021). https://doi.org/10.34808/tq2021/25.3/d
13. Merris, R.: Laplacian matrices of graphs: a survey. Linear Algebra Appl. **197**, 143–176 (1994). https://doi.org/10.1016/0024-3795(94)90486-3
14. Notarstefano, G., Parlangeli, G.: Controllability and observability of grid graphs via reduction and symmetries (2012). https://arxiv.org/abs/1203.0129
15. Ramachandran, R.K., Berman, S.: The effect of communication topology on scalar field estimation by networked robotic swarms (2016). https://arxiv.org/abs/1603.02381
16. Sevi, H., Jonckheere, M., Kalogeratos, A.: Generalized spectral clustering for directed and undirected graphs (2022). https://doi.org/10.48550/ARXIV.2203.03221. https://arxiv.org/abs/2203.03221
17. Spielman, D.: Specral graph theory and its applications incomplete draft, dated 4 December 2019. http://cs-www.cs.yale.edu/homes/spielman/sagt

18. Stankewicz, J.: On the gonality, treewidth, and orientable genus of a graph (2017). https://arxiv.org/abs/1704.06255
19. Tremblay, N., Puy, G., Gribonval, R., Vandergheynst, P.: Compressive spectral clustering. In: Proceedings of the 33rd International Conference on International Conference on Machine Learning, ICML 2016, vol. 48, pp. 1002–1011. JMLR.org (2016)
20. Tu, J., Mei, G., Picciallib, F.: An improved Nyström spectral graph clustering using k-core decomposition as a sampling strategy for large networks. J. King Saud Univ. Comput. Inf. Sci. (2022). https://doi.org/10.1016/j.jksuci.2022.04.009
21. Wierzchoń, S., Kłopotek, M.: Modern Clustering Algorithms. Studies in Big Data, vol. 34. Springer, Cham (2018). https://doi.org/10.1007/978-3-319-69308-8

Semi-supervised Malicious Domain Detection Based on Meta Pseudo Labeling

Yi Gao[1,2], Fangfang Yuan[1(✉)], Jinglin Yang[3], Dakui Wang[1], Cong Cao[1], and Yanbing Liu[1(✉)]

[1] Institute of Information Engineering, Chinese Academy of Sciences, Beijing, China
{gaoyi,yuanfangfang,wangdakui,caocong,liuyanbing}@iie.ac.cn
[2] School of Cyber Security, University of Chinese Academy of Sciences, Beijing, China
[3] National Computer network Emergency Response Technical Team/Coordination Center of China (CNCERT/CC), Beijing, China
yangjinglin@cert.org.cn

Abstract. The Domain Name System (DNS) is a crucial infrastructure of the Internet, yet it is also a primary medium for disseminating illicit content. Researchers have proposed numerous methods to detect malicious domains, with association-based approaches achieving relatively good performance. However, these methods encounter limitations in detecting malicious domains within isolated nodes and heavily relying on labeled data to improve performance. In this paper, we propose a semi-supervised malicious domain detection model named SemiDom, which is based on meta pseudo labeling. Firstly, we use associations among DNS entities to construct a semantically enriched domain association graph. In particular, we retain isolated nodes within the dataset that lack relationships with other entities. Secondly, a teacher network computes pseudo labels on the unlabeled nodes, which effectively augments the scarce labeled data. A student network utilizes these pseudo labels to transform both the structure and attribute features to domain labels. Finally, the teacher network is constantly optimized based on the student's performance feedback on the labeled nodes, enabling the generation of more precise pseudo labels. Extensive experiments on the real-world DNS dataset demonstrate that our proposed method outperforms the state-of-the-art methods.

Keywords: Malicious domain detection · Semi-supervised Learning · Meta Pseudo Labels

1 Introduction

Domain Name System (DNS) is a vital component of the Internet infrastructure, providing a crucial service of associating domains with IP addresses. Due to its

L. Franco et al. (Eds.): ICCS 2024, LNCS 14833, pp. 312–324, 2024.
https://doi.org/10.1007/978-3-031-63751-3_21

critical role in the Internet, DNS has emerged as an attack vector for cybercriminals, who leverage it for malicious activities such as spamming, phishing, and malware dissemination. Consequently, malicious domain detection is essential to maintaining cyberspace security. Early rule-based detection methods [11,18] rely on list filtering. With malicious domains multiplied, the size of the rule base grew quickly, making it increasingly difficult to maintain and causing a decline in detection performance. To overcome these limitations, feature-based methods were developed [7–9,19]. These methods extract domain features to train a detection model, but the effectiveness of detection is heavily dependent on feature selection, which requires expertise and is vulnerable to feature tampering by attackers. Recently, researchers [15,21–23] have proposed association-based methods that model the correlation between domains and utilize known domains to infer unknown ones. These methods are able to detect a greater number of concealed malicious domains and deliver outstanding detection outcomes. Nevertheless, association-based methods encounter the subsequent two challenges:

- As shown in Fig. 1, there are actually many isolated domain nodes in the real DNS traffic. The association-based methods ignore these nodes to prevent affecting knowledge propagation on the domain association graph (DAG). This operation not only causes a significant loss of domain data, but also loses the ability to detect malicious domains hidden within isolated nodes.
- Since the domain blacklist covers a limited portion of the domains, the labels of most domain nodes are unknown. Figure 1 depicts the training data used by association-based methods, which consists only of labeled domain data, ignoring the large amount of information in the unlabeled data.

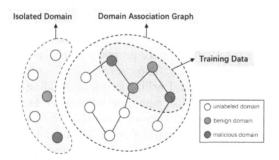

Fig. 1. A domain dataset constructed from real DNS traffic

To address the above challenges, we propose **SemiDom**, a semi-supervised malicious domain detection model based on meta pseudo labeling. In specific, we model the DNS scenario as a semantically rich DAG and preserves a large number of isolated domain nodes in the dataset. First, the *pseudo label generator* as a teacher network employs an adaptive label propagation algorithm to infer pseudo labels on unlabeled nodes in the DAG. Then, the *domain classifier*, which is a student network, evaluates the efficacy of the pseudo label generation and provides feedback to the *pseudo label generator*. Next, the *pseudo-label generator* adjusts the label propagation strategy by the feedback to infer more accurate pseudo labels. Finally, the pseudo-labeled data effectively augments the existing labeled data and helps the *domain classifier* to transform structural features as well as the node attribute features into domain labels.

In summary, the main contributions of this paper are as follows:

1. We utilize the association among domain, IP addresses, and clients to build a DAG, and we additionally preserve a large number of isolated domain nodes in the dataset that are not associated to other entities.
2. We propose SemiDom, a semi-supervised meta pseudo labeling framework that mines the rich information implicit in the unlabeled domain data for malicious domain detection.
3. We conduct extensive experiments on a dataset constructed from real DNS traffic, and the experimental results demonstrate the effectiveness of our proposed method.

2   Related Work

2.1   Malicious Domain Detection

As DNS Flexible technology advances, the size of malicious domains is growing, making rule-based malicious domain detection methods less effective. Researchers [6,8,9,20] have proposed feature-based detection methods, which train classifiers by extracting features from domain characters and DNS traffic. For example, Chin et al. [9] build a machine learning classifier to detect malicious domains using 27 features, including DNS records, average TTL, etc. However, attackers can modify the features of domains to evade the detection system. Recently, researchers [13,15,23,24] have proposed to utilize hard-to-fake associations to detect malicious domains. These association-based methods construct the DNS scenario as a graph and utilize graph embedding algorithms, graph neural networks, etc. to accomplish domain classification. For example, Peng et al. [15] construct a bipartite graph between domains and IP addresses, using RF and XGBoost to classify domains. Wang et al. [23] model the DNS scene as a heterogeneous graph consisting of domains, clients, and IP addresses, and use the HAN model to detect malicious domains. Association-based methods have achieved well detection results, but these methods are unable to detect malicious domains hidden in isolated domain nodes, and also heavily rely on labeled data to train the model. Different from existing works, we propose a semi-supervised malicious domain detection model based on meta pseudo labeling, which utilizes unlabeled data to augment labeled data and is capable of detecting isolated malicious domain nodes.

2.2   Meta Pseudo Labels

Meta Pseudo Labels (MPL) [17] is one of the state-of-the-art semi-supervised learning methods that adopts a Teacher-Student architecture. The teacher network generates pseudo labels for unlabeled data, and the student network learns knowledge on labeled data. In particular, the teacher in Meta Pseudo Labels is not fixed. It is constantly optimized according to the student's performance on labeled data to generate better pseudo labels for teaching students. Peng et al. [16] proposed a federated meta pseudo labeling framework SynFMPL, to address the challenges of limited labeled data and data heterogeneity in federated learning. Meta Pseudo Labels has also been widely used in the field of anomaly detection, Zhao et al. [25] proposed a meta pseudo labeling based anomaly detection framework, MPAD. This framework seeks to obtain valid pseudo anomalies from unlabeled samples to complement the observed anomaly set. In addition, Zhou et al. [26] improved the Meta Pseudo labels for recommendation attack detection by using an experienced teacher network to generate a set of student networks instead of only one student in the original Meta Pseudo Labels. To the best of our knowledge, there is no prior work that applies the Meta Pseudo Labels to malicious domain detection.

3   Preliminary

Definition 1. *Domain Association Graph*. *We define a Domain Association Graph (DAG) as $\mathcal{G} = (\mathcal{V}, \mathcal{E}, \mathcal{X}, \mathcal{Y})$, where $\mathcal{V}$ represents the set of domain nodes, and $\mathcal{E}$ represents the set of undirected edges. The feature vector of each domain is represented by $\mathcal{X} = (x_1, \ldots, x_n)$, and the corresponding label matrix is represented by $\mathcal{Y} = (y_1, \ldots, y_n)$.*

Definition 2. *Pseudo Label and Gold Label*. *In semi-supervised learning, human experts typically annotate a limited amount of unlabeled data. These manually annotated labels are **gold label**. In contrast, the model generates labels for the remaining unlabeled data. These labels generated through the model's predictions are **pseudo labels**.*

Definition 3. *Semi-supervised Malicious Domain Detection*. *The given domain dataset $\mathcal{D} = (\mathcal{G}, \mathcal{N})$ contains a DAG $\mathcal{G}$ and a set of isolated domain nodes $\mathcal{N}$ with gold labels. The node set in $\mathcal{G}$ is divided into an unlabeled node set $\mathcal{V}^u$ and a gold-labeled node set $\mathcal{V}^g$. Our goal is to learn a mapping function $\mathcal{U} : \mathcal{X} \to \mathcal{Y}$ that detects malicious domains using unlabeled data $\mathcal{V}^u$ and labeled data $(\mathcal{V}^g, \mathcal{N})$.*

4   Methodology

In this section, we describe SemiDom in detail. It consists of two parts: DAG construction and semi-supervised classification. The overall framework of SemiDom is shown in Fig. 2.

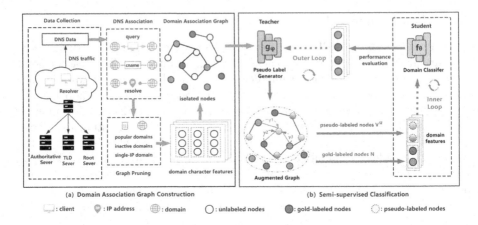

Fig. 2. The overall framework of SemiDom

4.1 Domain Association Graph Construction

Data Collection. The DNS traffic offers comprehensive information about communication exchanges that take place between clients, resolvers, and upper-level DNS servers. We extract the query association between clients and domains, the resolution association between domains and IP addresses, and the CNAME of domains from the DNS traffic. Based on these three types of relationships, we construct a DAG with rich edges as shown in Fig. 2(a). The rules for adding edges in the domain graph are specified as follows:

- **Query:** An edge between two domain nodes is constructed if they have both been requested by the same client. This is because attacked clients are more likely to query malicious domains, while normal clients typically query benign ones.
- **Resolve:** Construct edges between domain nodes that resolve to the same IP address. This is due to the fact that such domains are often registered by the same entity and belong to the same category.
- **CNAME:** Connects a domain to its corresponding domains in the CNAME record. This is because domains in CNAME records usually belong to the same category.

It is important to note that in actual DNS traffic, not all domains are linked based on the three associations mentioned above. In fact, most domains are isolated. Unlike other association-based malicious domain detection methods, we reserve many isolated domain nodes within the dataset to empower the model to detect malicious domains among isolated nodes.

Graph Pruning. Since there are many noisy nodes in the real DNS traffic, which are not beneficial for information propagation and increase the computational pressure, we use the following three strategies to prune the association graph:

- **Popular domains:** These domains, which are requested by more than $T\%$ of clients, tend to be benign domains. This is because these popular domains can be quickly detected by the security management system if they are maliciously attacked.
- **Inactive domains:** The times of visits to these domains are less than Q, producing very few edges in the DAG and lacking valuable information
- **Single-IP Domain:** These domains can only resolve to a single IP address, usually belong to less important, temporary, or testing domains.

Domain Features. We refer to FANCI [19] to extract a total of 21 features (each with a dimension of 41) as the initialization vector of domains. These features consist of three categories: structural, semantic and statistical features, such as domain character length, digit ratio and n-gram frequency distribution.

4.2 Semi-supervised Classification

Semi-supervised classification of malicious domains relies on a meta pseudo labeling framework which comprises a teacher network *pseudo label generator* and a student network *domain classifier*. The *pseudo label generator* infers pseudo labels on unlabeled nodes to teach the *domain classifier*, while it is constantly adapted by the feedback of the student's performance on the labeled nodes.

Pseudo Label Generator (Teacher). To adaptively balance the label information of each node from different neighborhoods, the *pseudo label generator* g_ϕ employs Adaptive Label Propagation (ALP) [10] algorithm to infer pseudo labels of unlabeled nodes. The goal of ALP is to get a an evenly smooth prediction matrix $\hat{\mathbf{Y}}$ through the label matrix $\mathbf{Y}$. Particularly, the propagation strategy of ALP can be described as the following equation:

$$\hat{\mathbf{Y}}_{i,:} = \sum_{k=0}^{K} \gamma_{ik} \mathbf{Y}_{i,:}^{(k)}, \quad \mathbf{Y}^{(k+1)} = \mathbf{T}\mathbf{Y}^{(k)}, \tag{1}$$

where $\mathbf{Y}^{(0)} = \mathbf{Y}$, $\mathbf{T}$ is the transition matrix, and $\mathbf{K}$ is the propagation step. γ_{ik} represents the influence of the k-hop neighborhood of node v_i, which is calculated using the attention mechanism described below:

$$\gamma_{ik} = \frac{\exp\left(\mathbf{a}^{\mathrm{T}} \mathrm{ReLU}\left(\mathbf{W}\mathbf{Y}_{i,:}^{(k)}\right)\right)}{\sum_{k'=0}^{K} \exp\left(\mathbf{a}^{\mathrm{T}} \mathrm{ReLU}\left(\mathbf{W}\mathbf{Y}_{i,:}^{(k')}\right)\right)}, \tag{2}$$

where $\mathbf{a}$ and $\mathbf{W}$ are the learnable attention vector and weight matrix, respectively. After $\mathbf{K}$ iterations, ALP learns a smooth predictive label matrix $\hat{\mathbf{Y}}$ that captures the label distribution of the k-hop neighborhood of node v_i. This matrix adjusts the impact of label propagation at each node while also capturing rich structural information on the graph.

Domain Classifier (Student). After encoding the structural knowledge of the DGA into pseudo labels, we constructed a *domain classifier* f_θ to transform the domain node features into node labels. The process of predicting labels can be formulated as follows:

$$\mathbf{P}_{i,:} = f_\theta\left(\mathbf{X}_{i,:}\right), \tag{3}$$

where f_θ is a multilayer perceptron with the addition of a softmax function. $\mathbf{X}_i$ and $\mathbf{P}_i$ are the feature and prediction label of the domain node v_i, respectively. The training data for the *domain classifier* consists of two parts: the set of unlabeled nodes $\mathcal{V}^u$ in the DAG and the set of gold-labeled isolated nodes $\mathcal{N}$.

Bi-level Optimization of Parameters. Ideally, generated pseudo labels should have the same contribution as the gold labels if they are accurate enough. Therefore, the optimization objective of SemiDom can be described as: *the generated pseudo labels should maximize the performance of the domain classifier at the gold-labeled nodes.* This objective implies a bi-level optimization problem with ϕ as the outer-loop parameters and θ as the inner-loop parameters.

Student (Inner-loop) Update: For gold-labeled nodes sampled from the isolated node set $\mathcal{N}$, we use their real labels as the ground truth. However, for the unlabeled nodes sampled from $\mathcal{V}^u$, we use the generated pseudo labels as the ground truth. The *domain classifier* updates θ according to the following equation:

$$\theta' = \theta - \eta_\theta \nabla_\theta J_{\text{pseudo}}\left(\theta, \phi\right), \tag{4}$$

where η_θ denotes the inner learning rate. $J_{\text{pseudo}}\left(\theta, \phi\right)$ denotes the loss of the *domain classifier* which is calculated on a batch of pseudo-labeled nodes and gold-labeled nodes.

Teacher (Outer-loop) Update: The parameters of the *pseudo label generator* are updated with the learning rate η_ϕ as follows:

$$\phi' = \phi - \eta_\phi \nabla_\phi J_{\text{gold}}\left(\theta'(\phi)\right), \tag{5}$$

where $J_{\text{gold}}\left(\theta'(\phi)\right)$ is the outer loop loss computed on gold-labeled nodes which is back-propagated to calculated the gradient for the *domain classifier*.

To compute the gradient of ϕ, we utilize chain rule to differentiate $J_{\text{gold}}\left(\theta'(\phi)\right)$ with respect to ϕ through intermediate function $\theta'(\phi)$. This is expressed in the following equation:

$$\nabla_\phi J_{\text{gold}}\left(\theta'(\phi)\right) \approx -\frac{\eta_\phi}{2\epsilon}\left[\nabla_\phi J_{\text{pseudo}}\left(\theta^+, \phi\right) - \nabla_\phi J_{\text{pseudo}}\left(\theta^-, \phi\right)\right], \tag{6}$$

where $\theta'(\phi) = \theta - \eta_\theta \nabla_\theta J_{\text{pseudo}}\left(\theta, \phi\right)$, $\theta^\pm = \theta \pm \epsilon \nabla_{\theta'} J_{\text{gold}}\left(\theta'(\phi)\right)$, and ϵ is a small scalar used for finite difference approximation in the computation of gradients.

In the above meta pseudo labeling framework, the *pseudo label generator* adjusts its label propagation strategy based on the *domain classifier*'s feedback to generate higher-quality pseudo labels. Using these improved pseudo labels, we can train a more precise and reliable *domain classifier*. The parameters of the *domain classifier* in the inner loop and the parameters of the *pseudo label generator* in the outer loop are updated alternatively. Algorithm 1 illustrates the complete detection algorithm.

Algorithm 1. The learning algorithm of SemiDom

Input: a DAG $\mathcal{G} = (\mathcal{V}, \mathcal{E}, \mathcal{X}, \mathcal{Y})$ with unlabeled node set $\mathcal{V}^u$ and gold-labeled node set $\mathcal{V}^g$, isolated node set $\mathcal{N}$, training epochs E, inner-loop learning rate η_θ and outer-loop learning rate η_ϕ.
Output: The well-trained *domain classifier*
1: Initialize the parameters ϕ and θ
2: **while** $e < E$ **do**
3:    Randomly sample a batch of nodes from $\mathcal{V}^u$ and $\mathcal{N}$.
4:    ▷ *Pseudo Label Generation*
5:    Compute the pseudo labels for sampled unlabeled nodes using the *pseudo label generator* g_ϕ.
6:    ▷ *Inner-loop update for θ.*
7:    Compute $J_{\text{pseudo}}(\theta, \phi)$ using the generated pseudo-labeled nodes and gold-labeled nodes.
8:    Update parameters θ of the *domain classifier* f_θ via Eq.(4).
9:    ▷ *Outer-loop update for ϕ.*
10:   Randomly sample a batch of nodes from $\mathcal{V}^g$ and $\mathcal{N}$
11:   Compute $J_{\text{gold}}(\theta'(\phi))$ on the labeled nodes using the updated *domain classifier*.
12:   Update parameters ϕ of the *pseudo label generator* g_ϕ via Eq.(5) and Eq.(6).
13: **end while**
14: **return** The well-trained *domain classifier*

---

5 Experiments

In this section, we evaluate the performance of SemiDom by conducting experiments on a dataset constructed from real DNS traffic. Further, we analyze the impact of the *pseudo label generator* and the sensitivity of hyper-parameters.

5.1 Dataset

To evaluate the performance of the model, we collected actual DNS traffic data from August 31, 2020 to September 13, 2020 for a total of two weeks. In order to build a highly connected DAG, we construct connecting edges between two domain nodes as long as any one of the three associations mentioned in Sect. 4.1 exists between these two nodes. It is worth emphasizing that domains collected from real traffic may not have any of the above associations, so we keep these isolated nodes in the dataset in order to detect malicious domains latent in them. In addition, we utilize the whitelist Alexa top 1M [1] to label benign domains, and the blacklists PhishTank [5], CoinBlockerLists [3], Malwaredomains [4] and AnudeepND [2] to label malicious domains. Finally, the dataset we constructed contains 101,023 isolated nodes, 104,583 associated nodes and 117,990 edges.

5.2   Baselines

In order to verify the effectiveness of our proposed SemiDom, we compared it with the following five baselines:

- **LP** [27]: LP (Label Propagation), a classical semi-supervised learning method, assumes that samples closely in the sample space are more similar. It ignores sample attribute features and uses only structural information.
- **GCN** [14]: GCN is a classical homogeneous graph neural network that uses the adjacency matrix and the node feature matrix to learn node representations through aggregation and convolution operations.
- **FANCI:** [19] FANCI is a feature-based malicious domain detection method. It analyzes the feature patterns of domains and classifies them using machine learning methods such as Support Vector Machine and Random Forest.
- **IpDom** [12]: This is a homogeneous graph-based method for malicious domain detection, which we refer to as IpDom for short. It utilizes the resolving relationship between domains and IP addresses to establish associations, and learns domain representations through an improved DeepWalk.
- **DeepDom** [22]: DeepDom is one of the state-of-the-art heterogeneous graph-based methods for malicious domain detection. It represents the DNS scenario as a heterogeneous graph, and uses short random walks based on meta-paths to guide the convolution operation.

5.3   Evaluation Metrics and Parameter Settings

To evaluate the models' performance, we use three standard metrics: precision (P), recall (R) and F1. Additionally, we draw ROC curves for each method, and the area under the curve (AUC) comprehensively evaluates the performance of the binary classification model. SemiDom is built on the PyTorch of version 1.9 and is executed for 200 iterations. The label propagation step K of the *pseudo label generator* is 10 and the learning rate η_ϕ is 0.0005. The *domain classifier* is configured with a 2-layer DNN and the learning rate η_θ is 0.001. The graph pruning strategy uses $T = 100$ and $Q = 5$. To ensure a fair comparison, the same number of isolated nodes are kept in the input data for all methods.

5.4   Performance Evaluation

Overall Performance. For each method, we conduct three separate sets of experiments using 10%, 30%, and 50% of the training set. Table 1 shows the experimental result with the best one marked in bold. Figure 3 illustrates the

ROC curves for different models under the label ratio of 10%. Based on this information, we can draw the following conclusions:

1. SemiDom significantly outperforms the structure-based LP and feature-based FANCI. This is because SemiDom more comprehensively considers both the attribute features and structural knowledge of domains.
2. SemiDom exhibits superior performance compared to the IpDom. The reason is that SemiDom retains isolated domain nodes in real DNS traffic and can detect malicious domains hidden in isolated nodes.
3. SemiDom outperforms GCN and DeepDom due to the fact that these two approaches rely on a large amount of labeled data to improve model performance, ignoring the rich information hidden in unlabeled data.

Table 1. Performance comparison of all methods under different label ratios

Label Ratio	10%			30%			50%		
Model	P	R	F1	P	R	F1	P	R	F1
LP	0.7327	0.7408	0.7366	0.7372	0.7402	0.7380	0.7401	0.7459	0.7423
GCN	0.8398	0.8427	0.8405	0.8440	0.8562	0.8499	0.8536	0.8619	0.8589
FANCI	0.8799	0.8650	0.8721	0.8802	0.8703	0.8780	0.8825	0.8754	0.8810
IpDom	0.7352	0.7388	0.7364	0.7423	0.7480	0.7454	0.7560	0.7594	0.7566
DeepDom	0.8982	0.9083	0.9010	0.9035	0.9165	0.9098	0.9156	0.9233	0.9189
SemiDom	**0.9519**	**0.9638**	**0.9553**	**0.9585**	**0.9780**	**0.9606**	**0.9610**	**0.9789**	**0.9676**

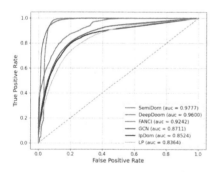

Fig. 3. ROC for each method   **Fig. 4.** Experimental results of SemiDom

SemiDom's Performance with Different Labeling Ratios. Figure 4 visualizes the experimental results of SemiDom at different label ratios. It can be observed that when the ratio of labeled data grows, SemiDom exhibits improved classification performance. The results indicate that SemiDom can achieve superior detection performance on datasets with an increased proportion of labeled data.

SemiDom at 50% Labeling vs. Baselines at 100% Labeling. We train all supervised learning methods in the baselines on the full training set, while SemiDom is trained on only 50% of the training set. As we can see in Table 2, SemiDom outperforms the other methods using only half of the labeled data, which fully validates the effectiveness of SemiDom.

Table 2. Performance: SemiDom at 50% Labeling vs. Baselines at 100% Labeling

Model	P	R	F1
GCN	0.8832	0.8901	0.8872
FANCI	0.8927	0.8990	0.8966
IpDom	0.7742	0.7787	0.7781
DeepDom	0.9339	0.9398	0.9361
SemiDom	**0.9610**	**0.9789**	**0.9676**

5.5   Model Analysis

Impact of the *Pseudo Label Generator*. In order to verify the enhancement of the *pseudo label generator* to the *domain classifier*, we designed **w/o ALP**, in which we removed SemiDom's *pseudo label generator* and used labeled domain nodes to train the *domain classifier*. In this experiment, the label ratio of SemiDom is set to 10%. As shown in Fig. 5, SemiDom has obvious advantages over **w/o ALP** when the number of labeled data is limited. This is because the *pseudo label generator* generates pseudo labels for unlabeled nodes, effectively augmenting the labeling data, whereas **w/o ALP** learns only limited knowledge from the labeled data. Moreover, the pseudo labels generated based on the labels of neighboring nodes encode the structural information of the DAG, while **w/o ALP** focuses only on attribute features.

Hyper-Parameter Sensitivity Study of K. The label propagation step K determines the number of times the label information is updated in the DAG. We study the hyper-parameter sensitivity of K in the *pseudo label generator*. As shown in Fig. 6, the performance of the model improves significantly as the number of iterations increases from $K = 2$ to $K = 10$. However, after $K = 10$ is reached, the performance gain from continuing to increase K diminishes and is accompanied by slower model convergence and more computational resource consumption. Considering the detection effect and training cost of the model, the number of steps for label propagation is set to $K = 10$.

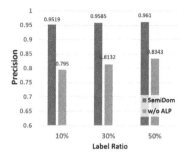

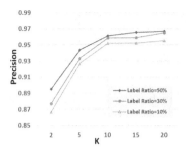

Fig. 5. Impact of pseudo label generator   **Fig. 6.** Hyper-parameter sensitivity study

6 Conclusion

In this paper, we propose SemiDom, a semi-supervised malicious domain detection model based on meta pseudo labeling. We first model the DNS scenario as a domain association graph and retain isolated nodes in the dataset. We then employ a meta pseudo labeling framework which contains a teacher network *pseudo label generator* and a student network *domain classifier*. The *pseudo label generator* infers pseudo labels on unlabeled nodes to teach the *domain classifier*. Meanwhile, it constantly optimizes the label propagation strategy by the feedback from the *domain classifier*'s performance on the labeled nodes. Extensive experiments show that SemiDom outperforms other state-of-the-art methods even with limited labeled data.

Acknowledgment. This work is supported by Xinjiang Uygur Autonomous Region key research and development program (No. 2022B03010).

References

1. alexa-top-sites (2022). https://aws.amazon.com/cn/alexa-top-sites/
2. Anudeepnd (2022). https://github.com/anudeepND/blacklist
3. Coinblockerlists (2022). https://gitlab.com/ZeroDot1/CoinBlockerLists
4. Malware domain block list (2022). http://www.malwaredomains.com/
5. Phishtank (2022). http://www.phishtank.com/
6. Anderson, H.S., Woodbridge, J., Filar, B.: Deepdga: adversarially-tuned domain generation and detection. In: Freeman, D.M., Mitrokotsa, A., Sinha, A. (eds.) Proceedings of the 2016 ACM Workshop on Artificial Intelligence and Security, pp. 13–21. ACM (2016)
7. Antonakakis, M., Perdisci, R., Dagon, D., Lee, W., Feamster, N.: Building a dynamic reputation system for DNS. In: 19th USENIX Security Symposium (USENIX Security 2010) (2010)
8. Bilge, L., Sen, S., Balzarotti, D., Kirda, E., Kruegel, C.: Exposure: a passive DNS analysis service to detect and report malicious domains. ACM Trans. Inf. Syst. Secur. (TISSEC) **16**(4), 1–28 (2014)

9. Chin, T., Xiong, K., Hu, C., Li, Y.: A machine learning framework for studying domain generation algorithm (DGA)-based malware. In: International Conference on Security and Privacy in Communication Systems (2018)

10. Ding, K., Wang, J., Caverlee, J., Liu, H.: Meta propagation networks for graph few-shot semi-supervised learning (2021)

11. Grill, M., Nikolaev, I., Valeros, V., Rehak, M.: Detecting DGA malware using netflow. In: 2015 IFIP/IEEE International Symposium on Integrated Network Management (IM), pp. 1304–1309. IEEE (2015)

12. He, W., Gou, G., Kang, C., Liu, C., Xiong, G.: Malicious domain detection via domain relationship and graph models. IEEE (2019)

13. Khalil, I., Yu, T., Guan, B.: Discovering malicious domains through passive DNS data graph analysis. In: Proceedings of the 11th ACM on Asia Conference on Computer and Communications Security, pp. 663–674 (2016)

14. Kipf, T.N., Welling, M.: Semi-supervised classification with graph convolutional networks. arXiv preprint arXiv:1609.02907 (2016)

15. Peng, C., Yun, X., Zhang, Y., Li, S.: Malshoot: shooting malicious domains through graph embedding on passive DNS data. In: Collaborative Computing (2018)

16. Peng, T., Chiu, T., Pang, A., Tail, W.: Synfmpl: a federated meta pseudo labeling framework with synergetic strategy. In: IEEE International Conference on Communications, ICC 2023, Rome, Italy, 28 May–1 June 2023 (2023)

17. Pham, H., Dai, Z., Xie, Q., Le, Q.V.: Meta pseudo labels. In: IEEE Conference on Computer Vision and Pattern Recognition, CVPR 2021, virtual, 19–25 June 2021 (2021)

18. Sato, K., Ishibashi, K., Toyono, T., Hasegawa, H., Yoshino, H.: Extending black domain name list by using co-occurrence relation between DNS queries. IEICE Trans. Commun. **95**(3), 794–802 (2012)

19. Schüppen, S., Teubert, D., Herrmann, P., Meyer, U.: {FANCI}: feature-based automated {NXDomain} classification and intelligence. In: 27th USENIX Security Symposium (USENIX Security 2018), pp. 1165–1181 (2018)

20. Shi, Y., Chen, G., Li, J.: Malicious domain name detection based on extreme machine learning. Neural Process. Lett. **48**(3), 1347–1357 (2018)

21. Sun, X., Tong, M., Yang, J., Xinran, L., Heng, L.: {HinDom}: a robust malicious domain detection system based on heterogeneous information network with transductive classification. In: 22nd International Symposium on Research in Attacks, Intrusions and Defenses (RAID 2019), pp. 399–412 (2019)

22. Sun, X., Wang, Z., Yang, J., Liu, X.: Deepdom: malicious domain detection with scalable and heterogeneous graph convolutional networks. Comput. Secur. **99**, 102057 (2020)

23. Wang, Q., et al.: Handom: heterogeneous attention network model for malicious domain detection. Comput. Secur. **125**, 103059 (2023)

24. Zhang, S., et al.: Attributed heterogeneous graph neural network for malicious domain detection. In: 2021 IEEE 24th International Conference on Computer Supported Cooperative Work in Design (CSCWD), pp. 397–403. IEEE (2021)

25. Zhao, S., Yu, Z., Wang, X., Marbach, T.G., Wang, G., Liu, X.: Meta pseudo labels for anomaly detection via partially observed anomalies. In: Database Systems for Advanced Applications - 28th International Conference, DASFAA 2023, Tianjin, China, 17–20 April 2023, Proceedings, Part IV (2023)

26. Zhou, Q., Li, K., Duan, L.: Recommendation attack detection based on improved meta pseudo labels. Knowl. Based Syst. **279**, 110931 (2023)

27. Zhu, X.: Learning from labeled and unlabeled data with label propagation. Tech Report (2002)

Elimination of Computing Singular Surface Integrals in the PIES Method Through Regularization for Three-Dimensional Potential Problems

Krzysztof Szerszeń$^{(\boxtimes)}$ (ID) and Eugeniusz Zieniuk (ID)

Faculty of Computer Science, University of Bialystok, Konstantego Ciołkowskiego 1M, 15-245 Białystok, Poland
{k.szerszen,e.zieniuk}@uwb.edu.pl

Abstract. We propose a technique to circumvent the direct computation of singular surface integrals in parametric integral equation system (PIES) employed for solving three-dimensional potential problems. It is based on the regularization of the original singular PIES formula, resulting in the simultaneous elimination of both strongly and weakly singular integrals. As a result, there is the possibility of numerically calculating the values of all integrals in the obtained formula using standard Gaussian quadrature. The evaluation of accuracy for the proposed approach is examined through an illustrative case, specifically focusing on the steady-state temperature field distribution problem.

Keywords: Regularized PIES · Singular Integrals · Three-Dimensional Boundary Value Problems · Bézier Surfaces

1 Introduction

Simulation studies typically involve formulating the considered problem as a boundary value problem (BVP) and describing it using partial differential equations (PDEs). Beyond selecting appropriate PDE, it becomes imperative to define the geometric configuration of the computational domain and establish boundary conditions. While formulating a boundary problem may seem relatively simple, obtaining a direct analytical solution is feasible only for a limited set of problems characterized by simple domain shapes and boundary conditions. In the case of practical problems with more complex geometry and complicated boundary conditions, obtaining a solution becomes a task that requires the application of numerical computational methods. Examples of such methods include the finite difference method (FDM) [1], the finite element method (FEM) [2], the boundary element method (BEM) [3] and meshless methods [4].

Parametric integral equatiom system (PIES) is a computational method for solving two and three-dimansional BVPs, where the necessity of subdividing the boundary and domain into conventional finite and boundary elements has been eliminated [5]. This is made possible through the analytical inclusion of the shape of the considered problem directly in the mathematical formula of PIES, along with the introduction of alternative methods to describe this geometry. Particularly promising is the description of the shape of three-dimensional boundary problems using parametric surface patches [6]. Consequently, instead of employing a mesh of boundary elements with declared nodes, which is characteristic of BEM, the shape of the boundary in the PIES can be defined using a smaller number of parametric surface patches, determined by a relatively small set of control points.

One of the primary challenges encountered in PIES is the computation of weakly and strongly singular integrals. Their presence results from the fact that PIES is an analytical modification of boundary integral equations (BIEs). This modification, previously employed in various differential equation contexts [5, 6], is designed to analytically account for the shape of the boundary problem without the necessity of using elements. PIES transforms the considered problem, originally described by its corresponding differential equation as a boundary-based problem in the case of BIEs, into a parameterized reference domain that maps this boundary. In 2D problems, this involves integrating along a parameterized straight line corresponding to the parameterized boundary contour, and in 3D problems, it requires integration over a two-dimensional parameterized plane representing the boundary surface.

Recognizing the importance of computing singular integrals, a substantial body of literature has been devoted to thoroughly addressing and advancing this subject. These methods are largely dedicated to BEM, among which we can mention adaptive element subdivision [7], distance transformation [8], variable transformation [9], polar coordinate transformation [10], analytical and semi-analytical methods [11, 12], and quadrature methods [13]. One of the most promising approaches is regularization methods [14–16].

The paper presents a regularization technique for PIES in 3D problems, aiming to eliminate both weak and strong singularities. This approach extends a previous regularization method developed for 2D problems [17–19]. The proposed method involves regularization of the original singular PIES formula by introducing an auxiliary regularization functions, incorporating unknown coefficients, and applying appropriate transformations. The effectiveness of this strategy is demonstrated through the analysis of the steady-state temperature field distribution governed by Laplace's equation.

2 PIES for the Laplace Equation in 3D and Its Numerical Solution

We consider the analysis of the steady-state temperature field within a three-dimensional domain Ω bounded by the boundary Γ. Mathematically, the problem is described as a boundary value problem for the Laplace equation

$$\frac{\partial^2 u}{\partial x_1^2} + \frac{\partial^2 u}{\partial x_2^2} + \frac{\partial^2 u}{\partial x_3^2} = 0, \tag{1}$$

with prescribed Dirichlet boundary conditions u_Γ and Neumann conditions p_Γ. In the case of practical problems related to more complex shapes of the domain, such as the one illustrated in Fig. 1a, computational methods are employed to determine the temperature field distribution. Figure 1b illustrates the boundary element modeling of the boundary Γ, commonly used in BEM. A similar discretization strategy applies to FEM, but it is confined to the domain defined by finite elements. While such modeling has gained popularity, it often results in a large number of elements and a substantial size of algebraic equations to be solved in practice.

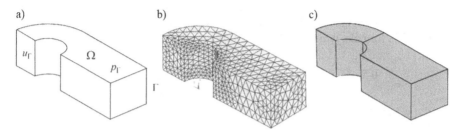

Fig. 1. Sample domain Ω with corresponding boundary conditions (a), discretization of the boundary using boundary elements in BEM (b), an alternative representation of the boundary with surface patches in PIES (c).

To overcome the limitations of FEM and BEM, the PIES method can be employed, enabling a mathematical simplification of the problem by one dimension. Analogous to BIE, the field within the domain Ω is determined by analyzing the boundary Γ of this domain. However, PIES introduces an analytical modification of BIE, transforming the problem from being directly defined on the boundary to one defined on a parameterized reference domain. In the case of 3D problems, this entails mapping the boundary of the

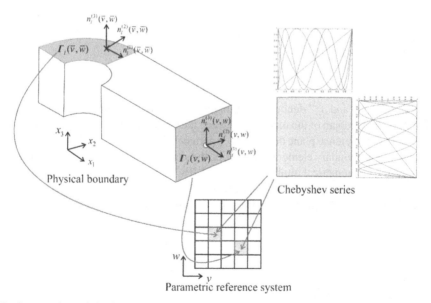

Fig. 2. Mapping of the boundary Γ onto a parameterized plane with the approximation of boundary functions Chebyshev series.

problem onto a parameterized plane. A visual representation of this process can be found in Fig. 2.

The presented approach allows for a general description of the boundary using various mathematical functions. By $\Gamma_j(v, w)$, we define a general parameterized function dependent on parameters v and w from the parameterized reference plane, describing the shape of the j-th segment of the boundary. These functions are analytically integrated into the PIES formula, which, for Laplace's equation, is expressed as [5]

$$0.5u_l(\bar{v}, \bar{w}) = \sum_{j=1}^{N} \int_{v_{j-1}}^{v_j} \int_{w_{j-1}}^{w_j} \{\bar{U}_{lj}^*(\bar{v}, \bar{w}, v, w)p_j(v, w)$$
$$- \bar{P}_{lj}^*(\bar{v}, \bar{w}, v, w)u_j(v, w)\}J_j(v, w)dvdw \tag{2}$$

where $v_{j-1} < \bar{v}, v < v_j, w_{j-1} < \bar{w}, w < w_j, l = 1,2,3,\ldots,n$.

Compared to BIE, the sub-integral functions $\bar{U}_{lj}^*(\bar{v}, \bar{w}, v, w)$ and $\bar{P}_{lj}^*(\bar{v}, \bar{w}, v, w)$ from formula (2) are not directly defined on the boundary but within the parameterized domain of $\Gamma_j(v, w)$. For the Laplace equation, they are defined as follows

$$\bar{U}_{lj}^*(\bar{v}, \bar{w}, v, w) = \frac{1}{4\pi} \frac{1}{\left(\eta_1^2 + \eta_2^2 + \eta_3^2\right)^{0.5,}} \tag{3}$$

$$\bar{P}_{lj}^*(\bar{v}, \bar{w}, v, w) = \frac{1}{4\pi} \frac{\eta_1 n_j^{(1)}(v, w) + \eta_2 n_j^{(2)}(v, w) + +\eta_3 n_j^{(3)}(v, w)}{\left(\eta_1^2 + \eta_2^2 + \eta_3^2\right)^{1.5}}, \tag{4}$$

where

$$\eta_1 = \Gamma_l^{(1)}(\bar{v}, \bar{w}) - \Gamma_j^{(1)}(v, w), \eta_2 = \Gamma_l^{(2)}(\bar{v}, \bar{w}) - \Gamma_j^{(2)}(v, w),$$
$$\eta_3 = \Gamma_l^{(2)}(\bar{v}, \bar{w}) - \Gamma_j^{(3)}(v, w) \tag{5}$$

where $\Gamma_j^{(1)}$, $\Gamma_j^{(2)}$ and $\Gamma_j^{(3)}$ are the scalar components of the function $\Gamma_j(v, w)$. Additionally, $n_j^{(1)}$, $n_j^{(2)}$ and $n_j^{(3)}$ denote the normal derivatives to the boundary, while $J_j(v, w)$ represents the Jacobian of the mapping between the Cartesian coordinate system and the parameterized reference plane dependent on parameters v and w. Formula (2) eliminates the need to use boundary elements to describe the shape of the boundary, as is the case in BEM, which is a numerical implementation of BIEs. In this work, we employ parametric Bézier surfaces for $\Gamma_j(v, w)$. Figure 1c illustrates the definition of the boundary, showcasing an example that utilizes six first-degree and seven third-degree Bézier surface patches.

PIES also allows for the separation of the boundary declaration, defined by $\Gamma_j(v, w)$, from the approximation on this boundary of the boundary functions denoted in Eq. (2) by $u_j(v, w)$ and $p_j(v, w)$. Boundary functions may take a different form than the functions $\Gamma_j(v, w)$. In this context, we assume that the boundary functions on each Bézier surface denoted by j are expressed as Chebyshev series, taking the following form

$$u_j(v, w) = \sum_{p=0}^{P-1} \sum_{r=0}^{R-1} u_j^{(pr)} T_j^{(p)}(v) T_j^{(r)}(w), \tag{6}$$

$$p_j(v, w) = \sum_{p=0}^{P-1} \sum_{r=0}^{R-1} p_j^{(pr)} T_j^{(p)}(v) T_j^{(r)}(w), \tag{7}$$

where $u_j^{(pr)}$ and $p_j^{(pr)}$ represent the values of successive coefficients in these series. Separating the approximation of boundary functions from the declaration of the boundary shape provides control over the accuracy of solutions obtained on the boundary without affecting the pre-defined functions $\Gamma_j(v, w)$ characterizing the boundary shape. When utilizing Chebyshev series, accuracy improvement is achieved by increasing the number of terms, denoted by P and R. This approach contrasts with BEM, where the same boundary elements both model the boundary shape and determine the field distribution on the boundary, necessitating a re-discretization of the boundary shape when increasing the number of elements to enhance accuracy in boundary solutions.

3 Elimination of Singulatities from PIES Through Regularization

In the case where $l = j$ and $v \to \bar{v}$, $w \to \bar{w}$, the sub-integral function $\bar{U}_{lj}^*(\bar{v}, \bar{w}, v, w)$ is weakly singular, while $\bar{P}_{lj}^*(\bar{v}, \bar{w}, v, w)$ is strongly singular. Utilizing Gaussian quadrature directly for computing integrals in formula (2), without isolating singular points, leads to significant errors that directly impact the accuracy of the solutions. The aim of this study is to eliminate singularities from this formula through regularization. To achieve this, in the first step of the proposed procedure, we rewrite (2) with a modified form of boundary functions, denoted as $\breve{u}_j(v, w)$ and $\breve{p}_j(v, w)$, which is presented as

$$0.5\breve{u}_l(\bar{v}, \bar{w}) = \sum_{j=1}^{N} \int_{v_{j-1}}^{v_j} \int_{w_{j-1}}^{w_j} \{\bar{U}_{lj}^*(\bar{v}, \bar{w}, v, w)\breve{p}_j(v, w) - \bar{P}_{lj}^*(\bar{v}, \bar{w}, v, w)\breve{u}_j(v, w)\} J_j(v, w) dv dw. \tag{8}$$

In our considerations, we assume that $\breve{u}_j(v, w)$ takes the following form

$$\breve{u}_j(v, w) = A_l(\bar{v}, \bar{w})\left[(\Gamma_l^{(1)}(\bar{v}, \bar{w}) - \Gamma_j^{(1)}(v, w) + \Gamma_l^{(2)}(\bar{v}, \bar{w}) - \Gamma_j^{(2)}(v, w) + \Gamma_l^{(3)}(\bar{v}, \bar{w}) - \Gamma_j^{(3)}(v, w)\right] + B_l(\bar{v}, \bar{w}), \tag{9}$$

along with its directional derivative with respect to the normal vector to the boundary, in the form

$$\breve{p}_j(v, w) = A_l(\bar{v}, \bar{w})\left[n_j^{(1)}(v, w) + n_j^{(2)}(v, w) + n_j^{(3)}(v, w)\right], \tag{10}$$

where $A_l(\bar{v}, \bar{w})$ and $B_l(\bar{v}, \bar{w})$ represent regularization coefficients defined as

$$A_l(\bar{v}, \bar{w}) = \frac{p_j(v, w)}{n_l^{(1)}(\bar{v}, \bar{w}) + n_l^{(2)}(\bar{v}, \bar{w}) + n_l^{(3)}(\bar{v}, \bar{w})}, \quad (11)$$

$$B_l(\bar{v}, \bar{w}) = u_l(\bar{v}, \bar{w}). \quad (12)$$

The functions (11, 12) are selected to satisfy the Laplace equation. By subtracting (8) from (2), the final formula for the regularized PIES is derived

$$\sum_{j=1}^{N} \left\{ \int_{v_{j-1}}^{v_j} \int_{w_{j-1}}^{w_j} \bar{U}_{lj}^*(\bar{v}, \bar{w}, v, w)[p_j(v, w) - d_{lj}(\bar{v}, \bar{w}, v, w)p_l(\bar{v}, \bar{w})] - \right.$$
$$\int_{v_{j-1}}^{v_j} \int_{w_{j-1}}^{w_j} \bar{P}_{lj}^*(\bar{v}, \bar{w}, v, w)[u_j(v, w) - u_l(\bar{v}, \bar{w}) - \quad (13)$$
$$\left. g_{lj}(\bar{v}, \bar{w}, v, w)p_l(\bar{v}, \bar{w})] \right\} J_j(v, w)dvdw = 0,$$

where

$$d_{lj}(\bar{v}, \bar{w}, v, w) = \frac{n_j^{(1)}(v, w) + n_j^{(2)}(v, w) + n_j^{(3)}(v, w)}{n_l^{(1)}(\bar{v}, \bar{w}) + n_l^{(2)}(\bar{v}, \bar{w}) + n_l^{(3)}(\bar{v}, \bar{w})}, \quad (14)$$

$$g_{lj}(\bar{v}, \bar{w}, v, w) = \frac{(\Gamma_j^{(1)}(v, w) - \Gamma_l^{(1)}(\bar{v}, \bar{w})) + (\Gamma_j^{(2)}(v, w) - \Gamma_l^{(2)}(\bar{v}, \bar{w})) + (\Gamma_j^{(3)}(v, w) - \Gamma_l^{(3)}(\bar{v}, \bar{w}))}{n_l^{(1)}(\bar{v}, \bar{w}) + n_l^{(2)}(\bar{v}, \bar{w}) + n_l^{(3)}(\bar{v}, \bar{w})}.$$
$$(15)$$

The use of functions $d_{lj}(\bar{v}, \bar{w}, v, w)$ and $g_{lj}(\bar{v}, \bar{w}, v, w)$ eliminates the presence of weak and strong singularities. To find the solution for the formula (13), the collocation method can be employed. Collocation points are be distributed in the parametric domain of PIES and defined by the parameter values $\bar{v}$ and $\bar{w}$. Evaluating (13) at the collocation points will yield a system of algebraic equations which can be expressed in general terms as

$$[G]\{u\} = [H]\{p\}. \quad (16)$$

Its size depends on the number of Bézier patches modeling the boundary and the number of terms in the Chebyshev approximation series. The off-diagonal elements in (16) are determined based on the following non-singular integrals

$$\left[g_{lj}^{(c,d,p,r)} \right] = \int_{v_{j-1}}^{v_j} \int_{w_{j-1}}^{w_j} \bar{P}_{lj}^*\left(\bar{v}^{(c,d)}, \bar{w}^{(c,d)}, v, w \right) T_j^{(p)}(v) T_j^{(r)}(w) J_j(v, w)dvdw, \quad (17)$$

$$\left[h_{lj}^{(c,d,p,r)} \right] = \int_{v_{j-1}}^{v_j} \int_{w_{j-1}}^{w_j} \bar{U}_{lj}^*\left(\bar{v}^{(c,d)}, \bar{w}^{(c,d)}, v, w \right) T_j^{(p)}(v) T_j^{(r)}(w) J_j(v, w)dvdw. \quad (18)$$

The practical application of the presented regularization relies on the introduction of auxiliary matrices $[\tilde{G}]$, $[\widehat{G}]$, $[\tilde{H}]$ incorporating regularization functions (14) and (15)

$$\left[G - diag\left\{ \sum\nolimits_{row} [\tilde{G}] \right\} \right] \{u\} = \left[H - diag\left\{ \sum\nolimits_{row} [\tilde{H}] \right\} - diag\left\{ \sum\nolimits_{row} [\widehat{G}] \right\} \right] \{p\}. \tag{19}$$

The elements of the matrices $[\tilde{G}]$, $[\widehat{G}]$, $[\tilde{H}]$ are as follows

$$\left[\tilde{g}_{ll}^{(c,d,p,r)} \right] = \int_{v_{j-1}}^{v_j} \int_{w_{j-1}}^{w_j} \bar{P}_{lj}^{*}\left(\bar{v}^{(c,d)}, \bar{w}^{(c,d)}, v, w \right) T_j^{(p)}(v) T_j^{(r)}(w) J_j(v, w) dv dw, \tag{20}$$

$$\left[\widehat{g}_{ll}^{(c,d,p,r)} \right] =$$

$$\int_{v_{j-1}}^{v_j} \int_{w_{j-1}}^{w_j} \frac{(\Gamma_j^{(1)}(v,w) - \Gamma_l^{(1)}(\bar{v},\bar{w})) + (\Gamma_j^{(2)}(v,w) - \Gamma_l^{(2)}(\bar{v},\bar{w})) + ((\Gamma_j^{(3)}(v,w) - \Gamma_l^{(3)}(\bar{v},\bar{w}))}{n_l^{(1)}\left(\bar{v}^{(c,d)}, \bar{w}^{(c,d)} \right) + n_l^{(2)}\left(\bar{v}^{(c,d)}, \bar{w}^{(c,d)} \right) + n_l^{(3)}\left(\bar{v}^{(c,d)}, \bar{w}^{(c,d)} \right)} *$$

$$\bar{P}_{jl}^{*}\left(\bar{v}^{(c,d)}, \bar{w}^{(c,d)}, v, w \right) T_j^{(p)}(v) T_j^{(r)}(w) J_j(v, w) dv dw, \tag{21}$$

$$\left[\widehat{h}_{ll}^{(c,d,p,r)} \right] = \int_{v_{j-1}}^{v_j} \int_{w_{j-1}}^{w_j} \frac{n_j^{(1)}(v,w) + n_j^{(2)}(v,w) + n_j^{(3)}(v,w)}{n_l^{(1)}\left(\bar{v}^{(c,d)}, \bar{w}^{(c,d)} \right) + n_l^{(2)}\left(\bar{v}^{(c,d)}, \bar{w}^{(c,d)} \right) + n_l^{(3)}\left(\bar{v}^{(c,d)}, \bar{w}^{(c,d)} \right)} *$$

$$\bar{U}_{jl}^{*}\left(\bar{v}^{(c,d)}, \bar{w}^{(c,d)}, v, w \right) T_j^{(p)}(v) T_j^{(r)}(w) J_j(v, w) dv dw$$

All integrals in (19) can be numerically computed through standard Gaussian quadrature. The comprehensive algorithm for solving the regularized PIES is presented below.

Algorithm for Regularized PIES

Read boundary input data (control points of n Bézier surfaces), **Read boundary conditions**

for $l \leftarrow 1, N$ do //loop over Bézier surfaces

for $j \leftarrow 1, N$ do

if $l == j$ then

for $p \leftarrow 0, P - 1$ do //loop over Chebyshev series

for $r \leftarrow 0, R - 1$ do

for $c \leftarrow 0, C - 1$ do //loop over collocation points

for $d \leftarrow 1, D - 1$ do

for $e \leftarrow 1, N$ do //loop over Bézier surfaces

$\left[g_{ll}^{(c,d,p,r)} \right] \leftarrow int[\bar{P}_{ll}^*(\bar{v}^{(c,d)}, \bar{w}^{(c,d)}, v, w)T_l^{(p)}(v)T_l^{(r)}(w)J_l(v,w) -$

$\bar{P}_{le}^*(\bar{v}^{(c,d)}, \bar{w}^{(c,d)}, v, w)T_e^{(p)}(v)T_e^{(r)}(w)J_e(v,w)]$

$\left[h_{ll}^{(c,d,p,r)} \right] \leftarrow int[\bar{U}_{ll}^*(\bar{v}^{(c,d)}, \bar{w}^{(c,d)}, v, w)T_l^{(p)}(v)T_l^{(r)}(w)J_l(v,w) -$

$\frac{\Gamma_e^{(1)}(v,w) - \Gamma_l^{(1)}(\bar{v},\bar{w}) + \Gamma_e^{(2)}(v,w) - \Gamma_l^{(2)}(\bar{v},\bar{w}) + \Gamma_e^{(3)}(v,w) - \Gamma_l^{(3)}(\bar{v},\bar{w})}{n_l^{(1)}(\bar{v}^{(c,d)}, \bar{w}^{(c,d)}) + n_l^{(2)}(\bar{v}^{(c,d)}, \bar{w}^{(c,d)}) + n_l^{(3)}(\bar{v}^{(c,d)}, \bar{w}^{(c,d)})} \bar{P}_{le}^*(\bar{v}^{(c,d)}, \bar{w}^{(c,d)}, v, w)T_e^{(p)}(v)T_e^{(r)}(w)J_e(v,w) -$

$\frac{n_e^{(1)}(v,w) + n_e^{(2)}(v,w) + n_e^{(3)}(v,w)}{n_l^{(1)}(\bar{v}^{(c,d)}, \bar{w}^{(c,d)}) + n_l^{(2)}(\bar{v}^{(c,d)}, \bar{w}^{(c,d)}) + n_l^{(3)}(\bar{v}^{(c,d)}, \bar{w}^{(c,d)})} \bar{U}_{le}^*(\bar{v}^{(c,d)}, \bar{w}^{(c,d)}, v, w)T_e^{(p)}(v)T_e^{(r)}(w)J_e(v,w)]$

end for

end for

end for

end for

end for

insert submatrix $\left[g_{ll}^{(c,d,p,r)} \right]$ to $\left[g_{ll} \right]$ and $\left[h_{ll}^{(c,d,p,r)} \right]$ to $\left[h_{ll} \right]$

else

for $p \leftarrow 0, P - 1$ do //loop over Chebyshev series

for $r \leftarrow 0, R - 1$ do

for $c \leftarrow 0, C - 1$ do // loop over collocation points

for $d \leftarrow 1, D - 1$ do

$\left[g_{lj}^{(c,d,p,r)} \right] \leftarrow int[\bar{P}_{lj}^*(\bar{v}^{(c,d)}, \bar{w}^{(c,d)}, v, w)T_j^{(p)}(v)T_j^{(r)}(w)J_j(v,w)]$

$\left[h_{lj}^{(c,d,p,r)} \right] \leftarrow int[\bar{U}_{lj}^*(\bar{v}^{(c,d)}, \bar{w}^{(c,d)}, v, w)T_j^{(p)}(v)T_j^{(r)}(w)J_j(v,w)]$

end for

end for

end for

end for

insert submatrix $\left[g_{lj}^{(c,d,p,r)} \right]$ to $\left[g_{lj} \right]$ and $\left[h_{lj}^{(c,d,p,r)} \right]$ to $\left[h_{lj} \right]$

end if

end for

end for

transform $[H]\{u\} = [G]\{p\}$ into $[A]\{x\} = \{b\}$, solve system of equations $[A]\{x\} = \{b\}$

---

4   Numerical Examples

The proposed approach is validated through examples featuring analytical solutions. Specifically, the investigation focuses on evaluating the impact of the relative positioning of collocation points and quadrature nodes on the accuracy of solutions.

4.1 Example 1

We consider the problem of temperature distribution in the domain depicted in Fig. 3, with the boundary modeled using 6 Bézier patches. Among them, 5 are degree-1 patches defined by 4 corner points each, and one is a degree-3 patch defined by 16 control points, allowing for the specification of the upper curvilinear part of the boundary. It is assumed that the expected temperature field distribution on the boundary and in the domain would be defined by the following analytical function, dependent on Cartesian coordinates and satisfying Laplace's equation

$$u(x_1, x_2, x_3) = x_1{}^2 + x_2 - x_3{}^2. \tag{23}$$

Based on this function, Dirichlet conditions are specified on each surface patch defining the boundary. Additionally, the normal derivative of this function with respect to the boundary in the form of

$$\frac{\partial u(x_1, x_2, x_3)}{\partial n} = 2x_1 n_1 + n_2 - 2x_3 n_3, \tag{24}$$

represents the analytical solution on the boundary.

To utilize Eqs. (2) and (13) for simulating a stationary temperature field, we apply the collocation method. Collocation points are positioned within the parametric domain of individual Bézier surfaces and are represented by points $\bar{v}$, $\bar{w}$. By expressing Eqs. (2) and (13) at these collocation points, we derive a system of algebraic equations that approximate the PIES. The size of this system is determined by the number of parametric surfaces modeling the boundary and the number of terms in the approximating series (6,7) on individual surface patches.

To find solutions on the boundary, 5×5 collocation points are specified, arranged according to the distribution of roots of Chebyshev polynomials of the second kind. In turn, for numerical integration, a Gauss quadrature of degree 25×25 are applied per Bézier parch, and its nodes are distributed in the same parametric domain. The parameterization of the boundary in PIES allows flexibility in positioning both collocation points and quadrature nodes, identified respectively by $\bar{v}$, $\bar{w}$ and v, w.

Below, we explore two algorithms for evaluating singular integrals in PIES:

- Isolation of singularities in weakly and strongly singular integrals using G-L quadratures, referred to as isolation;
- The proposed regularization, known as non-singular.

In all instances, non-singular integrals are computed using G-L quadrature. As previously mentioned, the proposed regularization employs standard G-L quadrature, with the quadrature nodes uniformly distributed along all boundary segments. This streamlines the node generation process, which may otherwise become recursive. While higher-degree quadrature can enhance precision, it comes at the cost of increased computational time, as demonstrated in the latter part of the example.

The second algorithm, which isolates singularities, also relies on G-L quadrature. However, it necessitates dividing the integration interval containing singular points into subintervals. This substantially complicates the integration process and necessitates two distinct quadrature distributions for boundary segments housing regular and singular integrals, respectively.

In the case of applying formula (2) to calculate the integral where the collocation point is treated as a singular point, it is common to isolate this point, as illustrated in Fig. 3a.

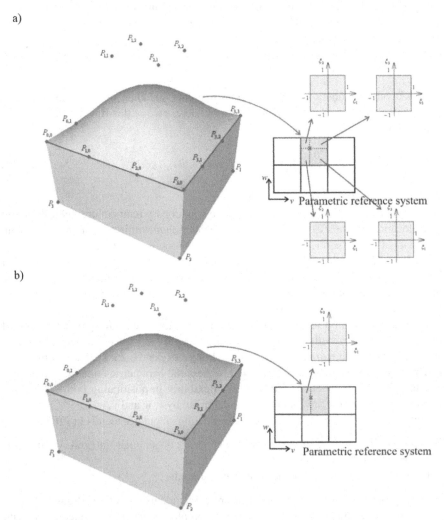

Fig. 3. Mapping of the boundary Γ of the problem onto a parameterized plane with isolating the singular point (a), in the context of the proposed regularization without the need for isolation (b).

In this case, for each Bézier surface, we need to isolate each colocation point by dividing the integration domain into four sections and applying independent Gaussian quadrature to each of them. The drawback of this approach is the necessity to modify the integration intervals for each collocation point. In Fig. 4a, b, two exemplary divisions are presented for two of the considered 5×5 collocation points, where 25×25 Gaussian quadratures were applied for each of the 4 integration intervals. As we can see in this case, we have an uneven distribution of Gaussian quadrature nodes. On the other hand, applying a varying number of quadrature points tailored to the sizes of the four subintervals requires additional intervention in the computational program.

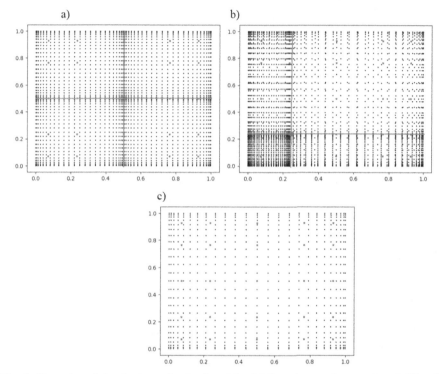

Fig. 4. Two selected distributions of nodes for a 25×25 Gaussian quadrature when dividing the integration domain into 4 parts during the isolation of collocation points $\bar{v}$, $\bar{w}$, with coordinates (0.5, 0.5) (a) and (0.27, 0.27) (b), in the context of the proposed regularization without the need for isolation (c).

When employing the regularized formula (13), isolation is unnecessary, as depicted in Fig. 3b. In this scenario, both collocation points and numerical integration quadrature nodes are within the same interval in the parametric domain of each Bézier patch. This observation is further illustrated in Fig. 4c, showcasing a representative configuration of 5×5 and Gauss quadrature 25×25, without the need to subdivide integration intervals. In Table 1, we investigate how the distance between collocation points and quadrature node points affects the accuracy of boundary solutions.

The results from Table 1 indicate the stability of obtained solutions at the boundary depending on the different distances between the positions of quadrature nodes and collocation points. The presented results refer to the vicinity of one selected point among

Table 1. Influence of the distance between collocation points and Gauss quadrature node on the accuracy of solutions at the boundary.

The coordinates of the central collocation point $\bar{v} = \bar{w}$	The distance between the collocation point and the nearest quadrature node	L_2 norm of the error of solutions at the boundary of the problem (compared to (24))
0.501	1e−3	0.029273
0.5001	1e−4	0.023153
0.50001	1e−5	0.023221
0.500001	1e−6	0.023234
0.5000001	1e−7	0.023197
0.50000001	1e−8	0.146456
0.500000001	1e−9	0.147884

the total of 25 defined. For the remaining points, the presented dependencies are analogous. The obtained results demonstrate the agreement with analytical solutions (24), confirming the effectiveness of the applied strategy.

4.2 Example 2

In the second example, we conducted an analysis of solution accuracy by varying the number of specified collocation points and Gauss quadrature points. The study focused on the domain depicted in Fig. 1a, with the boundary defined by 6 Bézier patches of degree 1 and 7 patches of degree 3. Specifying the shape of these patches involved specifying 112 control points. Dirichlet boundary conditions are imposed on the entire boundary and are defined based on the following function

$$u(x_1, x_2, x_3) = x_1^3 + 2x_2^3 + 3x_3^3 - 3x_1x_3^2 - 6x_2x_1^2 - 9x_3x_2^2. \qquad (25)$$

Meanwhile, the normal derivative of (25) represents the analytical solution at the boundary of this problem.

$$\frac{\partial u(x_1, x_2, x_3)}{\partial n} = n_1\left(3x_1^2 - 3x_3^2 - 12x_2x_1\right) + n_2\left(6x_2^2 - 6x_1^2 - 18x_3x_2\right)$$
$$+ n_3\left(9x_3^2 - 6x_1x_3 - 9x_2^2\right). \qquad (26)$$

In the analysis, the number of collocation points is varied from 2×2 to 5×5, while the number of Gauss quadrature nodes is set at 30×30 and 40×40 assigned to each Bézier patches. Similar to example 1, the application of formula (13) does not necessitate the isolation of a singular point. Figure 5 illustrates two exemplary combinations of collocation points and quadrature nodes that are under examination.

Figure 6 illustrates the L_2 norm of the error in solutions on the boundary as a function of the number of introduced collocation points and Gauss quadrature nodes.

a)   b)

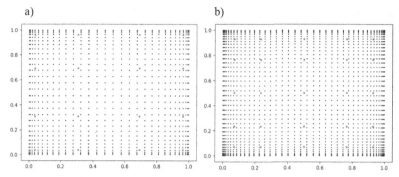

Fig. 5. Two exemplary analyzed combinations of collocation points (in red) and nodes (in black): 4 × 4 collocation points and 30 × 30 quadrature nodes (a), and 5 × 5 collocation points and 40 × 40 quadrature nodes (b). (Color figure online)

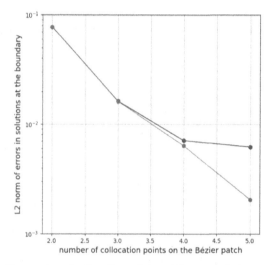

Fig. 6. Influence of the number of collocation points and Gauss quadrature nodes on the accuracy of solutions at the boundary of the problem.

The results again confirm the stability of solutions, even for a more complex domain compared to the first example.

5   Conclusions

The paper demonstrated the possibility of solving PIES without the need to calculate the values of singular integrals. The compiled results indicate a high accuracy of solutions even for a small number of solved algebraic equations. The presented approach can be applied to problems modeled by other differential equations, such as the Navier-Lame or Stokes equations.

References

1. Thomas, J.W.: Numerical partial differential equations: finite difference methods, vol. 22. Springer ScienceBusiness Media, New York 1995 (1998). https://doi.org/10.1007/978-1-4899-7278-1
2. Zienkiewicz, O.C., Taylor, R.L., Zhu, J.Z.: The finite element method: its basis and fundamentals. Elsevier (2005)
3. Brebbia, C.A., Telles, J.C.F., Wrobel, L.C.: Boundary element techniques: theory and applications in engineering. Springer Science & Business Media (2012)
4. Liu, G.R., Gu, Y.T.: An introduction to meshfree methods and their programming. Springer Science & Business Media (2005)
5. Zieniuk, E., Szerszen, K.: The PIES for solving 3D potential problems with domains bounded by rectangular Bézier patches. Eng. Comput. 31(4), 791–809 (2014)
6. Zieniuk, E., Szerszeń, K.: A separation of the boundary geometry from the boundary functions in PIES for 3D problems modeled by the Navier-Lamé equation. Comput. Math. Appl. 75(4), 1067–1094 (2018)
7. Telles, J.: A self-adaptive co-ordinate transformation for efficient numerical evaluation of general boundary element integrals. Int. J. Numer. Meth. Eng. 24(5), 959–973 (1987)
8. Miao, Y., Li, W., Lv, J.H., Long, X.H.: Distance transformation for the numerical evaluation of nearly singular integrals on triangular elements. Eng. Anal. Boundary Elem. 37(10), 1311–1317 (2013)
9. Cerrolaza, M., Alarcon, E.: A bi-cubic transformation for the numerical evaluation of the Cauchy principal value integrals in boundary methods. Int. J. Numer. Meth. Eng. 28(5), 987–999 (1989)
10. Sladek, V., Sladek, J., Tanaka, M.: Optimal transformations of the integration variables in computation of singular integrals in BEM. Int. J. Numer. Meth. Eng. 47(7), 1263–1283 (2000)
11. Zhou, H., Niu, Z., Cheng, C., Guan, Z.: Analytical integral algorithm applied to boundary layer effect and thin body effect in BEM for anisotropic potential problems. Comput. Struct. 86(15–16), 1656–1671 (2008)
12. Niu, Z., Wendland, W.L., Wang, X., Zhou, H.: A semi-analytical algorithm for the evaluation of the nearly singular integrals in three-dimensional boundary element methods. Comput. Methods Appl. Mech. Eng. 194(9–11), 1057–1074 (2005)
13. Hayami, K., Matsumoto, H.: A numerical quadrature for nearly singular boundary element integrals. Eng. Anal. Boundary Elem. 13(2), 143–154 (1994)
14. Shiah, Y.C., Shih, Y.S.: Regularization of nearly singular integrals in the boundary element analysis for interior anisotropic thermal field near the boundary. J. Chin. Inst. Eng. 30(2), 219–230 (2007)
15. Chen, H.B., Lu, P., Schnack, E.: Regularized algorithms for the calculation of values on and near boundaries in 2D elastic BEM. Eng. Anal. Boundary Elem. 25(10), 851–876 (2001)
16. Sladek, V., Sladek, J., Tanaka, M.: Regularization of hypersingular and nearly singular integrals in the potential theory and elasticity. Int. J. Numer. Meth. Eng. 36(10), 1609–1628 (1993)
17. Zieniuk, E., Szerszeń, K.: A regularization of the parametric integral equation system applied to 2D boundary problems for Laplace's equation with stability evaluation. J. Comput. Sci. 61, 101658 (2022)
18. Szerszeń, K., Zieniuk, E., Bołtuć, A., Kużelewski, A.: Comprehensive regularization of PIES for problems modeled by 2D Laplace's equation. In: International Conference on Computational Science, pp. 465–479. Springer International Publishing, Cham (2021). https://doi.org/10.1007/978-3-030-77961-0_38
19. Zieniuk, E., Szerszeń, K.: Near corner boundary regularization of the parametric integral equation system (PIES). Eng. Anal. Boundary Elem. 158, 51–67 (2024)

Application of Neural Graphics Primitives Models for 3D Representation of Devastation Caused by Russian Aggression in Ukraine

Illia Oholtsov[1]([✉])(ID), Yuri Gordienko[1](ID), Mariia Ladonia[1](ID), Sergii Telenyk[2](ID), Grzegorz Nowakowski[2](ID), and Sergii Stirenko[1](ID)

[1] National Technical University of Ukraine, "Igor Sikorsky Kyiv Polytechnic Institute", Kyiv, Ukraine
illia.oholtsov@gmail.com
[2] Cracow University of Technology, Cracow, Poland

Abstract. This work investigates the feasibility of applying Neural Radiance Fields (NeRFs) for reconstructing 3D representations of damaged structures caused by the ongoing aggression of Russia against Ukraine. The drone footage depicting the devastation was utilized and three NeRF models, Instant-NGP, Nerfacto, and SplatFacto, were employed. The models were evaluated across various damage levels (0: no damage, 4: high damage) using visual quality metrics like Structural Similarity Index Measure (SSIM), Learned Perceptual Image Patch Similarity (LPIPS), Peak Signal-to-Noise Ratio (PSNR) and rendering speed metrics like frames per second (FPS) and the number of rays per second (NRS). All input data (videos frames) and evaluation results (rendered visualizations) are available as a Kaggle dataset (http://tiny.cc/srasxz). No clear correlation was observed between damage level and reconstruction quality metrics, suggesting these metrics might not be reliable indicators of damage severity. SplatFacto consistently achieved the highest rendering speed (FPS, NRS) and exhibited the best visual quality (SSIM, PSNR, LPIPS) across all damage levels. The findings suggest that NeRFs, particularly SplatFacto, hold promise for rapid reconstruction and visualization of damaged structures, potentially aiding in damage assessment, documentation, and cultural heritage preservation efforts. Moreover, the study sheds light on the potential applications of such advanced modeling techniques in archiving and documenting conflict zones, providing a valuable resource for future investigations, humanitarian efforts, and historical documentation. However, further research is needed to explore the generalizability and robustness of NeRFs in diverse real-world scenarios.

Keywords: Artificial Intelligence · 3D representation · Neural Radiance Fields · Instant Neural Graphics Primitives · Damage Assessment · Ukraine

Supported in part by the National Research Foundation of Ukraine (NRFU) grant 2022.01/0199.

1   Introduction

Recent advancements in technology demonstrate the feasibility of restoring deteriorated artworks and cultural artifacts to their former glory. Employing advanced 3D reconstruction methodologies, these artifacts can be visually resurrected, even if they have been lost or irreparably damaged. Leveraging photographic documentation or 3D scanning data, it becomes possible to generate accurate three-dimensional models of cultural assets, thereby enabling viewers to engage with them once more. This restoration process extends to various forms of cultural heritage, including rediscovered ruins, damaged historical texts, ancient Egyptian and Greek temples, monumental structures, medieval frescoes, and other significant artifacts, all reconstructed with the aid of state-of-the-art techniques.

The ongoing aggression and war of Russia against Ukraine has left a profound mark on the landscape, particularly evident in the widespread destruction of cities [1,2]. The consequential impact of this aggression underscores the critical need to monitor, visualize, and document the extent of devastation with a depth that surpasses traditional methodologies.

Recently, several studies effectively demonstrated the potential of Neural Radiance Fields (NERFs) for reconstructing 3D representations [3,4]. NERF allows to create 3D models of objects and scenes from images of some scene from different angles. NERF "learns" the scene from these images ("remembering" the object's shape and details) and use this information to generate a realistic 3D model from any viewpoint, even one you didn't take images from before. This can be used for reconstructing damaged structures, capturing historical sites, or even creating immersive virtual experiences.

In this work, the feasibility of application of NERF models for reconstruction of 3D representations for damaged structures caused by Russian aggression in Ukraine is considered. Section 2 contains description of the state of the art, Sect. 3 describes dataset, models, experiments, and the whole workflow, Sect. 4 gives the results obtained during the experiments, Sect. 5 contains discussions of results and resumes them in Sect. 6.

2   Background and Related Work

Neural Radiance Fields (NeRFs) have revolutionized the field of 3D scene representation by offering a novel deep learning approach. NeRFs operate under a paradigm of learning a continuous volumetric representation of a scene from a collection of 2D images captured from various viewpoints [3]. This representation allows for the synthesis of novel views, enabling visualization of the scene from any desired perspective, even those not present in the original capture set.

NeRFs have experienced remarkable progress since their inception [3], effectively addressing initial limitations and paving the way for promising future applications. The Mip-NeRF approach employs a multiscale representation, significantly improving image detail and rendering speed while mitigating aliasing artifacts [5]. Mip-NeRF 360 tackles the challenge of unbounded scenes by enabling the reconstruction of large and intricate environments with 360-degree camera views [6]. Instant Neural Graphics Primitives utilizes multiresolution hash encoding for efficient training and real-time rendering, significantly improving the speed and efficiency of NeRF models [4]. High-Fidelity Reconstruction combines multi-resolution 3D hash grids with rendering, achieving superior surface reconstruction detail, leading to high-fidelity scene representations [7]. K-Planes presents a "white-box" model using k-planes to represent dynamic scenes like videos [8]. This approach enables efficient optimization and incorporation of specific priors, making it suitable for dynamic scene representation. 3D Gaussian Splatting leverages pre-computed 3D Gaussians and visibility-aware rendering, achieving real-time rendering at high quality, opening doors for interactive applications [9].

These advancements collectively demonstrate the rapid development of NeRF technology. By addressing initial limitations, these innovations push the capabilities of NeRFs towards real-time applications, detailed scene reconstruction, and handling complex scenes with diverse material properties and dynamics.

While NeRFs have seen significant advancements, several key challenges remain. For example, training NeRFs on large datasets with high resolution can be computationally expensive and time-consuming. Ongoing research explores techniques for improving training efficiency and reducing memory footprints. NeRFs can struggle with scenes containing complex materials with non-diffuse BRDF (Bidirectional Reflectance Distribution Function) properties or challenging lighting conditions that deviate from the typical assumptions made during training. Real-world data often contains noise, occlusions, or missing information. Improving the robustness of NeRFs to handle such data remains an ongoing area of research that can potentially improve generalization and reconstruction quality for integrating NeRFs into real-world applications. Addressing these challenges is crucial for further advancing the practical applications and capabilities of NeRF technology. That is why the main of the work is to investigate the feasibility of application of NERF models for reconstruction of 3D representations for damaged structures caused by Russian aggression in Ukraine.

3   Methodology

3.1   Data

The data for this project consisted of drone footage depicting the aftermath of Russian aggression in Ukraine. Due to the ongoing conflict, the initial step involved collecting existing videos from various sources, primarily YouTube. While this approach provided a starting point, the preferred method for future data acquisition would be to utilize drone cameras directly. This would enable high-resolution video capture and ensure comprehensive coverage of all key details and angles. All input data (videos frames) and evaluation results (rendered visualizations) are available as a Kaggle dataset[1].

---

[1] http://tiny.cc/srasxz.

To effectively analyze and categorize the scenes within the dataset, a method of differentiation based on "damage levels" was employed. This categorization was crucial for understanding the varying degrees of destruction observed in each scene. Annotators were instructed to assign damage level scores ranging from 0 to 5 to each scene, representing the extent of destruction. This scoring system enabled a quantitative representation of the severity of damage, ensuring consistency and accuracy in the evaluation process. Initially, the damage level classification was based on the existing dataset, which provided a range of damage levels. However, it is important to note that this classification system can be extended to encompass a broader range of destruction scenarios with the inclusion of new data. For instance, level 5 classifications could represent scenes where structures are completely obliterated, leaving only ruins with no discernible remnants of their original form. Incorporating such detailed classifications aims to provide a more nuanced understanding of the extent of devastation caused by the conflict. These efforts contribute to the comprehensive analysis and visualization of the aftermath captured in the drone footage.

Borodyanka Residential Area, Damage Level 4. This dataset focuses on the aftermath of conflict in Borodyanka, characterized by a lack of varied perspectives (Fig. 1). The footage reveals a high level of destruction, providing intricate details and figures. The camera path is complex, offering a nuanced view of the affected area. However, the presence of many background objects introduces a challenge, as limited information is available about these elements, potentially impacting the overall interpretability of the dataset.

Fig. 1. Examples of drone imagery for Borodyanka, damage level 4.

Chernihiv Hotel "Ukraine", Damage Level 3. This dataset features drone footages capturing the aftermath of Russian aggression on the Chernihiv hotel "Ukraine" (Fig. 2). The camera follows a 360-degree path, capturing the scene from all angles and perspectives. The medium to high level of destruction is emphasized, with minimum background objects, focusing primarily on the hotel.

Chernihiv Residential Area, Damage Level 2. This dataset captures drone footages of a residential area in Chernihiv (Fig. 3). The footage follows 360-degree

Fig. 2. Examples of drone imagery for Chernihiv Hotel "Ukraine", damage level 3.

camera path but with several jumps. The level of destruction is medium to low, and the main scene contains numerous objects. There are many background objects, which may impact the clarity of the footage.

Fig. 3. Examples of drone imagery for Chernihiv residential area, damage level 2.

Chernihiv Bridge, Damage Level 1,3. The Chernihiv Bridge dataset show-cases drone footage with a forward-flying camera path, providing a singular perspective of the scene (Fig. 4). The scene exhibits a low level of destruction on one half and a high level on the other. There are minimum background objects, with the main focus on the Bridge.

Fig. 4. Examples of drone imagery for Chernihiv Bridge, damage level 'mixed'.

Chernihiv Cathedral, Damage Level 0. This dataset showcases drone footage captured using a 360-degree camera path, featuring the Chernihiv Cathedral (Fig. 5) with occasional zoom adjustments. The scene focuses primarily on the cathedral and does not depict any destruction.

Fig. 5. Examples of drone imagery for Chernihiv cathedral, damage level 0.

3.2 Models

This section outlines the NeRF models employed in our research, trained and evaluated using the user-friendly NerfStudio framework [10]. NerfStudio provides various implementations of popular NeRF-based methods, including those directly relevant to this work[2].

SplatFacto. The SplatFacto implementation within NerfStudio was used to explore the 3D Gaussian Splatting (3D GS) approach [9]. The model uses default parameters to achieve a balance between speed, rendering quality, and the size of the generated splat file. One noteworthy parameter is cull-alpha-thresh (0.1) that defines the opacity threshold used for culling Gaussians, effectively removing those with minimal contribution to the final rendering.

Instant-NGP. For comparison purposes, NerfStudio's implementation of the Instant Neural Graphics Primitives (Instant-NGP) model [4] was used also and utilizes the following key parameters: grid-resolution (128), resolution of the grid used for the field; grid-levels (4), levels of the grid used for the field; and max-res (2048), maximum resolution of the hashmap for the base multi-layer perceptron (MLP).

Nerfacto. Nerfacto is a unified approach within NerfStudio that combines elements from various research papers to create a fast and high-quality rendering method [11]. It incorporates several key components:

- Pose refinement: This improves the accuracy of the camera poses, leading to a more realistic rendered scene.
- Piecewise sampler: This sampling strategy allocates samples throughout the scene, prioritizing areas with more objects and detail.
- Proposal sampler: This sampler further refines the sample locations by focusing on areas that significantly impact the final image. It utilizes a density function, implemented with a small fused MLP and hash encoding, to identify these crucial regions.

---

[2] https://github.com/nerfstudio-project/nerfstudio/.

- Density field: This field represents a coarse estimation of the scene's density, guiding the sampling process. It is also implemented with a small fused MLP and hash encoding.

Similar to the previous models, Nerfacto utilizes the following common parameters: grid-resolution (128), resolution of the grid used for the field; grid-levels (4), the number of levels within the hierarchical grid structur; and max-res (2048), maximum resolution of the hashmap for the base MLP.

All models share additional default parameters for training:

- the number of steps between saves (default: 2000), i.e. frequency at which the model checkpoints are saved during training,
- the maximum number of iterations (default: 30000), i.e. training iterations,
- the number of steps between batches (default: 500), i.e. the number of training steps before updating the model with a new batch of data.

3.3   Metrics

The paper presents various metrics to evaluate the performance of NERF models in reconstructing 3D representations of damaged buildings. These metrics include:

- SSIM (Structural Similarity Index Measure): measures the visual quality of the reconstructed scene compared to the ground truth. A higher SSIM value indicates better visual quality.
- PSNR (Peak Signal-to-Noise Ratio): measures the peak signal power relative to the noise power, with higher values indicating better fidelity to the original image.
- LPIPS (Learned Perceptual Image Patch Similarity): measures the perceptual similarity between the reconstructed image and the ground truth. A low LPIPS score means that image patches are perceptual similar.
- FPS (Frames Per Second): represents the rendering speed of the model, indicating how many frames can be generated per second.
- NRS (Number of Rays per Second): indicates the computational efficiency of the model, capturing the number of rays the model can process per second.

3.4   Workflow

The initial step involved collecting drone footage capturing the aftermath of Russian aggression in Ukraine. Due to the ongoing conflict, the primary method for acquiring this data was utilizing existing videos from diverse sources, mainly from YouTube. However, in practical scenarios, capturing high-resolution videos using drone cameras would be preferable to ensure comprehensive coverage of all relevant angles and details.

Once the video data is acquired, frames are extracted at a predetermined frame rate. In this work, a frame rate of 2 frames per second (fps) was chosen to achieve a balance between capturing sufficient scene information and maintaining dataset size within manageable limits.

While all extracted frames contribute to the dataset, they vary in informational value. To optimize representation and computational efficiency, frames capturing unique viewpoints or revealing crucial scene details were prioritized. Prioritized frames were those that maintained clarity and focus on the main object, discarding blurred frames or those not primarily focused on the main objects. This selection process aimed to ensure the dataset's quality and relevance for subsequent analyses and model training. Additionally, frames from multiple videos were combined to increase dataset size and viewpoint diversity, enhancing model robustness. This comprehensive approach aimed to create a rich and diverse dataset capable of supporting various analyses and training tasks effectively.

The next step involved processing the prepared frames with the COLMAP which is an open-source general-purpose Structure-from-Motion (SfM) and Multi-View Stereo (MVS) pipeline with a graphical and command-line interface[3]. It serves two primary purposes:

1. Camera Parameter Estimation: COLMAP helped estimate the intrinsic and extrinsic camera parameters associated with each extracted frame. These parameters are crucial for NeRF training, as they relate pixel locations within the images to their corresponding 3D world coordinates.
2. Feature Extraction and Matching: COLMAP was also employed to extract distinctive features from each frame and establish robust matches between corresponding features across different viewpoints. These feature matches provide valuable information about the scene's geometry and facilitate the subsequent NeRF training process.

Though COLMAP scripts usually require model-specific adjustments, NeRF-Studio allows a single script for all models within the framework, improving efficiency and simplifying data preparation. Following COLMAP processing, the NeRFStudio framework was used for training. This framework offers various NeRF model implementations, facilitating the exploration of different approaches. Three specific models were employed in this work:

- Instant-NGP. This model was chosen for its fast training speed, achieved through efficient multi-resolution representations.
- SplatFacto. This model emphasizes rendering efficiency, utilizing pre-computed 3D Gaussians for real-time rendering capabilities.
- NeRFacto. This model combines multi-resolution encoding with neural surface rendering, aiming to achieve high-fidelity reconstructions.

Each of these models was trained on all prepared datasets within the NeRF-Studio environment. This multi-model approach allowed the comparison of performance and characteristics of each method in the context of reconstructing the specific scene captured in the drone footage.

---

[3] https://colmap.github.io/.

The tools provided within the NeRFStudio framework were leveraged to evaluate the performance of trained models. Before initiating training, each dataset was split into dedicated training and testing sets. This ensured a fair evaluation process where models were assessed on previously unseen frames.

Various evaluation metrics are provided by NeRFStudio and calculated on a per-frame basis. These metrics typically include standard image quality measures like PSNR, SSIM, and LPIPS, offering different perspectives on reconstruction accuracy and visual fidelity. After calculating these metrics for all frames, the mean and standard deviation are computed across the entire test set. This aggregation provides a summary of the overall model performance, indicating both the central tendency and the variability across frames. The training and testing trials were performed in Google Colab environment[4] with access to a graphics processing unit NVIDIA Tesla T4 with 16 GB of video random access memory and 12 GB of system random access memory.

4 Results

After several attempts the mean and standard deviation values of metrics were measured for Instant-NGP (Table 1), Nerfacto (Table 2), and SplatFacto (Table 3) models and visualized in the plots below (Fig. 6 and 7). The maximal values for SSIM, PSNR, LPIPS, FPS, NRS, and minimal values for LPIPS are emphasized by **bold** font.

Based on these results (Table 1, 2, and 3), one can make the following observations regarding the correlations between metrics and damage levels. There is no clear trend between the damage level and the values of SSIM, PSNR, LPIPS, FPS, or NRS across the different objects. While there might be slight variations in the means for each damage level, the standard deviations suggest significant overlap between different damage categories for each metric. This indicates that reconstruction quality metrics alone might not be reliable indicators of damage severity.

Table 1. Mean and standard deviation values of metrics for Instant-NGP model.

Object	Damage	SSIM	PSNR	LPIPS	FPS	NRS
Borodyanka	4	0.38 ± 0.09	15.4 ± 1.8	0.79 ± 0.06	0.10 ± 0.01	$9.3E4 \pm 1.1E4$
Hotel	3	$\mathbf{0.66 \pm 0.08}$	20.8 ± 2.1	$\mathbf{0.31 \pm 0.07}$	0.07 ± 0.01	$6.7E4 \pm 0.8E4$
Residential	2	0.45 ± 0.09	18.2 ± 1.5	0.55 ± 0.10	0.064 ± 0.004	$5.9E4 \pm 0.4E4$
Bridge	mixed	0.49 ± 0.12	$\mathbf{21.6 \pm 1.5}$	0.56 ± 0.06	0.08 ± 0.02	$7.5E4 \pm 1.9E4$
Cathedral	0	0.63 ± 0.08	20.6 ± 2.3	0.44 ± 0.10	$\mathbf{0.13 \pm 0.02}$	$\mathbf{11.9E4 \pm 1.7E4}$

---

[4] https://colab.research.google.com/.

Table 2. Mean and standard deviation values of metrics for Nerfacto model.

Object	Damage	SSIM	PSNR	LPIPS	FPS	NRS
Borodyanka	4	0.38 ± 0.13	15.4 ± 2.4	0.50 ± 0.14	0.13 ± 0.01	11.4E4 ± 8.3E4
Hotel	3	**0.55 ± 0.07**	**18.8 ± 1.4**	**0.28 ± 0.06**	**0.21 ± 0.02**	**19.2E4 ± 1.9E4**
Residential	2	0.38 ± 0.09	16.9 ± 1.8	0.51 ± 0.11	0.13 ± 0.01	11.6E4 ± 6.6E4
Bridge	mixed	0.37 ± 0.14	**18.8 ± 1.6**	0.56 ± 0.05	0.13 ± 0.01	11.9E4 ± 7.3E4
Cathedral	0	0.50 ± 0.10	18.5 ± 2.2	0.44 ± 0.11	0.12 ± 0.01	11.4E4 ± 9.6E4

Table 3. Mean and standard deviation values of metrics for SplatFacto model.

Object	Damage	SSIM	PSNR	LPIPS	FPS	NRS
Borodyanka	4	0.42 ± 0.11	16.1 ± 2.4	0.50 ± 0.07	0.55 ± 0.09	49.0E4 ± 8.8E4
Hotel	3	**0.93 ± 0.01**	**29.0 ± 1.4**	**0.06 ± 0.01**	**0.63 ± 0.08**	**57.6E4 ± 7.5E4**
Residential	2	0.75 ± 0.18	23.3 ± 4.1	0.18 ± 0.11	0.55 ± 0.08	50.8E4 ± 8.0E4
Bridge	mixed	0.78 ± 0.08	23.9 ± 3.2	0.28 ± 0.14	0.56 ± 0.10	51.6E4 ± 9.1E4
Cathedral	0	0.76 ± 0.13	23.1 ± 3.6	0.24 ± 0.09	0.60 ± 0.10	55.0E4 ± 8.6E4

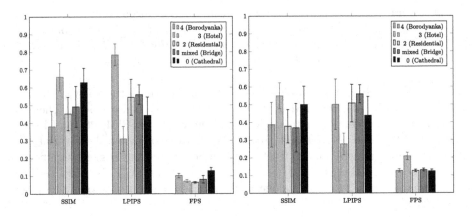

Fig. 6. Graphic visualization of some metrics for the objects with various damage levels for Instant-NGP model from Table 1 (left) and Nerfacto model from Table 2 (right).

5   Discussion

This chapter delves deeper into the performance of the three rendering models: Splatfacto, Nerfacto, and Instant-NGP. The models were assessed under varying object damage levels (0: no damage, 4: high damage) using both visual quality metrics (SSIM, LPIPS, PSNR) and rendering speed metrics (FPS, rays per second).

Based on the results (Table 1, 2, and 3), one can make the following observations regarding the correlations between metrics and damage levels for various models that are summarized in Tables 4, 5, 6, 7, 8 and visualizations (Fig. 8, 9 and 10).

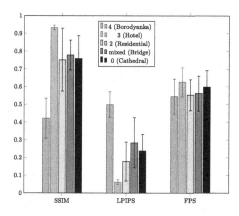

Fig. 7. Graphic visualization of some metrics for the objects with various damage levels for SplatFacto model from Table 3.

Table 4. Mean and standard deviation values of metrics for damage level 0 ("No Damage").

Model	SSIM	PSNR	LPIPS	FPS	NRS
Instant-NGP	0.63 ± 0.08	20.6 ± 2.3	0.44 ± 0.10	0.13 ± 0.02	11.8E4 ± 1.7E4
Nerfacto	0.50 ± 0.10	18.5 ± 2.2	0.44 ± 0.11	0.12 ± 0.01	11.4E4 ± 9.6E4
SplatFacto	**0.76 ± 0.13**	**23.1 ± 3.6**	**0.24 ± 0.09**	**0.60 ± 0.09**	**55.0E4 ± 8.6E4**

Table 5. Mean and standard deviation values of metrics for the damage level 2.

Model	SSIM	PSNR	LPIPS	FPS	NRS
Instant-NGP	0.45 ± 0.09	18.2 ± 1.5	0.54 ± 0.10	0.064 ± 0.004	5.9E4 ± 3.9E4
Nerfacto	0.38 ± 0.09	16.9 ± 1.8	0.51 ± 0.11	0.126 ± 0.007	11.6E4 ± 6.6E4
SplatFacto	**0.75 ± 0.18**	**23.3 ± 4.1**	**0.18 ± 0.11**	**0.55 ± 0.09**	**50.9E4 ± 8.0E4**

Table 6. Mean and standard deviation values of metrics for the damage level 3.

Model	SSIM	PSNR	LPIPS	FPS	NRS
Instant-NGP	0.66 ± 0.08	20.8 ± 2.1	0.31 ± 0.07	0.07 ± 0.01	6.7E4 ± 7.9E4
Nerfacto	0.55 ± 0.07	18.8 ± 1.4	0.27 ± 0.06	0.21 ± 0.02	19.2E4 ± 1.9E4
SplatFacto	**0.93 ± 0.01**	**29.0 ± 1.4**	**0.06 ± 0.01**	**0.63 ± 0.08**	**57.6E4 ± 7.5E4**

SplatFacto achieves the highest SSIM and PSNR scores for all damage levels. This indicates that SplatFacto generally produces reconstructions that best preserve the structural similarity between the original scene and the reconstructed one. Nerfacto is the worst one and Instant-NGP is intermediate between Splat-Facto and Nerfacto.

Table 7. Mean and standard deviation values of metrics for the damage level 4.

Model	SSIM	PSNR	LPIPS	FPS	NRS
Instant-NGP	0.38 ± 0.09	15.5 ± 1.8	0.79 ± 0.06	0.10 ± 0.01	$9.3\text{E}4 \pm 1.1\text{E}4$
Nerfacto	0.39 ± 0.13	15.4 ± 2.4	0.50 ± 0.14	0.13 ± 0.01	$11.4\text{E}4 \pm 8.3\text{E}4$
SplatFacto	$\mathbf{0.42 \pm 0.11}$	$\mathbf{16.1 \pm 2.4}$	$\mathbf{0.50 \pm 0.07}$	$\mathbf{0.55 \pm 0.01}$	$\mathbf{49.0\text{E}4 \pm 8.9\text{E}4}$

Table 8. Mean and standard deviation values of metrics for the damage level "mixed".

Model	SSIM	PSNR	LPIPS	FPS	NRS
Instant-NGP	0.49 ± 0.12	21.6 ± 1.5	0.56 ± 0.06	0.08 ± 0.02	$7.5\text{E}4 \pm 1.9\text{E}4$
Nerfacto	0.37 ± 0.14	18.8 ± 1.6	0.56 ± 0.05	0.13 ± 0.01	$11.9\text{E}4 \pm 7.3\text{E}4$
SplatFacto	$\mathbf{0.78 \pm 0.08}$	$\mathbf{23.9 \pm 3.2}$	$\mathbf{0.28 \pm 0.14}$	$\mathbf{0.56 \pm 0.10}$	$\mathbf{51.7\text{E}4 \pm 9.1\text{E}4}$

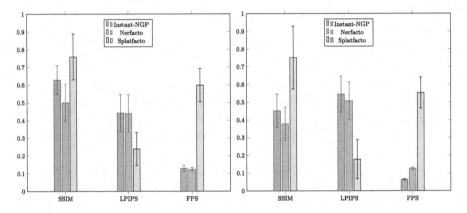

Fig. 8. Graphic visualization of some metrics for various models for the damage level "No Damage" (left) from Table 4 and 2 (right) Table 5.

SplatFacto demonstrates the most consistent and best performance in terms of the lowest LPIPS (perceptual similarity) scores across all damage levels. This indicates that for all damage scenarios SplatFacto generally produces reconstructions with the highest perceptual similarity to the ground truth.

FPS performance of models across damage levels demonstrate that SplatFacto achieves the highest FPS consistently across all damage levels. This indicates that SplatFacto can generate reconstructions at a significantly faster rate compared to other models. Nerfacto has the lowest FPS across all damage levels and it might be computationally expensive and unsuitable for real-time applications. Instant-NGP falls between SplatFacto and Nerfacto and shows a moderate improvement over Instant-NGP but doesn't reach the speed of SplatFacto.

Finally, SplatFacto consistently outperforms other models, Instant-NGP and Nerfacto, in both visual quality and rendering speed across all damage levels. It

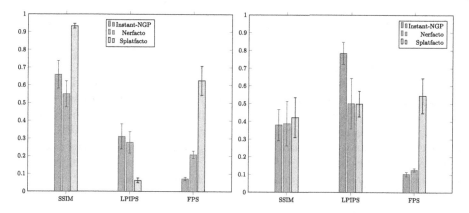

Fig. 9. Graphic visualization of some metrics for various models for the damage level 3 (left) from Table 6 and the damage level 4 (right) from Table 7.

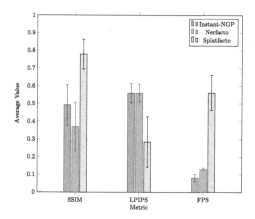

Fig. 10. Graphic visualization of some metrics from Table 8 for various models for the damage level "mixed".

excels at preserving fine details and achieves impressive rendering speeds compared to Nerfacto and Instant-NGP.

6   Conclusions

The study extensively tests and compares three modern NERF models on a dataset of drone footages capturing the aftermath of Russian attacks in Ukraine. Each model is tasked with reconstructing 3D representations of destroyed buildings, offering insights into the spatial and structural aspects of the damage. The study successfully demonstrates the potential of NERF models for 3D reconstruction of damaged structures in conflict zones. The findings suggest that Splatfacto offers a compelling combination of high-quality reconstructions (measured by

SSIM, PSNR, and LPIPS), fast rendering speed and computational efficiency. The paper acknowledges that the current research is in its initial stages, and further development is needed to improve the accuracy and efficiency of NERF models for real-world applications.

The research highlights several potential areas for future research. The impact of training data variations should be investigated, especially, sensitivity of the model performance with different datasets capturing diverse damage scenarios, lighting conditions, and viewpoints. The usage of various advanced NERF models, such as K-planes [8], TensoRF [12], Neuralangelo [7], and others, could be explored to understand how different model architectures could improve the reconstruction quality while maintaining efficiency. Also additional modalities (auxiliary data such as depth maps or semantic segmentation) should be considered for incorporation to enhance the reconstruction accuracy. The very important aspect is related to integration of this approach in real-time reconstruction systems, especially, in the context of optimisation of NERF models for real-time applications, enabling faster generation of 3D representations for immediate damage assessment or monitoring purposes.

By addressing these future research directions, one can further refine NERF technology for comprehensive documentation and analysis of conflict zones, potentially aiding humanitarian efforts and historical preservation. Overall, the study highlights the promising potential of NERF technology for revolutionizing the way one can use to document and analyze conflict zones, offering valuable insights for future research and real-world applications.

References

1. Rawtani, D., Gupta, G., Khatri, N., Rao, P.K., Hussain, C.M.: Environmental damages due to war in Ukraine: a perspective. Sci. Total Environ. **850**, 157932 (2022)
2. Pereira, P., Bašić, F., Bogunovic, I., Barcelo, D.: Russian-Ukrainian war impacts the total environment. Sci. Total Environ. **837**, 155865 (2022)
3. Mildenhall, B., Srinivasan, P.P., Tancik, M., Barron, J.T., Ramamoorthi, R., Ng, R.: NeRF: representing scenes as neural radiance fields for view synthesis. Commun. ACM **65**(1), 99–106 (2021)
4. Müller, T., Evans, A., Schied, C., Keller, A.: Instant neural graphics primitives with a multiresolution hash encoding. ACM Trans. Graph. **41**(4), 102:1–102:15 (2022)
5. Yu, T.-H., Feng, Y., Lai, S., Li, M., Zhou, J., Bao, H.: Mip-NeRF: a multi-scale representation for anti-aliasing neural radiance fields. In: Proceedings of the IEEE/CVF International Conference on Computer Vision, pp. 12868–12877 (2021)
6. Barron, J.T., Mildenhall, B., Verbin, D., Srinivasan, P.P., Hedman, P.: Mip-NeRF 360: unbounded anti-aliased neural radiance fields. In: Proceedings of the IEEE/CVF Conference on Computer Vision and Pattern Recognition, pp. 5470–5479 (2022)
7. Li, Z., et al.: Neuralangelo: high-fidelity neural surface reconstruction. In: Proceedings of the IEEE/CVF Conference on Computer Vision and Pattern Recognition, pp. 8456–8465 (2023)

8. Fridovich-Keil, S., Meanti, G., Warburg, F.R., Recht, B., Kanazawa, A.: K-planes: explicit radiance fields in space, time, and appearance. In: Proceedings of the IEEE/CVF Conference on Computer Vision and Pattern Recognition, pp. 12479–12488 (2023)

9. Kerbl, B., Kopanas, G., Leimkühler, T., Drettakis, G.: 3D gaussian splatting for real-time radiance field rendering. ACM Trans. Graph. **42**(4), 1–14 (2023)

10. Tancik, M., et al.: Nerfstudio: a modular framework for neural radiance field development. In: ACM SIGGRAPH 2023 Conference Proceedings, pp. 1–12 (2023)

11. Nerfstudio Team. Nerfacto (2022). https://docs.nerf.studio/nerfology/methods/nerfacto.html

12. Chen, A., Xu, Z., Geiger, A., Yu, J., Su, H.: TensoRF: tensorial radiance fields. In: Avidan, S., Brostow, G., Cissé, M., Farinella, G.M., Hassner, T. (eds.) ECCV 2022. LNCS, vol. 13692, pp. 333–350. Springer, Cham (2022). https://doi.org/10.1007/978-3-031-19824-3_20

GPU-Accelerated FDTD Solver
for Electromagnetic Differential Equations

MohammadReza HoseinyFarahabady$^{(\boxtimes)}$ [ID] and Albert Y. Zomaya [ID]

The University of Sydney, School of Computer Science, Center for Distributed
and High Performance Computing, Sydney, NSW, Australia
{reza.hoseiny,albert.zomaya}@sydney.edu.au

Abstract. Computational electromagnetics plays a crucial role across
diverse domains, notably in fields such as antenna design and radar sig-
nature prediction, owing to the omnipresence of electromagnetic phenomena. Numerical methods have replaced traditional experimental
approaches, expediting design iterations and scenario characterization.
The emergence of GPU accelerators offers an efficient implementation
of numerical methods that can significantly enhance the computational
capabilities of partial differential equations (PDE) solvers with specific
boundary-value conditions. This paper explores parallelization strategies for implementing a Finite-Difference Time-Domain (FDTD) solver
on GPUs, leveraging shared memory and optimizing memory access patterns to achieve performance gains. One notable innovation presented
in this research involves utilizing strategies such as exploiting temporal locality and avoiding misaligned global memory accesses to enhance
data processing efficiency. Additionally, we break down the computation
process into multiple kernels, each focusing on computing different electromagnetic (EM) field components, to enhance shared memory utilization and GPU cache efficiency. We implement crucial design optimizations to exploit GPU's parallel processing capabilities fully. These include
maintaining consistent block sizes, analyzing optimal configurations for
field-updating kernels, and optimizing memory access patterns for CUDA
threads within warps. Our experimental analysis verifies the effectiveness
of these strategies, resulting in improvements in both reducing execution
time and enhancing the GPU's effective memory bandwidth. Throughput evaluation demonstrates performance gains, with our CUDA implementation achieving up to 17 times higher throughput than CPU-based
methods. Speedup gains and throughput comparisons illustrate the scalability and efficiency of our approach, showcasing its potential for developing large-scale electromagnetic simulations on GPUs.

Keywords: Numerical Computational Electromagnetics · GPU
Accelerators · Finite-Difference Time-Domain Solver · Partial
Differential Equations · Geometric Discretization

© The Author(s), under exclusive license to Springer Nature Switzerland AG 2024
L. Franco et al. (Eds.): ICCS 2024, LNCS 14833, pp. 354–367, 2024.
https://doi.org/10.1007/978-3-031-63751-3_24

1   Introduction

Electromagnetics, the study of electrical and magnetic fields and their interaction, has been a cornerstone technology since the twentieth century and continues to be vital in the twenty-first. Maxwell's equations are at the heart of modern electromagnetic engineering, typically solved using computational electromagnetics (CEM) [19]. CEM has undergone significant advancement in the last decade, enabling highly accurate predictions for various electromagnetic phenomena such as wave scattering analysis, radar target scattering and the precise design of antennas and microwave devices, and analysis of electromagnetic interference (EMI) and electromagnetic compatibility (EMC) issues in electronic devices [11]. Commonly used CEM methods fall into two categories: those based on *differential equation* (DE) methods and those based on *integral equation* (IE) methods, leveraging Maxwell's equations and appropriate boundary conditions. IE methods typically offer approximations using finite sums, while DE methods employ finite differences. Previously, numerical EM analysis primarily occurred in the frequency domain due to its suitability for obtaining analytical solutions and limited experimental hardware. However, recent advancements in computational resources have led to a shift towards more advanced *time-domain* CEM models, mainly focusing on DE time-domain approaches like the *finite-difference time-domain* (FDTD) method [8,13]. The FDTD method solves problems in time while providing frequency-domain responses via Fourier transform and is applicable across a wide range of fields [4,5,15].

Implementing the Finite-Difference Time-Domain (FDTD) algorithm on Graphics Processing Units (GPUs) offers a promising avenue for accelerating numerical simulations [20]. GPUs, originally designed for graphics rendering, excel at parallel computation, making them well-suited for tasks like CEM models solver that involve heavy computational loads. By leveraging CUDA (Compute Unified Device Architecture), NVIDIA's parallel computing programming model, developers can harness the massive parallelism of various NVIDIA GPUs (e.g., Tesla architecture [12]) to significantly speed up FDTD computations. The implementation typically involves partitioning the space domain into smaller cells assigned to individual GPU threads. Each thread performs calculations for a specific portion of the domain in parallel with other threads. By carefully optimizing memory access patterns, developers can exploit the GPU's architecture to achieve high throughput and efficiency. One key advantage of GPU-accelerated FDTD computations is its ability to handle larger and more complex geometries with finer spatial resolution in a reasonable time frame. Such scalability is particularly beneficial for large-scale applications such as antenna design, electromagnetic compatibility analysis, and photonics research, where intricate geometries and high-fidelity computations are standard requirements. As a result, GPU-based FDTD implementations offer researchers and engineers a powerful tool for exploring and analyzing electromagnetic phenomena with unprecedented speed and accuracy.

This paper introduces an innovative approach for GPU-accelerated FDTD implementation, focusing on leveraging shared memory, optimizing memory access patterns, and minimizing divergence paths. We ensure streamlined execution by maintaining consistent block sizes and employing coalesced memory access to maximize floating-point operations per second. Additionally, we introduce key strategies such as exploiting temporal locality and breaking down computation into multiple kernels. Through rigorous experimentation, we identify optimal configurations for updating electromagnetic fields and determine the number of cells assigned to each GPU thread. This optimization effectively balances execution time reduction and GPU memory bandwidth enhancement. Furthermore, we compare the performance of our GPU implementation with a CPU-based approach using the Meep software package. Our extensive experimental analysis across various simulation sizes and configurations demonstrates substantial speedup gains and throughput improvements compared to CPU methods.

The remainder of this paper is structured as follows: Sect. 2 provides an overview of the background concepts relevant to our study, including Finite-Difference Time-Domain (FDTD) methods and GPU acceleration techniques. Following this, Sect. 3 details the methodology adopted in our study, encompassing the design and implementation of our GPU-accelerated FDTD solver. In Sect. 4, we outline the experimental setup used to evaluate the performance of our implementation, discussing hardware, software, parameter configurations, benchmark scenarios, and the results of our experiments. Finally, Sect. 5 concludes the paper, highlighting avenues for future research.

2   FDTD Framework for Numerical CEM

The FDTD method, introduced by Yee in 1966, discretizes Maxwell's equations in both space and time, enabling their solution within the time domain. Electric and magnetic field components are positioned at discrete spatial points, progressing through discrete time steps by approximating derivatives to model field evolution. In FDTD, fields are sampled at discrete time intervals, with electric and magnetic components sampled at distinct intervals, offset by $\Delta t/2$. The FDTD algorithm starts with Maxwell's time-domain equations, which are discretized using second-order accurate central difference formulas. The 3D geometry is divided into cells, forming a grid with rectangular Yee cells for stepped surface approximation. Field components within Yee cells are positioned with electric vectors at edge centers and magnetic vectors at face centers, representing Faraday's and Ampere's laws, respectively. Four electric field vectors surround each magnetic field vector, and vice versa, depicting law simulations. Field sampling in FDTD occurs at discrete time intervals, with electric components sampled at integer time intervals and magnetic components at half-integer intervals, offset by $\Delta t/2$. This necessitates spatial and temporal indices to differentiate components.

The foundation of constructing an FDTD-based CEM algorithm lies in Maxwell's time-domain equations. These differential equations describe the field behavior over time. The equations are as follows [7]:

$$\nabla \cdot \mathbf{D} = \rho_e$$
$$\nabla \cdot \mathbf{B} = \rho_m = 0$$
$$\nabla \times \mathbf{E} = -\frac{\partial \mathbf{B}}{\partial t} - \mathbf{M}$$
$$\nabla \times \mathbf{H} = \mathbf{J} + \frac{\partial \mathbf{D}}{\partial t}$$

Here, $\mathbf{E}$ represents the electric field strength vector (in Volts per meter), $\mathbf{D}$ represents the electric displacement vector (in Coulombs per square meter), $\mathbf{H}$ represents the magnetic field strength vector (in Amperes per meter), $\mathbf{B}$ represents the magnetic flux density vector (in Webers per square meter), $\mathbf{J}$ represents the electric current density vector (in Amperes per square meter), $\mathbf{M}$ represents the magnetic current density vector (in Volts per square meter), ρ_e represents the electric charge density (in Coulombs per cubic meter), and ρ_m represents the magnetic charge density (in Webers per cubic meter), equal to zero everywhere. Additionally, constitutive relations complement Maxwell's equations to characterize the material media [7]. Constitutive relations for linear, isotropic, and non-dispersive materials can be expressed as $\mathbf{D} = \epsilon \mathbf{E}$ and $\mathbf{B} = \mu \mathbf{H}$, where ϵ and μ represent the permittivity (in Farad/meter) and the permeability (in Henry/meter) of the material.

When deriving Finite-Difference Time-Domain (FDTD) equations, we can focus on the curl equations as the divergence equations can be fulfilled by the developed FDTD updating equations [7,19]. The electric current density $\mathbf{J}$ comprises the sum of the conduction current density $\mathbf{J}_c = \sigma_e \mathbf{E}$ and the impressed current density $\mathbf{J}_i$ such that $\mathbf{J} = \mathbf{J}_c + \mathbf{J}_i$. For the magnetic current density $\mathbf{M}$, we have $\mathbf{M} = \mathbf{M}_c + \mathbf{M}_i$, where $\mathbf{M}_c = \sigma_m \mathbf{H}$. Here, σ_e represents the electric conductivity in Siemens per meter, and σ_m represents the magnetic conductivity in Ohms per meter. By decomposing the current densities into conduction and impressed components and employing the constitutive relations, we can rewrite Maxwell's curl equations as:

$$\epsilon \frac{\partial \mathbf{E}}{\partial t} = \nabla \times \mathbf{H} - \sigma_e \mathbf{E} - \mathbf{J}_i,$$
$$\mu \frac{\partial \mathbf{H}}{\partial t} = -\nabla \times \mathbf{E} - \sigma_m \mathbf{H} - \mathbf{M}_i.$$

This formulation treats only the electromagnetic fields $\mathbf{E}$ and $\mathbf{H}$. All four constitutive parameters ε, μ, σ_e, and σ_m are the input parameters so that any linear isotropic material can be specified. Treatment of electric and magnetic sources is included through the impressed currents. Each vector equation can be further decomposed into three scalar equations for three-dimensional space, i.e., $\mathbf{E} = (\mathbf{E}_x, \mathbf{E}_y, \mathbf{E}_z)$ and $\mathbf{H} = (\mathbf{H}_x, \mathbf{H}_y, \mathbf{H}_z)$.

Central Difference Approximation Schemes

The central difference formula approximates derivatives through finite differences and is a foundational technique within numerical analysis, especially concerning the resolution of differential equations. These formulas estimate a function's derivative at a specific point by assessing the function at neighboring points. Specifically, for a first-order derivative, the central difference formula calculates the disparity between function values at symmetrically positioned points around the point of interest and divides by the spacing between these points. This method offers several benefits, including straightforward implementation and relatively high accuracy compared to alternative finite difference approaches. Nonetheless, it's crucial to recognize that the selection of grid spacing and the order of the difference formula can significantly influence the accuracy and stability of the numerical solution.

The central difference formula to approximate the derivative $f'(x)$ involves averaging the forward and backward difference formulas. This technique yields:

$$f'(x) = \frac{f(x + \Delta x) - f(x - \Delta x)}{2\Delta x} - \frac{(\Delta x)^2}{6} f''(x) + O(\Delta x^2)$$

Therefore, the central difference formula for $f'(x)$ exhibits second-order accuracy. This level of accuracy implies that the dominant term in the error introduced by a second-order accurate formula is proportional to the square of the sampling period. For example, halving the sampling period reduces the error by a factor of four. Consequently, a second-order accurate formula like the central difference formula provides greater precision than a first-order accurate formula.

Updating Procedures in FDTD Method

The FDTD method, introduced by Yee in 1966 [21], discretizes Maxwell's equations in both space and time, enabling their solution in the time domain. It places electric and magnetic field components at discrete spatial points within a grid, advancing through discrete time steps to simulate field evolution. In FDTD, fields are sampled at discrete time instants, with electric and magnetic components sampled at different intervals offset by $\Delta t/2$. While the algorithm calculates the fields at discrete time points, electric and magnetic components are not sampled simultaneously. Electric field components are sampled at integer time steps $(0, \Delta t, 2\Delta t, \ldots)$, while magnetic field components are sampled at half-integer time steps $(\frac{1}{2}\Delta t, (1 + \frac{1}{2}) \Delta t, \ldots)$, introducing an offset of $\Delta t/2$ between them. Figure 1 showcases a single Yee cell within the grid utilized in the FDTD method.

As outlined previously, the FDTD updating process approximates derivatives using the central difference formula. Here, field components are referenced by spatial and temporal indices, denoted with superscript notation. For example, $\mathbf{E}_z^n(i, j, k)$ represents the z component of an electric field vector sampled at time instant $n\Delta t$, positioned at $((i - 1)\Delta_x, (j - 1)\Delta_y, (k - 0.5)\Delta_z)$. Similarly, $\mathbf{H}_y^{n+\frac{1}{2}}(i, j, k)$ represents the y component of a magnetic field vector positioned at $((i - 0.5)\Delta_x, (j - 1)\Delta_y, (k - 0.5)\Delta_z)$ and sampled at time instant $(n + \frac{1}{2}) \Delta t$. Remarkably, for a computational domain with the number of cells (M_x, M_y, M_z), the total spatial problem size is $M = M_x \times M_y \times M_z$. Hence, The storage space required for this is approximately $24 \times M$ bytes for 32-bit precision and $48 \times M$ bytes for 64-bit precision [1].

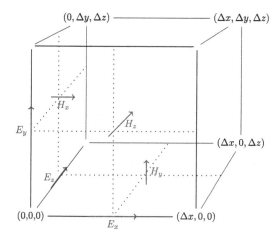

Fig. 1. Arrangement of electric and magnetic field vector components within a single Yee cell (Inspired by the diagrams in [19,22]).

3   Our Proposed Implementation of the FDTD Method

Recent advancements in GPU accelerators promise efficient numerical method implementations, significantly boosting the computational efficiency of EM solvers. Ongoing progress in GPU technology, with increased computational power, memory bandwidth, and specialized hardware features like tensor cores, continues to advance numerical simulations. This section emphasizes designing and implementing a GPU-accelerated FDTD solver, showcasing its effectiveness in solving EM equations with enhanced computational speed. This paper primarily focuses on NVIDIA's GPU platform [17] and the associated CUDA programming model [6,16]. While a similar approach can be developed for GPUs supporting the OpenCL standard, we center our discussion on NVIDIA's technology. In the following section, we provide readers with a concise overview of modern GPU processor architecture, including its memory hierarchy and the programming model utilized to map thread blocks onto the GPU streaming processors effectively. This foundational knowledge is essential for readers to grasp the design considerations before implementing the FDTD method on modern GPUs.

GPU Architecture and CUDA Programming Model

A GPU comprises Streaming Multiprocessors (SMs) containing processing cores and shared low-latency memory. SMs execute kernel functions in parallel with variable core counts. Global DRAM memory is partitioned and managed by an interconnection network for read/write requests. The speed difference between shared and global memory motivates minimizing data transfers for efficiency. Thread blocks can use local/shared memory as a manual cache, compensating for the lack of automatic caching in the older models of NVIDIA GPUs. Accessing shared memory is faster than global memory, mitigating latency with numerous threads emphasizing arithmetic operations.

CUDA, developed by NVIDIA, allows C/C++ code to run directly on GPUs, exploiting parallel processing. Programs consist of host code on the CPU and device code on the GPU, organized into kernel functions executed by multiple threads in parallel. CUDA manages parallelism and memory, with barrier instructions ensuring synchronized operations among threads. CUDA streams allow asynchronous execution of commands on NVIDIA GPUs, enabling concurrent operations and facilitating overlap of computation with data transfers [16]. Conditional branching significantly impacts stream performance, particularly in kernels with numerous threads intended to hide global memory access latency, which can range from 400 to 600 cycles [2].

Parallelization Strategies of FDTD Implementation

The field-updating kernel's parallelization in CUDA is feasible since the update equation is independently applied at each Yee's cell at every time step. In traditional CPU-based computations, the sequential nature of processing limits the speed and scalability of simulations involving large-scale electromagnetic problems. However, the parallelization of the field-updating kernel in CUDA is highly advantageous due to the inherent independence of the update equation at each Yee cell and time step. This independence allows for efficient exploitation of parallel processing capabilities offered by NVIDIA GPUs. CUDA enables the execution of thousands of threads in parallel on GPU cores, thereby significantly accelerating the computation process.

CUDA harnesses the massive parallelism of GPU architectures to update multiple Yee cells simultaneously at each time step by distributing the workload across multiple threads. This strategy significantly reduces the computational time needed for complex electromagnetic simulations. The CUDA programming model also offers developers fine-grained control over thread management and memory allocation. This level of control guarantees efficient parallelization of the field-updating kernel, leading to substantial performance enhancements compared to traditional CPU-based methods.

To optimize performance on GPUs, it is crucial to reduce global memory access latency. This entails following specific architectural guidelines, such as leveraging shared memory to minimize latency. We employ strategies like exploiting temporal locality and avoiding misaligned global memory accesses to reduce additional memory transactions. Additionally, minimizing divergence paths, where threads within a warp take different control flow paths, ensures efficient data processing. Furthermore, breaking down the computation process into multiple kernels, each focusing on computing different components of the EM field, can enhance shared memory utilization and GPU cache efficiency, which is the approach we have adopted in this work.

As highlighted below, we have implemented several crucial design decisions and optimizations to exploit the GPU's parallel processing capabilities fully.

- For executing the field-updating kernels, we maintain a consistent number of blocks per grid. This approach ensures that the workload is evenly distributed across the GPU cores, preventing the overloading of individual GPU cores and minimizing idle resources. This approach also simplifies memory management and synchronization, as each block operates independently without dependencies on other blocks. This enables seamless coordination between threads within the same block, facilitating efficient execution of the field-updating kernels.

- We conduct a thorough experimental analysis across various block sizes to ascertain the optimal configurations for all kernels that update the electric and magnetic fields. This evaluation involves systematically varying the number of threads per block and the total number of blocks per grid to explore the performance impact on different GPU architectures and computational workloads. Our experimental methodology includes profiling each kernel's memory access patterns and computational intensity to gain insights into their performance characteristics under varying block configurations. This analysis helps us understand how different block sizes affect memory bandwidth usage, register pressure, and arithmetic throughput, enabling us to uncover the most efficient block configurations that balance computational efficiency, memory bandwidth utilization, and resource utilization on the GPU.

- The most effective performance is consistently achieved with a block size of $B_1 \times B_2$, where B_1 is set to a relatively large value, such as greater than 256, on our GeForce 4070 Ti GPU. Meanwhile, B_2 remains relatively small, for example, less than 8. This configuration optimizes the utilization of the GPU's resources while minimizing potential overhead. A large value for B_1 allows for a high degree of parallelism within each block, enabling efficient utilization of the GPU's streaming multiprocessors (SMs) and maximizing the number of threads running concurrently. On the other hand, keeping B_2 small helps mitigate potential memory contention and resource conflicts within each block. With a smaller B_2, the threads within a block have access to a more localized and efficient shared memory space, reducing the likelihood of memory access conflicts and improving overall memory access latency.

- Such configurations consistently yield the shortest execution times for field-updating kernels, thereby maximizing the number of floating-point operations per second (FLOPS) for each kernel. By meticulously optimizing the block size parameters, we ensure that the GPU's computational resources are fully utilized, enabling the kernels to achieve peak performance. The reduction in execution times directly translates to a higher throughput of FLOPS, as more operations can be completed within a given time frame. This enhanced efficiency allows for faster simulation runs and enables us to tackle larger and more complex electromagnetic problems within a reasonable timeframe.

- In our CUDA implementation, data arrays are typically accessed from global memory, while integer constants are accessed from constant memory.

To optimize the usage of global memory, we employ coalesced access strategy. Coalesced access involves ensuring that consecutive threads within a thread block access consecutive memory addresses when reading or writing data. This approach allows for the efficient transfer of whole data sets in a single transaction, maximizing memory bandwidth utilization and reducing memory access latency. By arranging threads to access contiguous memory locations, we can streamline memory transfers and minimize the number of transactions required to load or store data in shared memory. This, in turn, enhances memory access efficiency and the overall performance of memory-bound kernels.

- In our CUDA implementation, we adopt a one-dimensional (1D) indexing scheme to define block and thread numbers, aiming to optimize memory access patterns and enhance overall performance. By organizing blocks and threads in a 1D manner, we facilitate memory coalescing and streamline access operations, both for reading data from and writing data to global memory. The decision to use 1D indexing is rooted in its effectiveness in maximizing memory coalescing. With 1D indexing, consecutive threads within a block access contiguous memory locations, enabling efficient data transfer in a single transaction and minimizing memory access latency. Furthermore, *column-major* order ensures that data elements along the same column are stored contiguously in memory, aligning with the memory access patterns of CUDA kernels. This organization is particularly advantageous during computations involving a time-consuming partial FDTD solver, where efficient memory access is essential for performance optimization.

- The strategy we have adopted involves assigning multiple Yee's cells to each thread, enabling direct access from shared memory for neighboring values within each warp. This approach minimizes memory access overhead by reducing the frequency of global memory accesses per thread. By allowing inner threads of a warp block to read from shared memory instead of global memory, our strategy eliminates the need for additional global memory reads within the warp. Moreover, when data is arranged in global memory, our strategy leverages the on-chip caching mechanism in the latest NVIDIA GPUs, further optimizing memory access and reducing latency. To determine the most effective performance, we conducted extensive experimental analysis across various sizes of cells per thread computation. For example, in our in-house system, we have identified that the optimal value for the cell size per thread computation is 8. This optimal configuration maximizes the efficiency of thread utilization and memory access, resulting in improved performance and overall computational throughput.

4   Results

This section presents the results of the electromagnetic differential equation solutions analysis conducted using the developed GPU-accelerated FDTD method. We begin by showcasing the experimental results designed to validate the implemented FDTD method through two simple scenarios. Subsequently, we present the performance evaluation results, comparing them to the outcomes of other FDTD implementations running on the CPU.

Platform Setup: The numerical simulations described in this study were conducted on a desktop computer featuring an Intel Core i7-13700 CPU, boasting 24 cores and 64 GB of memory, alongside a single NVIDIA GTX 4070 Ti Graphics card equipped with *8 GB* of global memory. The desktop operates on Ubuntu 22.04, and the kernels were developed using the CUDA Toolkit v12.2.

Accuracy Assessment

To validate the numerical solver, we compared our GPU-implemented FDTD method and other FDTD implementations primarily running on the CPU. This evaluation was performed using two distinct scenarios, as outlined below.

Scenario 1: In our evaluation, we simulated the reflection coefficient of a frequency-selective surface composed of a dipole array at normal incidence. Using a 3D FDTD method within free space, the dipole array was arranged in the XY plane with a size of 8 mm and evenly spaced intervals of 30 mm in all X, Y, and Z directions. The FDTD simulation was executed twice: once on the CPU using the Meep software package [18] – an open-source FDTD implementation for electromagnetism simulation on CPU many-core processors – and once on the GPU using our proposed implementation. Both simulations employed mesh sizes of 0.1 mm in each direction. To ensure accurate results, we applied boundary conditions at the regional boundary. Additionally, we set the time step to $\Delta t = 0.004$ ns to maintain the stability and convergence of the FDTD method. This approach enabled us to assess the performance and effectiveness of our implementation precisely. We calculated the 250-time step E and H fields in each direction under identical parameter settings by comparing the results obtained from both the Meep implementation and our proposed method. The simulation results revealed consistency between the two methods, with a maximum error of 0.8% in computational differences.

Scenario 2: We further conducted a second experimental evaluation to calculate the Radar Cross Section (RCS) of a Perfect Electric Conductor (PEC) sphere, which serves as a fundamental benchmark for electromagnetic simulation methods due to its well-understood analytical solution. Using our proposed implementation of the FDTD method, we simulated the electromagnetic scattering from a sphere to compare the computed RCS values against the analytically derived RCS values. The simulation setup consisted of a cubic domain with a PEC sphere at its center, spanning 25 cells in radius within a calculation domain of $600 \times 600 \times 600$ cells. The domain was filled with free-space, and the temporal resolution was set to $\Delta t = 0.004$ ns. The sphere, with a radius of $r = 0.2\lambda$, was subjected to an incident electric field at $0°$ in the x direction and had

PEC boundaries with Perfect Magnetic Conductor (PMC) boundary conditions. Throughout the simulation of 300 steps, we computed the absorption efficiency of the sphere and meticulously compared it with analytical results. We observed a degree of agreement between the two methods exceeding 99.4%.

Efficiency Evaluation of GPU-implemented FDTD Method

In this section, we compare the performance of our proposed GPU massive parallel implementation of the FDTD method with the parallel vectorized CPU implementation using the Meep software package for the two scenarios mentioned above. While there are many other GPU implementations of the FDTD method in the literature, such as those referenced in [3,9,10,14,20], the source code for various other GPU implementations is not publicly available. Hence, direct comparisons of performance evaluations between our method and these implementations are not feasible in this study.

Speedup Gain: Figure 2 (a) plots the speedup of our implementation compared to the Meep implementation on CPU cores. The GPU version's speedup demonstrates our approach's potential and the consistency of such speedup as the simulation size increases. It is evident that GPU implementation is more efficient when more extensive simulations are accommodated in GPU global memory. However, when the simulation space exceeds the capacity of the GPU global memory (*i.e.,* more than 8GB in our setup), the speedup declines. Such a decline is attributed to the challenges in hiding the context switching with a small number of active warps, compounded by the mandatory data transfer between the main memory in the host and the GPU global memory per each simulation step.

Furthermore, maximizing the concurrent execution of a large number of threads on the GPU proves to be more efficient, given that the primary bottleneck for FDTD computation lies in the availability of data in the global/shared memory of the GPU. Additionally, the cache memory of the GPU plays a nonsignificant role in influencing speedup performance. For smaller simulation sizes, even where data fits within the SM caches, the speedup can reach values near 8x, as for smaller simulation sizes; the time costs of this memory management become more significant with respect to the whole execution time as the communication time of data movement between host and device is dominated. Similarly, as the simulation size increases beyond the capacity of GPU global memory, the occurrence of communication time between host and device becomes more significant again. This results in heightened communication time between host and device memory, as well as the global and shared memory within the GPU, consequently leading to severe degradation in performance. Consequently, the speedup diminishes significantly to as low as 2x when the simulation size surpasses the GPU's global memory capacity by one order of magnitude.

Throughput: Throughput refers to the number of finite difference cells updated per second. Figure 2 (b) plots the average performance throughput (in Millions of Cells per Second) and the throughput gain of our implementation compared to the CPU-based Meep implementation as the side length of the cubic domain increases from 200 to 1000. This calculation excludes the wasted communication time transferring data from the host's main memory to GPU global memory. Results show that the throughput of the CPU-based Meep solver of the FDTD code remains constant at approximately 50 Mega Cells/s, while our CUDA implementation ranges from 130 to 877 Mega Cells/s, maintaining around 100 when the simulation size exceeds the capacity of the GPU's global memory by one order of magnitude. Furthermore, our method demonstrates a 17x higher throughput performance than the CPU-based method when the entire simulation data fits into the GPU global memory. The consistent speedup and throughput gain trend observed in Scenario 2 experiments across varying simulation sizes aligns with earlier reported results. Therefore, we omit the repetition of these findings.

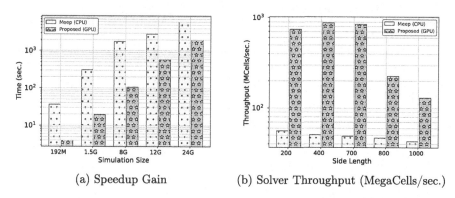

(a) Speedup Gain   (b) Solver Throughput (MegaCells/sec.)

Fig. 2. Comparative analysis of (a) speedup gain, and (b) throughput with increasing simulation size: our GPU implementation versus Meep (CPU) method. Evaluation conducted with varying model sizes and a consistent time step in FDTD simulations for Scenario 1.

5 Conclusions

This paper illuminates the promising potential of employing GPU-accelerated FDTD methods to implement large-scale PDE solvers. By harnessing advanced features of the CUDA framework, such as CUDA streams, we have developed a GPU-accelerated FDTD solver and extensively evaluated its performance across NVIDIA's GPUs. Comparative analyses against the parallelized CPU solver have revealed substantial performance advantages. Notably, our GPU solver achieved throughput rates of up to 877 Mega Cells per second, marking up to a 17-fold improvement over the open-source Meep CPU solver on standard desktops. This affordability renders cutting-edge GPU technology accessible to a wide range of individuals utilizing commodity workstations in partial differential equations solver. While GPUs inherently offer superior computational performance compared to traditional CPUs, owing to their heightened floating-point capabilities and memory bandwidth, future investigations will explore alternative FDTD implementations tailored for diverse contexts alongside their scalability on multi-GPU cluster platforms.

Acknowledgments. Professor Zomaya would like to acknowledge the support of the Australian Research Council Discovery Project (DP200103494). Dr. M. Reza HoseinyFarahabady acknowledges the continued support of The Center for Distributed and High-Performance Computing at The University of Sydney for giving access to advanced high-performance computing and cloud facilities, digital platforms, and necessary tools.

References

1. Andersson, U.: Time-Domain Methods for the Maxwell Equations. Doctoral dissertation, Royal Institute of Technology, Stockholm (2001)
2. Ansorge, R.: Programming in Parallel with CUDA: A Practical Guide. Cambridge University Press (2022)
3. Baumeister, P.F., Hater, T., Kraus, J., Pleiter, D., Wahl, P.: A performance model for gpu-accelerated fdtd applications. In: 2015 IEEE 22nd International Conference on High Performance Computing (HiPC), pp. 185–193 (2015). https://doi.org/10.1109/HiPC.2015.24
4. Carcione, J.M., Valle, S., Lenzi, G.: GPR modelling by the Fourier method: improvement of the algorithm. Geophys. Prospect. **47**(6), 1015–1029 (1999)
5. Cassidy, N.J., Millington, T.M.: The application of finite-difference time-domain modelling for the assessment of GPR in magnetically lossy materials. J. Appl. Geophys. **67**(4), 296–308 (2009)
6. Cook, S.: CUDA Programming: A Developer's Guide to Parallel Computing with GPUs. Morgan Kaufmann (Nov 2012)
7. Elsherbeni, A., Demir, V.: The Finite-difference Time-domain for Electromagnetics: With MATLAB Simulations. ACES series on computational electromagnetics and engineering, Institution of Engineering and Technology (2016)
8. Feng, D.S., Dai, Q.W.: GPR numerical simulation of full wave field based on UPML boundary condition of ADI-FDTD. NDT E Int. **44**(6), 495–504 (2011)
9. Francés, J., Bleda, S., Neipp, C., Márquez, A., Pascual, I., Beléndez, A.: Performance analysis of the FDTD method applied to holographic volume gratings: Multi-core CPU versus GPU computing. Comput. Phys. Commun. **184**(3), 469–479 (2013)
10. Francés, J., Otero, B., Bleda, S., Gallego, S., Neipp, C., Márquez, A., Beléndez, A.: Multi-GPU and multi-CPU accelerated FDTD scheme for vibroacoustic applications. Comput. Phys. Commun. **191**, 43–51 (2015)
11. Giannakis, I., Giannopoulos, A., Warren, C.: A realistic FDTD numerical modeling framework of ground penetrating radar for landmine detection. IEEE J. Selected Topics Appl. Earth Observ. Remote Sens. **9**(1), 37–51 (2016)
12. Lindholm, E., Nickolls, J., Oberman, S., Montrym, J.: Nvidia tesla: a unified graphics and computing architecture. IEEE Micro **28**(2), 39–55 (2008)
13. Liu, J., Shen, L.: Numerical simulation of subsurface radar for detecting buried pipes. IEEE Trans. Geosci. Remote Sens. **29**(5), 795–798 (1991)

14. Livesey, M., Stack, J.F., Jr., Costen, F., Nanri, T., Nakashima, N., Fujino, S.: Development of a CUDA implementation of the 3d FDTD method. IEEE Antennas Propag. Mag. **54**(5), 186–195 (2012)
15. Lopez, J., Carnicero, D., Ferrando, N., Escolano, J.: Parallelization of the finite-difference time-domain method for room acoustics modelling based on CUDA. Math. Comput. Model. **57**(7), 1822–1831 (2013)
16. NVIDIA Corporation: NVIDIA CUDA Toolkit Documentation, CUDA Streams: Asynchronous Concurrent Execution (2023). Accessed Nov 2023
17. NVIDIA Corporation: NVIDIA's GPU Platform (2023). Accessed Nov 2023
18. Oskooi, A., Roundy, D., Ibanescu, M., Bermel, P.A., Joannopoulos, J.: Meep: a flexible free-software package for electromagnetic simulations by the FDTD method. Comput. Phys. Commun. **181**(3), 687–702 (2010)
19. Taflove, A.: Computational Electrodynamics: The Finite-Difference Time-Domain Method, 3rd edn. Artech House, Norwood, MA, USA (2005)
20. Warren, C., Giannopoulos, A., Gray, A., et al.: A CUDA-based GPU engine for gprMax: Open source FDTD electromagnetic simulation software. Comput. Phys. Commun. **237**, 208–218 (2019)
21. Yee, K.S.: Numerical solution of initial boundary value problems involving Maxwell's equations in isotropic media. IEEE Trans. Antennas Propagat. **14**(3), 302–307 (1966)
22. Yu, W., Yang, X., Liu, Y., Mittra, R., Muto, A.: Advanced FDTD Method: Parallelization, Acceleration, and Engineering Applications. Artech House, Norwood, MA (2011)

Computational Aspects of Homogeneous Approximations of Nonlinear Systems

Marcin Korzeń[1] ⓘ, Grigory Sklyar[1] ⓘ, Svetlana Ignatovich[2] ⓘ,
and Jarosław Woźniak[1](✉) ⓘ

[1] Faculty of Computer Science and Information Technology, West Pomeranian
University of Technology in Szczecin, Żołnierska str. 49, 71-210 Szczecin, Poland
{mkorzen,gsklyar,jwozniak}@zut.edu.pl
[2] Department of Applied Mathematics, V. N. Karazin Kharkiv National University,
Svobody sqr., 4, Kharkiv 61022, Ukraine
s.ignatovich@karazin.ua
http://www.wi.zut.edu.pl, https://start.karazin.ua/i/school/mathematics

Abstract. The objective of the paper is to describe computational
methods and techniques of investigation of certain algebraic structures
needed in order to apply the results in concrete problems in mathematical control theory of nonlinear systems. Contemporary theoretic research
requires more and more sophisticated tools for a possible application of
the results. In the paper we propose computational tools and techniques
for a certain type of simplification of driftless control systems. Such simplification still preserve most crucial properties of the original ones like
controllability, but the simplified system have a special feedforward form
that is much easier to integrate or allows to solve other problems in control theory. We present the computational procedure and foundations of
the library as the extension of existing software libraries in Python language. The approach is illustrated with some numerical experiments and
simulations. We conclude with a discussion about related computational
issues.

Keywords: nonlinear system · nonlinear approximation ·
homogeneous approximation · computational procedures

1 Introduction

We consider control systems – nonlinear with respect to state and linear with
respect to controls – namely (driftless) systems of the form

$$\dot{x} = \sum_{i=1}^{m} X_i(x) u_i, \tag{1}$$

where $X_i(x)$ are real analytic vector fields in the neighborhood of the origin in
$\mathbb{R}^n$. It is one of class of systems widely considered in modern control theory.
In general, control theory is devoted to studying dynamical processes that are

L. Franco et al. (Eds.): ICCS 2024, LNCS 14833, pp. 368–382, 2024.
https://doi.org/10.1007/978-3-031-63751-3_25

described in most cases as systems of difference or differential equations (usually with additional constraints) where some parameters – controls – can be externally changed at any moment of time during the process, as desired. One of most important problems arising is construct a proper method of designing controls which transfer a system from a given state to some other preassigned target state. This desired control should usually also satisfy additional requirements: it should be, for example, bounded or optimal in a certain sense.

An important feature of the modern control theory is that the studied problems are mostly nonlinear and, moreover, their linear approximations often lose the structural properties of original systems thus cannot be loosely applied. This means that, among nonlinear systems, the simplest systems should be chosen, in particular, to approximate other nonlinear systems of more complicated structure. This direction is represented by a homogeneous approximation problem which was actively developed during several decades [1,2,4,6,10,16,17].

Along with the differential-geometric methods which are most commonly applied, algebraic methods proved to be useful in such problems. The first idea was given by M. Fliess [8]; he proposed to consider a series in the free associative algebra instead of the control system. One of the follow-ups was proving that the free algebraic interpretation of concepts related to homogeneous approximation allows more exact description of local properties of nonlinear affine control systems [18]. This approach, based mainly on algebraic constructions, is well suitable for numerical implementation.

In this paper we continue this research direction depicting the computational approach to the control synthesis problem of a class of non-linear dynamical systems. Firstly we introduce the needed notions, then we describe the numerical procedures for obtaining homogeneous approximations of our systems. In the end we provide numerical experiments for three systems – two artificially constructed and one physical – illustrating our approach.

2   Theoretical Background

Thus, we consider the control system (1), where $X_i(x)$ are real analytic vector fields in the neighborhood of the origin. We are interested in trajectories of this system starting at the origin and corresponding to a control $u(t) = (u_1(t), \dots, u_m(t))$, which is supposed to be measured and bounded. In this paper we assume that the system is locally controllable in a neighborhood of the origin. Then, for some t_f, the trajectory $x(t)$ of the system exists for $t \in [0, t_f]$. Let us denote by $\mathcal{E}_{X_1,\dots,X_m}(t_f, u)$ the "end-point map" which takes $u(t)$ to the end point of the trajectory $x(t_f)$. Linearity in the control and analyticity allows expressing the end-point map as a series

$$\mathcal{E}_{X_1,\dots,X_m}(t_f, u) = \sum_{k=1}^{\infty} \sum_{1 \le i_1,\dots,i_k \le m} c_{i_1 \dots i_k} \eta_{i_1 \dots i_k}(t_f, u), \qquad (2)$$

where $\eta_{i_1 \ldots i_k}(t_f, u)$ are "iterated integrals" of the form

$$\eta_{i_1 \ldots i_k} = \eta_{i_1 \ldots i_k}(t_f, u) = \int_0^{t_f} \int_0^{\tau_1} \cdots \int_0^{\tau_{k-1}} u_{i_1}(\tau_1) \cdots u_{i_k}(\tau_k) d\tau_k \cdots d\tau_1.$$

Coefficients $c_{i_1 \ldots i_k} \in \mathbb{R}^n$ are defined by the system as

$$c_{i_1 \ldots i_k} = X_{i_k} X_{i_{k-1}} \cdots X_{i_1} E(0), \tag{3}$$

where $E(x) = x$ is the identity map and X_i are understood as differential operators acting as $X_i \varphi = D\varphi(x) \cdot X_i(x)$, $i = 1, \ldots, m$.

If controls are such that $|u_i(t)| \le 1$, $t \in [0, t_f]$, iterated integrals satisfy the estimate $|\eta_{i_1 \ldots i_k}(t_f, u)| \le \frac{1}{k!} t_f^k$. Hence, for small t_f, it is natural to truncate the series in (2) and to consider an approximation that is described by a finite number of iterated integrals. If such a truncated series actually *is* an end-point map for a certain system, then such a system can be regarded as a reasonable approximation of the initial system (1).

It turns out that there exist such coordinates in which one easily finds an appropriate truncation. Suppose we want to write the system (1) in the new coordinates $y = F(x)$; then coefficients of the series for the end-point map in the new coordinates equal $X_{i_k} X_{i_{k-1}} \cdots X_{i_1} F(0)$. This leads to studying the operators $X_{i_k} X_{i_{k-1}} \cdots X_{i_1}$ (so-called nonholonomic derivatives); the goal is to find the convenient *privileged coordinates*: if the system is written in these coordinates then the end-point map takes the componentwise form

$$(\mathcal{E}_{X_1, \ldots, X_m})_j(t_f, u) = (\mathcal{E}_{\widehat{X}_1, \ldots, \widehat{X}_m})_j(t_f, u) + \rho_j(t_f, u),$$

where $\mathcal{E}_{\widehat{X}_1, \ldots, \widehat{X}_m}$ is the end-point map of the system

$$\dot{z} = \sum_{i=1}^m \widehat{X}_i(z) u_i, \tag{4}$$

which is locally controllable in a neighborhood of the origin and is such that $(\mathcal{E}_{\widehat{X}_1, \ldots, \widehat{X}_m})_j$ contains iterated integrals of length w_j and ρ_j contains iterated integrals of length greater than w_j. Then the system (4) is called *a homogeneous approximation* of the system (1). Due to homogeneity (more specifically, quasi-homogeneity), a homogeneous approximation is the simplest approximation of the nonlinear system (1). Numbers w_j are interpreted as weights of coordinates; for convenience assume $w_1 \le \cdots \le w_n$.

The question is how to construct a homogeneous approximation. In [2], the general method is described, which suggests to find privileged coordinates applying successive polynomial change of coordinates. After a finite number of steps, the system (4) can be constructed. In practical implementation we deal with vector fields, therefore, symbolic computations are needed.

Another way for finding a homogeneous approximation is as follows. Let us start with the series representation (2), where we assume that coefficients $c_{i_1 \ldots i_k}$ are *fixed vectors*. Then we turn our attention to the iterated integrals.

As was noticed by M. Fliess [8], for a fixed $t_f > 0$, iterated integrals form a free associative algebra $\mathcal{F}$ over $\mathbb{R}$ with the algebraic operation $\eta_{i_1 \ldots i_k} \eta_{j_1 \ldots j_p} = \eta_{i_1 \ldots i_k j_1 \ldots j_p}$. We introduce the inner product in $\mathcal{F}$ assuming the basis $\eta_{i_1 \ldots i_k}$ to be orthonormal. Notice that $\eta_1, \ldots, \eta_m$ can be considered as the free generators of $\mathcal{F}$. A natural grading in $\mathcal{F}$ is defined as

$$\mathcal{F} = \sum_{k=1}^{\infty} \mathcal{F}^k, \quad \mathcal{F}^k = \mathrm{Lin}\{\eta_{i_1 \ldots i_k} : i_1, \ldots, i_k \in \{1, \ldots m\}\}.$$

We write $\mathrm{ord}(a) = k$ if $a \in \mathcal{F}^k$. Let $\mathcal{L}$ be a free graded Lie algebra generated by $\eta_1, \ldots, \eta_m$ with the Lie brackets operation $[\ell_1, \ell_2] = \ell_1 \ell_2 - \ell_2 \ell_1$ and grading $\mathcal{L} = \sum_{k=1}^{\infty} \mathcal{L}^k$, where $\mathcal{L}^k = \mathcal{L} \cap \mathcal{F}^k$.

Now, with the series (2) we associate a linear map $c : \mathcal{F} \to \mathbb{R}^n$ defined on the basis as $c(\eta_{i_1 \ldots i_k}) = c_{i_1 \ldots i_k}$. Suppose an analytic change of variables $y = F(x)$ is applied; then in the new coordinates the end-point map has a series representation $F(\mathcal{E}_{X_1, \ldots, X_m})$; it should be written in the form of a series of iterated integrals with vector coefficients. While transforming, products of iterated integrals arise, which should be represented as linear combinations of iterated integrals. In algebraic terms, the product of iterated integrals corresponds to the "shuffle product" defined recursively as

$$\eta_{i_1 \ldots i_k} \sqcup\!\sqcup \eta_{j_1 \ldots j_p} = \eta_{i_1}(\eta_{i_2 \ldots i_k} \sqcup\!\sqcup \eta_{j_1 \ldots j_p}) + \eta_{j_1}(\eta_{i_1 \ldots i_k} \sqcup\!\sqcup \eta_{j_2 \ldots j_p}).$$

The shuffle product in $\mathcal{F}$ and the Lie algebra $\mathcal{L}$ are surprisingly connected: by R. Ree's theorem, an element from $\mathcal{F}$ belongs to $\mathcal{L}$ if and only if it is orthogonal to shuffle product of any two elements from $\mathcal{F}$.

We use this property in order to define the homogeneous approximation. Namely, let us consider the following linear subspace in $\mathcal{L}$,

$$\mathcal{L}_{X_1, \ldots, X_m} = \sum_{k=1}^{\infty} \mathcal{P}^k, \quad \text{where } \mathcal{P}^k = \{\ell \in \mathcal{L}^k : c(\ell) \in c(\mathcal{L}^1 + \cdots + \mathcal{L}^{k-1})\}, \ k \geq 1.$$

$$(5)$$

One can show that $\mathcal{L}_{X_1, \ldots, X_m}$ is a graded subalgebra of $\mathcal{L}$ of codimension n; we call it a *core Lie subalgebra* of the system (1). Let $\ell_1, \ldots, \ell_n \in \mathcal{L}$ be (homogeneous) elements such that $\mathcal{L} = \mathrm{Lin}\{\ell_1, \ldots, \ell_n\} + \mathcal{L}_{X_1, \ldots, X_m}$, and let $\{\ell_j\}_{j=n+1}^{\infty}$ be a (homogeneous) basis of $\mathcal{L}_{X_1, \ldots, X_m}$. Without loss of generality we assume that $\ell_i \in \mathcal{L}^{w_i}$, $i = 1, \ldots, n$, where $w_1 \leq \cdots \leq w_n$. Due to the Poincaré-Birkhoff-Witt theorem, elements $\ell_{i_1}^{q_1} \cdots \ell_{i_k}^{q_k}$ form a basis of $\mathcal{F}$, where $i_1 < \cdots < i_k$ and ℓ^q means the q-th power of ℓ. Then there exists a biorthogonal basis, and, due to the Melançone-Reutenauer theorem, its elements are of the form

$$\frac{1}{q_1! \cdots q_k!} d_{i_1}^{\sqcup\!\sqcup q_1} \sqcup\!\sqcup \cdots \sqcup\!\sqcup d_{i_k}^{\sqcup\!\sqcup q_k},$$

where $d^{\sqcup\!\sqcup q}$ means the q-th shuffle power of d and d_i are elements of the biorthogonal basis orthogonal to all elements of the Poincaré-Birkhoff-Witt basis except ℓ_i and such that the inner product of d_i and ℓ_i equals 1.

The main result about homogeneous approximation is as follows.

- There exists the system (4) (homogeneous approximation of (1)) whose end-point map equals

$$(\mathcal{E}_{\widehat{X}_1,\ldots,\widehat{X}_m})_j = d_j, \quad j = 1, \ldots, n. \tag{6}$$

- There exists a change of coordinates $y = F(x)$ in the system (1) which reduces it to the form $\dot{y} = \sum_{i=1}^{m} Y_i(y) u_i$ such that

$$(\mathcal{E}_{Y_1,\ldots,Y_m})_j = d_j + \rho_j, \quad j = 1, \ldots, n, \tag{7}$$

where ρ_j contains iterated integrals of length greater than w_j.

We emphasize that it may be more convenient to consider the system (4) in other coordinates for which the end-point map equals

$$(\mathcal{E}_{\widehat{X}_1,\ldots,\widehat{X}_m})_j = d_j + P_j(d_1, \ldots, d_{j-1}), \quad j = 1, \ldots, n, \tag{8}$$

where P_j are shuffle polynomials containing elements from $\mathcal{F}^{w_j}$ only.

3 Procedure for Obtaining Homogeneous Approximations

Thus, the rough plan for finding a homogeneous approximation using the results described in the previous section is as follows.

1. Find the core Lie subalgebra and the elements $d_1, \ldots, d_n$ of the biorthogonal basis.
2. Reconstruct the system (4) having the end-point map (6).
3. Find a change of variables $y = F(x)$ reducing the end-point map of the system (1) to the form (7).

In [15], the first attempt was taken to implement these steps (for a slightly different class of systems). Now we present a much better implementation, which allows us to perform computer experiments. We discuss the algorithm in more detail.

In step 1, we first of all find the core Lie subalgebra (5). To use its definition, we perform the following steps:

1.1. Find a basis in the free Lie algebra $\mathcal{L}$.
1.2. Find coefficients $c_{i_1 \ldots i_k}$ of the series (2).
1.3. Find bases for subspaces $\mathcal{P}^k$.

Actually, we need to find $c(b_j)$, where b_j are basis elements in $\mathcal{L}$. Taking into account the grading, we find bases of subspaces $\mathcal{L}^1, \mathcal{L}^2, \ldots$ successfully until $\mathcal{L}^r$ such that $c(\mathcal{L}^1 + \cdots + \mathcal{L}^r) = \mathbb{R}^n$. More specifically, for any k, we find a basis b_j of $\mathcal{L}^k$ and the coefficients $c_{i_1 \ldots i_k}$; then we form the vectors $c(b_j)$ and find the subspace $\mathcal{P}^k$.

Coefficients $c_{i_1...i_k}$ are defined by (3). For a sufficiently general form of vector fields X_i, this step is of a large complexity, where symbolic and numerical approaches should be combined. However, some simplification can be implemented: for example, if the coefficient c_j vanishes then all $c_{i_1...i_q j}$ vanish as well.

Finding a basis in the free Lie algebra is a standard problem. However we observe that the dimension of $\mathcal{L}^k$ grow rapidly, so the complexity here is related to the depth of singularity r rather than the dimension of the system n.

Finally, for step 1 two more sub-steps are needed:

1.4. Find elements of the Poincaré-Birkhoff-Witt basis.

1.5. Find elements of the biorthogonal basis $d_1, \ldots, d_n$.

As above, we consider each subspace $\mathcal{F}^k$ separately and restrict ourselves by $k \le r$ and solve linear algebra problems, possible with sparse matrices.

The algorithm for step 2 is given in [18]. Namely, step 2 consists of two sub-steps applied to each of d_j, $j = 1, \ldots, n$:

2.1. Represent d_j as a sum $d_j = \sum_{k=1}^{m} \eta_k a_k$. If $w_j = 1$, then $a_k \in \mathbb{R}$. If $w_j \ge 2$, then $a_k \in \mathcal{F}^{w_j - 1}$ and, moreover, a_k equals a shuffle polynomial of $d_1, \ldots, d_{j-1}$, i.e.,

$$a_k = \sum \alpha^k_{q_1 \ldots q_{j-1}} d_1^{\sqcup\!\sqcup q_1} \sqcup\!\sqcup \cdots \sqcup\!\sqcup d_{j-1}^{\sqcup\!\sqcup q_{j-1}}, \quad \text{where } \alpha^k_{q_1 \ldots q_{j-1}} \in \mathbb{R}.$$

2.2. For any $k = 1, \ldots, n$, define the j-th component of $\widehat{X}_k$ as follows: if $w_j = 1$, then put $(\widehat{X}_k(z))_j = a_k$; if $w_j \ge 2$, then $(\widehat{X}_k(z))_j = \sum \alpha^k_{q_1 \ldots q_{j-1}} z_1^{q_1} \cdots z_{j-1}^{q_{j-1}}$.

Concerning step 3, we note that such a change of variables is not unique. In particular, one can find $F(x)$ as a polynomial [17,18] in the following way:

3.1. Consider the polynomial mapping $\Phi : \mathbb{R}^n \to \mathbb{R}^n$ of the form

$$\Phi(z) = \sum_{q_1 w_1 + \cdots + q_n w_n \le w_n} \frac{1}{q_1! \cdots q_n!} c(\ell_1^{q_1} \cdots \ell_n^{q_n}) z_1^{q_1} \cdots z_n^{q_n}.$$

3.2. Find a mapping $y = F(x)$ that reduces $\Phi(z)$ to the "upper triangular form", i.e., such that component-wise

$$(F(\Phi(z)))_j = z_j + p_j(z_1, \ldots, z_{j-1}),$$

where p_j is a polynomial that contains terms of the form $z_1^{q_1} \cdots z_n^{q_n}$ such that $q_1 w_1 + \cdots + q_n w_n \ge w_j + 1$.

Our general goal to provide a fully automatic procedure for any system in the form (1). The presented method is effective, but computational complexity may depend on complexity of the system, of length and order of the endpoint series expansion, and of the number of gaps in the resulting Lie core algebra. Computational issues of the presented procedure consist of combinatorial, symbolic, and numerical algebra procedures. The main difficulty is the need to combine symbolic and numeric computations. There are two separate parts where symbolic-based computations are required: high-order differentiation for computations of Lie brackets and Lie elements and finding nonlinear mapping and change of variables. Currently, we use sympy [13] and lambdification, but at least for differentiation, we see the possibility of using modern tools of automatic differentiation like JAX/Autograd [5].

4   Numerical Experiments

Systems and Transformations. In the experimental part we consider three control systems:

1. Artificial system 1:

$$
\begin{pmatrix} \dot{x}_0 \\ \dot{x}_1 \\ \dot{x}_2 \\ \dot{x}_3 \end{pmatrix} = \begin{pmatrix} 1 \\ 0 \\ 0 \\ \sin(x_2) \end{pmatrix} u_0 + \begin{pmatrix} x_2 \\ 1 \\ x_1(1-x_3) \\ 1 \end{pmatrix} u_1; \tag{9}
$$

2. Artificial system 2:

$$
\begin{pmatrix} \dot{x}_0 \\ \dot{x}_1 \\ \dot{x}_2 \\ \dot{x}_3 \end{pmatrix} = \begin{pmatrix} 1 \\ 1-x_1^2 \\ 0 \\ x_3(1-x_0) \end{pmatrix} u_0 + \begin{pmatrix} x_1(1-x_3) \\ 0 \\ x_1(1-x_2)+x_3 \\ 1 \end{pmatrix} u_1; \tag{10}
$$

3. Vehicle (truck/lorry) linked to with two trailers:

$$
\begin{pmatrix} \dot{x}_0 \\ \dot{x}_1 \\ \dot{x}_2 \\ \dot{x}_3 \\ \dot{x}_4 \end{pmatrix} = \begin{pmatrix} \cos(x_2) \\ \sin(x_2) \\ 0 \\ \sin(x_2 - x_3) \\ \sin(x_3 - x_4)\cos(x_2 - x_3) \end{pmatrix} u_0 + \begin{pmatrix} 0 \\ 0 \\ 1 \\ 0 \\ 0 \end{pmatrix} u_1, \tag{11}
$$

where x_0, x_1 are car position and x_2 is the angle between the car direction and the Ox_0 axis.

For each system, we aim to find the control signal that moves the system from a given initial point to zero.

First, we write the solution of the considered system in the form of the series (2). Using the procedure described in the previous section, we find the transformation that allows the truncation of the series to the homogeneous form. System (9) has the following homogeneous approximation:

$$
\begin{pmatrix} \dot{z}_0 \\ \dot{z}_1 \\ \dot{z}_2 \\ \dot{z}_3 \end{pmatrix} = \begin{pmatrix} 1 \\ 0 \\ 0 \\ 0 \end{pmatrix} u_0 + \begin{pmatrix} 0 \\ 1 \\ -z_0 z_1 \\ -\frac{1}{6} z_0 z_1^3 \end{pmatrix} u_1, \tag{12}
$$

and the transformation $y = F(x)$ can be chosen as

$$y = F(x) = \begin{pmatrix} x_0 \\ x_1 \\ x_3 - x_1 - \frac{1}{2}x_0 x_1^2 \\ \frac{1}{6}x_2 - \frac{1}{12}x_1^2 - \frac{1}{36}x_1^3 - \frac{1}{24}x_0 x_1^4 + \frac{1}{12}x_1^2 x_3 \end{pmatrix}. \tag{13}$$

Homogeneous approximation of the systems (10) is the following:

$$\begin{pmatrix} \dot{z}_0 \\ \dot{z}_1 \\ \dot{z}_2 \\ \dot{z}_3 \end{pmatrix} = \begin{pmatrix} 1 \\ 0 \\ 0 \\ 0 \end{pmatrix} u_0 + \begin{pmatrix} 0 \\ 1 \\ z_0 \\ -z_0 z_1 \end{pmatrix} u_1, \tag{14}$$

and transformation for the system (10) can be

$$y = F(x) = \begin{pmatrix} x_1 \\ x_3 \\ x_2 - \frac{1}{2}x_3^2 \\ \frac{1}{3}x_0 - \frac{1}{3}x_1 - \frac{1}{3}x_2 - \frac{1}{6}x_1^2 + \frac{1}{6}x_3^2 - \frac{1}{9}x_1^3 - \frac{1}{3}x_1 x_3^2 \end{pmatrix}. \tag{15}$$

The homogeneous approximation for the car–trailers system (11) is given by:

$$\begin{pmatrix} \dot{z}_0 \\ \dot{z}_1 \\ \dot{z}_2 \\ \dot{z}_3 \\ \dot{z}_4 \end{pmatrix} = \begin{pmatrix} 1 \\ 0 \\ 0 \\ 0 \\ 0 \end{pmatrix} u_0 + \begin{pmatrix} 0 \\ 1 \\ z_0 \\ -\frac{1}{2}z_0^2 \\ \frac{1}{6}z_0^3 \end{pmatrix} u_1, \tag{16}$$

and the transformation can be

$$y = F(x) = \begin{pmatrix} x_0 \\ x_2 \\ -x_1 + x_0 x_2 \\ -x_4 + x_0 x_1 - \frac{1}{2}x_0^2 x_2 \\ x_3 - x_1 + x_4 - \frac{1}{2}x_0^2 x_1 + x_0 x_4 + \frac{1}{6}x_0^3 x_2 \end{pmatrix}. \tag{17}$$

As we see, the homogeneous systems (12), (14) and (16) are much simpler than the original systems (9), (10), (11). Please note that the transformations $y = F(x)$ are effectively invertible. The linear part is easy to invert, and the nonlinear part has the feedforward form. As was mentioned above, choosing a homogeneous approximation in a slightly different form (8) could be more convenient, for example, if we are interested in simplifying the change of variables $y = F(x)$.

The homogeneous approximation is a certain kind of simplification of a nonlinear control system, that makes the system easier to being integrated and also allows easier solutions of various controllability problems. A homogeneous approximation simplifies the given system and maintains its main properties like controllability. Both systems are equivalent up to a nonlinear transformation. Having the approximation and transformation, we can compare the trajectories of both systems with the same control signals. We can also find the control for the approximated homogeneous system in a simpler transformed space, and we apply the same control to the original system in the transformed space. In order to compare both trajectories, we must transform the homogeneous system trajectory to the original space using inverse transformations. In our experiments we inverse the time in the initial system and in its homogeneous approximation and consider the origin as a goal point.

Results. Comparing the system trajectories with the same control signals, we would like to show and briefly discuss the quality of such approximations. One of objectives of the presented research is to review how good the homogeneous approximation is with regard to the original system.

Finding the optimal control u_i was not the main task in this paper. Nevertheless, because the approximation is valid close to the origin, we should find some control signal for the approximated system that moves the system from a given initial point to the origin (point 0) or close to the origin. In our experiment we chose the sequence of radii of increasing size and randomly selected ten initial points per radius. Each initial point in chosen from an original space is transformed to z's space i.e. $z^0 = F(x^0)$. Next, we try to find a control signal that moves the transformed homogeneous system from the initial point z^0 to the final point 0. Finally, we inspect trajectories of the original and approximated systems in the original space. To transform the trajectory of the homogeneous system to the original space, we have to apply the transformation $x = F^{-1}(z)$. As the quality measure, we use two following measures: the distance between final positions,

$$e_{t_f} = \|x(t_f) - x_h(t_f)\|, \tag{18}$$

and summed up distance between trajectories,

$$e_1 = \int_0^{t_f} \|x(t) - x_h(t)\| dt, \tag{19}$$

where x denotes the original systems' trajectories, and x_h the trajectories of their homogeneous approximations transformed back to the original coordinates (Fig. 1).

Detailed results are summarized in Table 1. Figures 2 and 3 present sample trajectories for different initial positions. It is worth to notice that in the car-trailers system (11) our simulation shows that the behaviour of the approximated system is better than the theory suggests. Here the trajectory (namely state variables x_3 and x_4) at some point goes out of the region of applicability but it comes back in later, and in the end the trajectory arrives very close to the end-point of the original system. One can observe this quantitatively in Table 1 where the total (integral) error e_1 may be much larger than e_{t_f} which measures the deviation of the end-point only.

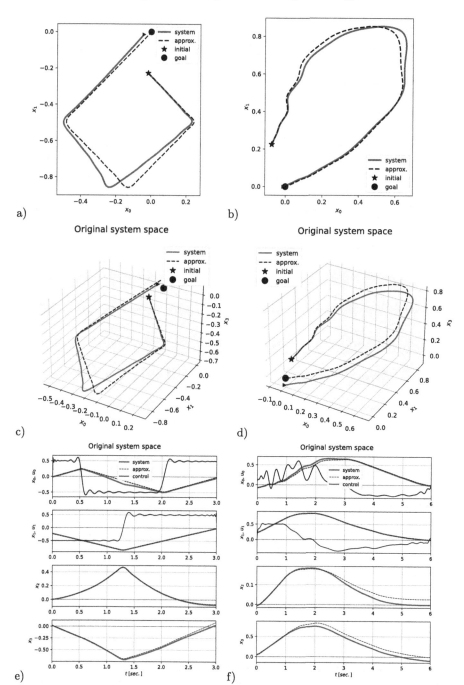

Fig. 1. Illustration of the proposed approach with the system (9); a–d) projection of system's trajectories to 2D and 3D subspaces starting from different initial points. e), f) control signals u_0, u_1 (solid black) and systems' trajectories in separate coordinates. Red solid lines – original system trajectories, black dashed lines represent behavior of approximated system in original system's coordinates (Color figure online)

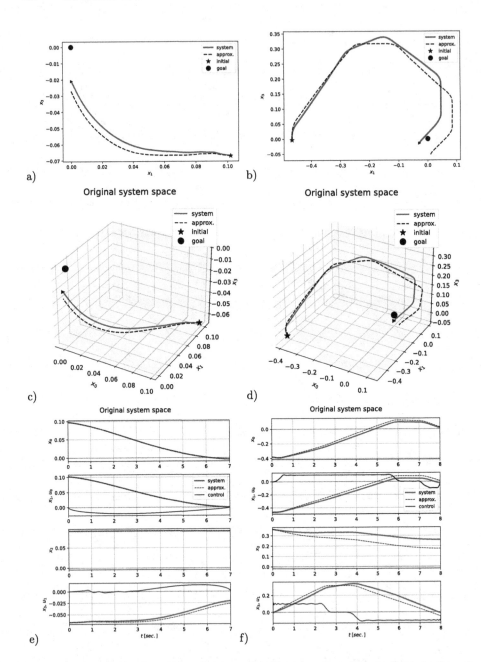

Fig. 2. Illustration of the proposed approach with the system (10); a–d) projection of system's trajectories to 2D and 3D subspaces starting from different initial points. e), f) control signals u_0, u_1 (solid black) and systems' trajectories in separate coordinates. Red solid lines – original system trajectories, black dashed lines represent behavior of approximated system in original system's coordinates (Color figure online)

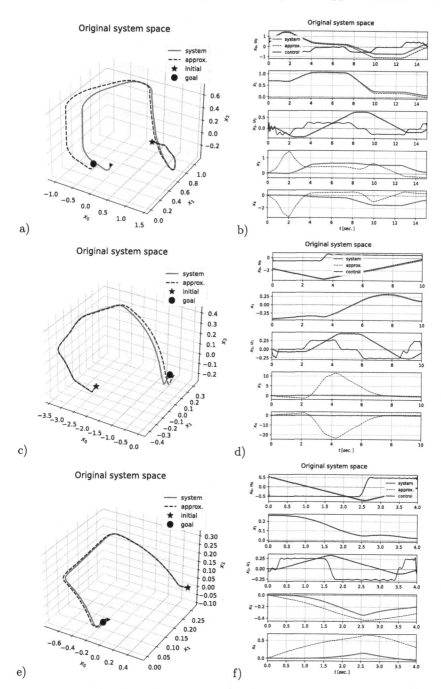

Fig. 3. Illustration of the proposed approach with the car-trailers system (11); a), c), e) projection of system's trajectory to 3D subspace ($x_0 = x$, $x_1 = y$, $x_2 = \alpha$). b), d), f) systems' trajectories in separate coordinates. Black-dashed lines represent behavior of approximated system in original system's coordinates, red lines are original system trajectories. Control signals u_0, u_1 are presented in first and third subplots (Color figure online).

Table 1. Comparison of accuracy of homogeneous approximation. Here $\|X(0)\|$ denotes the norm of initial point, $\|X_h(t_f)\|$ – the norm of the final point (i.e. closeness of the solution to the goal), e_{t_f}, e_1 – global measures of quality of approximation defined by (18) and (19). Each trial represents the average of ten repetitions of the experiment

Model	Trial	$\|X(0)\|$	$\|X_h(t_f)\|$	e_{t_f}	e_1	$e_{t_f}/\|X_0\|$	$e_1/\|X_0\|$
art1	0	0.209	0.082	0.012	0.001	0.061	0.006
	1	0.410	0.095	0.052	0.013	0.128	0.031
	2	0.729	0.296	0.136	0.113	0.163	0.114
	3	1.225	0.134	0.272	0.550	0.187	0.353
art2	0	0.277	0.248	0.008	0.000	0.028	0.001
	1	0.517	0.568	0.036	0.008	0.057	0.011
	2	1.133	0.787	0.162	0.187	0.129	0.141
	3	1.649	1.314	0.634	1.074	0.373	0.608
car	0	0.301	0.172	0.027	0.019	0.088	0.068
	1	0.517	0.400	0.069	0.441	0.125	0.786
	2	1.271	0.411	0.104	28.000	0.094	17.138
	3	1.650	0.377	0.211	9.954	0.133	7.103

Numerical Details. Now we state some remarks about numerical details of the experiments. A proper numerical representation for control signal and state variables is an important issue. In our experiments we arbitrary choose representations based on Chebyshev polynomials. First, we assume that the control signal is a polynomial of degree n written in the Chebyshev basis of the first kind T_i, $i = 0, \ldots, n$

$$f_n(t) = \frac{1}{2}a_0 + \sum_{i=1}^{n} a_i T_i(t).$$

We do not use it directly, but we use the barycentric interpolation on the Chebyshev points of the second kind (or the Chebyshev extreme points) instead [3]. Interpolation on Chebyshev nodes is equivalent to the Chebyshev expansion. Both representations are widely used in practice, e.g. packages Chebfun [7] and PaCal [11] are based on barycentric interpolation, package numpy.polynomial.chebyshev provides common operations on Chebyshev expansion. The transformation between both representations (called Chebyshev transformation), i.e. transition from the interpolation using Chebyshev nodes to coefficients of expansion in the Chebyshev basis, can be effectively performed using Fast Fourier Transform [12, ch. 4.]. Representation using interpolation is convenient for most operations like algebraic operations of function evaluation. However, the cumulative integration is better to perform using Chebyshev expansion [12, ch. 2.], where we have a direct formula:

$$\int_0^t f_n(\tau)d\tau = \frac{1}{2}c_0 + \sum_{i=1}^{n} c_i T_i(t),$$

where constant c_0 is determined from the initial condition.

Recalling that the homogeneous system has a special feedforward form, we can easily solve it directly using a presented Chebyshev framework. Under the assumption that control is polynomial in Chebyshev expansion, we use only basic algebraic operations (addition, subtraction and multiplication), cumulative integrals, and feedforward substitutions to solve subsequent equations. Each step of procedure is closed with respect to Chebyshev representation used. Computations with this representation are fast and numerically very stable, accuracy is typically close to the machine precision.

To find the control, we use the optimization procedure with goal function:

$$g(u_{01}, \ldots, u_{1n}; u_{11}, \ldots, u_{1n}) = \|x_h(t_f) - 0\|_2 + \sum_{j=0}^{n} \alpha \|x_j\|_\infty$$

where u_{ij} are values of control signals $u_i, i = 0, 1$ in Chebyshev nodes $t_j, j = 0, 1, \ldots, n_c$ scaled to the interval $[0, t_f]$, t_f is a fixed time of controls, α is a (small) parameter. As the optimization procedure we use l-bfgs-b from scipy [14], with additional restrictions on maximal values of nodes of the control signals.

To compare the quality of the homogeneous approximation we apply the same control signals to the original system. To solve the original system we have to use a general ODE solver. In our case we use odeint function from scipy. For the homogeneous system direct integration using Chebyshev framework is more accurate and much faster. In the experiments we use interpolators with $n_c = 50$ nodes to represent control, in such case scipy's ODE solver was about 4 times slower. For higher degrees this difference will be even bigger.

We prepared experimental part using Python language with numpy [9] libraries mainly for arrays, scipy [19] for optimization and ODE solvers, and sympy [13] for symbolic computing – mainly differentiation and computing of Lie brackets.

5   Summary

In this paper we presented the procedure of determining of a homogeneous approximation from a computational point of view, and we provide the numerical experiments with some nonlinear control systems and their homogeneous approximations. After comparing the system trajectories, we briefly discussed the quality of such approximations. The experiments confirmed that the theoretical results concerning homogeneous approximations of nonlinear systems can be used in practice, and the need to construct suitable software libraries to be used in possible applications – e.g. in control design – is evident.

Disclosure of Interests. The authors have no competing interests to declare that are relevant to the content of this article.

References

1. Agrachev, A., Marigo, A.: Nonholonomic tangent spaces: intrinsic construction and rigid dimensions. Electron. Res. Announc. Amer. Math. Soc. **9**, 111–120 (2003)
2. Bellaïche, A.: The tangent space in sub-Riemannian geometry. Progr. Math. **144**, 4–78 (1996)
3. Berrut, J.P., Trefethen, L.N.: Barycentric Lagrange interpolation. SIAM Rev. **46**(3), 501–517 (2004)
4. Bianchini, R., Stefani, G.: Graded approximation and controllability along a trajectory. SIAM J. Control Optimiz. **28**, 903–924 (1990)
5. Bradbury, J., et al.: JAX: composable transformations of Python+NumPy programs (2018). http://github.com/google/jax
6. Crouch, P.E.: Solvable approximations to control systems. SIAM J. Control Optimiz. **22**, 40–54 (1984)
7. Driscoll, T.A., Hale, N., Trefethen, L.N.: Chebfun Guide. Pafnuty Publications (2014). http://www.chebfun.org/docs/guide/
8. Fliess, M.: Fonctionnelles causales non linéaires et indéterminées non commutatives. Bull. Soc. Math. France **109**, 3–40 (1981)
9. Harris, C.R., et al.: Array programming with NumPy. Nature **585**(7825), 357–362 (2020)
10. Hermes, H.: Nilpotent and high-order approximations of vector field systems. SIAM Rev. **33**, 238–264 (1991)
11. Jaroszewicz, S., Korzeń, M.: Arithmetic operations on independent random variables: a numerical approach. SIAM J. Sci. Comp. **34**(4), A1241–A1265 (2012)
12. Mason, J., Handscomb, D.: Chebyshev Polynomials. CRC Press (2002)
13. Meurer, A., et al.: Sympy: symbolic computing in Python. PeerJ Comput. Sci. **3**, e103 (2017)
14. Morales, J.L., Nocedal, J.: Remark on "algorithm 778: L-bfgs-b: Fortran subroutines for large-scale bound constrained optimization". ACM Trans. Math. Softw. **38**(1) (Dec 2011)
15. Sklyar, G., Barkhayev, P., Ignatovich, S., Rusakov, V.: Implementation of the algorithm for constructing homogeneous approximations of nonlinear control systems. Math. Control Signals Syst. **34**(4), 883–907 (2022)
16. Sklyar, G.M., Ignatovich, S.Y.: Approximation of time-optimal control problems via nonlinear power moment min-problems. SIAM J. Control Optimiz. **42**, 1325–1346 (2003)
17. Sklyar, G.M., Ignatovich, S.Y.: Description of all privileged coordinates in the homogeneous approximation problem for nonlinear control systems. C. R. Math. Acad. Sci. Paris **344**, 109–114 (2007)
18. Sklyar, G.M., Ignatovich, S.Y.: Free algebras and noncommutative power series in the analysis of nonlinear control systems: an application to approximation problems. Dissertationes Math. (Rozprawy Mat.) **2014**, 1–88 (2014)
19. Virtanen, P., et al.: SciPy 1.0: Fundamental algorithms for scientific computing in Python. Nature Methods **17**, 261–272 (2020). https://doi.org/10.1038/s41592-019-0686-2

An Asymptotic Parallel Linear Solver and Its Application to Direct Numerical Simulation for Compressible Turbulence

Mitsuo Yokokawa[1]([✉])[iD], Taiki Matsumoto[1], Ryo Takegami[1], Yukiya Sugiura[2], Naoki Watanabe[3], Yoshiki Sakurai[4][iD], Takashi Ishihara[5][iD], Kazuhiko Komatsu[6][iD], and Hiroaki Kobayashi[7][iD]

[1] Graduate School of System Informatics, Kobe University, Kobe, Japan
yokokawa@port.kobe-u.ac.jp
[2] Graduate School of Informatics, Kyoto University, Kyoto, Japan
[3] Mizuho Research & Technologies, Ltd., Chiyoda-ku, Tokyo, Japan
[4] Graduate School of Environment and Information Sciences, Yokohama National University, Yokohama, Japan
[5] Faculty of Environmental, Life, Natural Science and Technology, Okayama University, Okayama, Japan
[6] Cyberscience Center, Tohoku University, Sendai, Japan
[7] Graduate School of Information Sciences, Tohoku University, Sendai, Japan

Abstract. When solving numerically partial differential equations such as the Navier-Stokes equations, higher-order finite difference schemes are occasionally applied for spacial descretization. Compact finite difference schemes are one of the finite difference schemes and can be used to compute the first-order derivative values with smaller number of stencil grid points, however, a linear system of equations with a tridiagonal or pentadiagonal matrix derived from the schemes have to be solved. In this paper, an asymptotic parallel solver for a reduce matrix, that obtained from the Mattor's method in a computation of the first-order derivatives with an eighth-order compact difference scheme under a periodic boundary condition, is proposed. The asymptotic solver can be applied as long as the number of grid points of each Cartesian coordinate in the parallelized subdomain is 64 or more, and its computational cost is lower than that of the Mattor's method. A direct numerical simulation code has also been developed using the two solvers for compressible turbulent flows under isothermal conditions, and optimized on the vector supercomputer SX-Aurora TSUBASA. The optimized code is 1.7 times faster than the original one for a DNS with 2048^3 grid points and the asymptotic solver achieves approximately a 4-fold speedup compared to the Mattor's solver. The code exhibits excellent weak scalability.

Keywords: Asymptotic parallel linear solver · Eighth-order compact difference scheme · Direct numerical simulation · Finite difference method · Vector system · SX-Aurora TSUBASA

L. Franco et al. (Eds.): ICCS 2024, LNCS 14833, pp. 383–397, 2024.
https://doi.org/10.1007/978-3-031-63751-3_26

1   Introduction

Turbulence is a core physical phenomenon in various natural phenomena and problems in science and technology, and it is important to understand and elucidate its universal nature. However, the Navier-Stokes equations, which are the governing equations for fluid flows, are known to be highly nonlinear and difficult to solve analytically. Therefore, it is effective to study turbulent flows by numerical simulations using supercomputers. Among numerical simulations of turbulent flows, a direct numerical simulation (DNS) is widely used for clarifying the universal properties of turbulence because it solves the governing equations directly without modeling and resolves eddy motions at the smallest scale [2].

Large-scale DNSs for incompressible turbulence using a Fourier spectral method with up to 12288^3 grid points have been performed to study the statistical properties in the inertial subrange of turbulence at the Taylor-microscale Reynolds number up to 2300 [3]. For compressible turbulence, on the other hand, such large-scale DNSs have not yet been performed, and the statistical properties in the inertial subrange of compressible turbulence at high Reynolds numbers have not yet been fully studied. Recently, to the best of our knowledge, the largest DNSs for compressible isothermal turbulence using finite difference methods up to 4096^3 grid points have been performed [8]. However, the Reynolds numbers achieved by these DNSs are not high enough to study inertial subrange properties. To obtain high-resolution results in DNSs for compressible turbulence using finite difference methods requires a large number of grid points and a large amount of computational resources [8]. Compact finite difference schemes are often used to discretize the convection terms in the Navier-Stokes equations. Though a high resolution solutions can be obtained compared to usual finite difference schemes, linear systems have to be solved. Therefore, it is necessary to develop a fast solver of the systems and a parallel DNS code using the solvers.

In parallel computations, a computation domain is divided into several subdomains, calculations of which are assigned to parallel tasks and executed. In this case, the number of parallel tasks and the method of dividing the computational domain, which can change the amount of computation and communication in each task, must be chosen appropriately to obtain results fast. In general, when a sufficiently large number of grid points is taken so as to calculate the flow fields precisely, a three-dimensional domain decomposition (cuboid decomposition) can increase the amount of parallelism compared to a two-dimensional domain decomposition (pencil decomposition), and large-scale computation with massively parallel supercomputers is expected.

In this paper, we propose an asymptotic parallel linear solver for the linear system of equations obtained when first-order derivatives in space is discretized by an eighth-order compact difference method. The solver is then used to develop a finite difference DNS code of three-dimensional homogeneous isotropic compressible isothermal turbulence in a box with a periodic boundary condition. The code is parallelized by the cuboid decomposition to increase the number of parallel tasks. The code is optimized and its performance is evaluated on the supercomputer SX-Aurora TSUBASA installed at Cyberscience Center, Tohoku University.

The remainder of this paper is organized as follows. Section 2 outlines a direct numerical simulation code for compressible turbulent flows under isothermal conditions. Section 3 describes parallel solutions of the linear systems of equations obtained by applying an eighth-order compact scheme to the first-order derivative calculations. Section 4 gives optimization and performance evaluation results of the code on SX-Aurora TSUBASA. Section 5 concludes the paper.

2   Direct Numerical Simulation Code

We consider the three-dimensional compressible turbulent flows with isothermal conditions in a cube with a side length of 2π subject to periodic boundary conditions that obey the following equations;

$$\frac{\partial \rho}{\partial t} + \nabla \cdot (\rho \boldsymbol{u}) = 0, \tag{1}$$

$$\frac{\partial (\rho \boldsymbol{u})}{\partial t} + \nabla \cdot (\rho \boldsymbol{u}\boldsymbol{u} + p\mathbf{I}) = \nabla \cdot \boldsymbol{\tau} + \boldsymbol{F}, \tag{2}$$

$$p = \rho c^2, \tag{3}$$

where t is the time, ρ the density, $\boldsymbol{u}$ the velocity, p the pressure, $\boldsymbol{\tau}$ the viscous stress tensor, and c the speed of sound. $\boldsymbol{F}$ is the forcing terms to maintain the flows in statistically quasi-steady states. $\mathbf{I}$ is the identity matrix.

Here, Eqs. (1) and (2) are discretized at gird points that equally divide the computational region (the cube), those are $(x_i, y_j, z_k) = (i\Delta, j\Delta, k\Delta)$ $(0 \leq i,j,k \leq N)$ and $\Delta = 2\pi/N$, where N is the number of division in each Cartesian coordinate. Then, N^3 ordinary differential equations with respect to $\rho_{i,j,k}$ and $(\rho \boldsymbol{u})_{i,j,k}$ are obtained, where $\rho_{i,j,k} = \rho(x_i, y_j, z_k)$ and $(\rho \boldsymbol{u})_{i,j,k} = \rho(x_i, y_j, z_k)\boldsymbol{u}(x_i, y_j, z_k)$. In disetizing the equations in space, an eighth-order compact difference scheme (CD8) [5] for the convection terms and an eighth-order central finite difference scheme (FD8) for the viscous terms are used. This is because the amount of contribution from the viscous terms is small for high Reynolds number flows, and a higher precision scheme is not necessary for the viscous terms. Describing a physical value $f(x, y, z)$ and the first-order partial derivative in the x-direction $\frac{\partial f(x,y,z)}{\partial x}$ at the point (x_i, y_j, z_k) by $f_{i,j,k}$ and $f'_{i,j,k}$, respectively, the CD8 and FD8 schemes are written as follows,

$$\alpha f'_{i-1,j,k} + f'_{i,j,k} + \alpha f'_{i+1,j,k} =$$
$$a\frac{f_{i+1,j,k} - f_{i-1,j,k}}{2\Delta} + b\frac{f_{i+2,j,k} - f_{i-2,j,k}}{4\Delta} + c\frac{f_{i+3,j,k} - f_{i-3,j,k}}{6\Delta},$$
$$\alpha = \frac{3}{8}, \quad a = \frac{25}{16}, \quad b = \frac{1}{5}, \text{ and } c = -\frac{1}{80}, \tag{4}$$

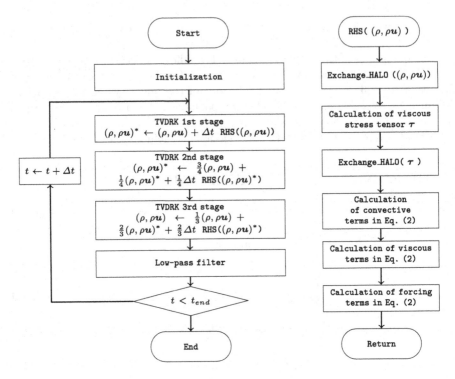

Fig. 1. DNS Code Flow

and

$$f'_{i,j,k} = a\frac{f_{i+1,j,k} - f_{i-1,j,k}}{\Delta} + b\frac{f_{i+2,j,k} - f_{i-2,j,k}}{\Delta}$$
$$+ c\frac{f_{i+3,j,k} - f_{i-3,j,k}}{\Delta} + d\frac{f_{i+4,j,k} - f_{i-4,j,k}}{\Delta},$$
$$a = \frac{4}{5}, \quad b = -\frac{1}{5}, \quad c = \frac{4}{105}, \text{ and } d = -\frac{1}{280}. \quad (5)$$

The first-order partial derivative in each y- and z- direction is also written by the similar formula. To maintain a statistically quasi-steady state of turbulence, the same forcing terms as Petersen and Livescu [7] are incorporated. That is, the Fourier coefficients of the external force F is not zero only at low wavenumbers $|k| = \sqrt{k_1^2 + k_2^2 + k_3^2} < 3$, where k_1, k_2, and k_3 are wavenumbers corresponding to each direction of the Cartesian coordinates. Since the number of non-zero Fourier coefficients is at most 100 out of N^3, not fast Fourier transforms but discrete Fourier transforms (DFT) are used to calculate them to decrease computational costs.

The third-order total variation diminishing Runge-Kutta (TVDRK) scheme [1] with a constant time step is used for a temporal integration of the equations of $\rho_{i,j,k}$ and $(\rho u)_{i,j,k}$ over time. Additionally, an eighth-order low-pass filter scheme is applied to remove unphysical numerical oscillations at high wavenumbers. The code flow is illustrated in Fig. 1.

A uniform grid of dimensions $N \times N \times N$ discretizing the cube is partitioned into some sets of grid points using a standard cuboid decomposition and those are assigned to parallel tasks with layout $n_{px} \times n_{py} \times n_{pz}$. The number of grid points in a subregion assigned to a task is $(N/n_{px}) \times (N/n_{py}) \times (N/n_{pz})$. Halo regions are added around each subregion to perform the CD8 and FD8 calculations within a task. The number of grids of halo regions that are extended outward in the direction of each coordinate of the calculation region is 4 so that the computation of FD8 is possible. Physical values at the grid points in the halo regions between adjacent tasks are exchanged in the procedure Exchange_HALO in the code flow, and then the CD8 and FD8 values can be calculated using Eqs. (4) and (5) independent of other tasks. To calculate the viscous terms in Eq. (2), each element of the viscous stress tensor τ is first calculated using the FD8, and then the derivative of each element is computed by the FD8 again.

3   Calculation of the System Obtained by the Eighth-Order Compact Scheme

3.1   Linear System Obtained by the Eighth-Order Compact Scheme

We consider here a solution for $N \times N$ linear systems of equations to calculate the first partial derivatives in the x-direction written as a matrix-vector form of Eq. (4) below:

$$
\begin{bmatrix}
1 & \alpha & & \cdots & & \alpha \\
\alpha & 1 & \alpha & & & \\
 & \alpha & 1 & \alpha & & \vdots \\
 & & \ddots & \ddots & \ddots & \\
\vdots & & & \alpha & 1 & \alpha \\
\alpha & \cdots & \cdots & & \alpha & 1
\end{bmatrix}
\begin{bmatrix}
f'_{1,j,k} \\
f'_{2,j,k} \\
\vdots \\
\vdots \\
f'_{N-1,j,k} \\
f'_{N,j,k}
\end{bmatrix}
=
\begin{bmatrix}
\Delta_x f_{1,j,k} \\
\Delta_x f_{2,j,k} \\
\vdots \\
\vdots \\
\Delta_x f_{N-1,j,k} \\
\Delta_x f_{N,j,k}
\end{bmatrix}
\quad (1 \le j, k \le N), \quad (6)
$$

where

$$
\Delta_x f_{i,j,k} = a\frac{f_{i+1,j,k} - f_{i-1,j,k}}{2\Delta} + b\frac{f_{i+2,j,k} - f_{i-2,j,k}}{4\Delta} + c\frac{f_{i+3,j,k} - f_{i-3,j,k}}{6\Delta}
$$

The matrix is a slightly perturbed form of the tridiagonal matrix with α added to the upper-right and lower-left elements due to the periodic boundary conditions. Similar linear systems for each y- and z-direction are obtained.

3.2   Parallel Solution by the Mattor Method

To simplify the computational procedure, we denote the linear system as $Ax = b$ as follows,

$$
A = \begin{bmatrix} 1 & \alpha & \cdots & & & \alpha \\ \alpha & 1 & \alpha & & & \vdots \\ & \alpha & 1 & \alpha & & \vdots \\ & & \ddots & \ddots & \ddots & \\ \vdots & & & \alpha & 1 & \alpha \\ \alpha & \cdots & \cdots & & \alpha & 1 \end{bmatrix} \in R^{N \times N}, x = \begin{bmatrix} x_1 \\ x_2 \\ \vdots \\ \vdots \\ x_{N-1} \\ x_N \end{bmatrix} \in R^N, b = \begin{bmatrix} b_1 \\ b_2 \\ \vdots \\ \vdots \\ b_{N-1} \\ b_N \end{bmatrix} \in R^N.
$$

$$(7)$$

Consider the system $Ax = b$ is solved by P parallel tasks, where $N = MP$ and M is the number of grid points to be solved in a task. The parallel solver of a tridiagonal matrix system proposed by Mattor et al. [6] is applied to the linear system (7). The matrix is divided into $P \times P$ blocks and the vectors are divided into P subvectors as

$$
\begin{bmatrix} A_M & \alpha e_M e_1^T & & \cdots & & \alpha e_1 e_M^T \\ \alpha e_1 e_M^T & A_M & \alpha e_M e_1^T & & & \vdots \\ & \alpha e_1 e_M^T & A_M & \alpha e_M e_1^T & & \vdots \\ & & \ddots & \ddots & \ddots & \\ & & & \alpha e_1 e_M^T & A_M & \alpha e_M e_1^T \\ \alpha e_M e_1^T & \cdots & & \cdots & \alpha e_1 e_M^T & A_M \end{bmatrix} \begin{bmatrix} x_1 \\ x_2 \\ \vdots \\ \vdots \\ \vdots \\ x_{P-1} \\ x_P \end{bmatrix} = \begin{bmatrix} b_1 \\ b_2 \\ \vdots \\ \vdots \\ \vdots \\ b_{P-1} \\ b_P \end{bmatrix}, \quad (8)
$$

where

$$
A_M = \begin{bmatrix} 1 & \alpha & & & & \\ \alpha & 1 & \alpha & & \mathbf{0} & \\ & \alpha & 1 & \alpha & & \\ & & \ddots & \ddots & \ddots & \\ & \mathbf{0} & & \alpha & 1 & \alpha \\ & & & & \alpha & 1 \end{bmatrix} \in R^{M \times M}, e_1 = \begin{bmatrix} 1 \\ 0 \\ \vdots \\ \vdots \\ 0 \end{bmatrix} \in R^M, e_M = \begin{bmatrix} 0 \\ \vdots \\ \vdots \\ 0 \\ 1 \end{bmatrix} \in R^M, \quad (9)
$$

$$
x_p \text{ and } b_p \in R^M \quad (p = 1, \cdots, P).
$$

x^T stands for a transpose vector of the vector x. The p-th block system is written as

$$
\alpha e_1 e_M^T x_{p-1} + A_M x_p + \alpha e_M e_1^T x_{p+1} = b_p. \quad (10)
$$

Following the Mattor method, $\boldsymbol{x}_p$ is written by three vectors $\boldsymbol{x}_p^R$, $\boldsymbol{x}_p^U$ and $\boldsymbol{x}_p^L$, and two scalars s_p and t_p as

$$\boldsymbol{x}_p = \boldsymbol{x}_p^R - t_p\boldsymbol{x}_p^U - s_p\boldsymbol{x}_p^L \quad (p = 1, \cdots, P), \tag{11}$$

$$A_M\boldsymbol{x}_p^R = \boldsymbol{b}_p, \quad A_M\boldsymbol{x}_p^U = \begin{bmatrix} 0 \\ \vdots \\ 0 \\ \alpha \end{bmatrix}, \quad \text{and } A_M\boldsymbol{x}_p^L = \begin{bmatrix} \alpha \\ 0 \\ \vdots \\ 0 \end{bmatrix}. \tag{12}$$

Note that $\boldsymbol{x}_p^U = F\boldsymbol{x}_p^L$ with a flip matrix F, i.e. $x_{p,j}^U = x_{p,M-j+1}^L (j = 1, \cdots, M)$, where $x_{p,j}$ stands for the j-th element of vector $\boldsymbol{x}_p$. By substituting $\boldsymbol{x}_p$ in Eq. (10) by Eq. (11) and considering Eq. (12), $2P$ equations below are obtained.

$$(x_{p-1,M}^R - t_{p-1}x_{p-1,M}^U - s_{p-1}x_{p-1,M}^L) - s_p = 0, \tag{13}$$
$$(x_{p+1,1}^R - t_{p+1}x_{p+1,1}^U - s_{p+1}x_{p+1,1}^L) - t_p = 0 \quad (p = 1, \cdots P),$$

where $x_{0,M}^R = x_{P,M}^R$, $x_{P+1,M}^R = x_{1,M}^R$, and so forth in the first and the P-th block system. Since $x_{p,j}^U = x_{p,M-j+1}^L$, we obtain a following reduced system (14) whose matrix size is $2P \times 2P$, and solve this system to get the coefficients s_p and t_p $(p = 1, \cdots, P)$.

$$\begin{bmatrix} x_{1,M}^U & x_{1,1}^U & & & & & 1 \\ x_{1,1}^U & x_{1,M}^U & 1 & & & & \vdots \\ & 1 & x_{2,M}^U & x_{2,1}^U & & & \\ & & x_{2,1}^U & x_{2,M}^U & & & \\ & & & & \ddots & \ddots & \ddots & \\ \vdots & & & & \ddots & \ddots & \ddots & \\ & & & & 1 & x_{P,M}^U & x_{P,1}^U \\ 1 & & \cdots & & & x_{P,1}^U & x_{P,M}^U \end{bmatrix} \begin{bmatrix} s_1 \\ t_1 \\ s_2 \\ t_2 \\ \vdots \\ \vdots \\ s_P \\ t_P \end{bmatrix} = \begin{bmatrix} x_{1,1}^R \\ x_{1,M}^R \\ x_{2,1}^R \\ x_{2,M}^R \\ \vdots \\ \vdots \\ x_{P,1}^R \\ x_{P,M}^R \end{bmatrix}. \tag{14}$$

The final solution is computed by Eq. (11). Note that the reduced matrix is organized by the first and M-th elements of the vector $\boldsymbol{x}_p^U$ $(p = 1, \cdots, P)$ and the reduced matrix can be assembled once at the initialization phase of the code by solving the second linear system in Eq. (12). Hereafter, the procedure is referred to as a normal solver and is shown in **Algorithm 1**.

3.3 Asymptotic Property of the Reduced Matrix

Here, let us derive the asymptotic property of the system $A_M\boldsymbol{x}_p^U = \alpha e_M^T$, i.e.

Algorithm 1. (Normal Solver)

Initialization

1: Compute the Cholesky decomposition of the matrix A_M in all tasks.
2: Solve the equation $A_M x_p^U = [0, \cdots, 0, \alpha]^T$ and assemble the reduced matrix in all tasks concurrently.

In time advancement loop

1: Solve the equation $A_M x_p^R = b_p$ in each task after computing the vector b_p.
2: Assemble the right hand side of the reduced system (14) by using **MPI_Allgather** function in all tasks.
3: Solve the system in each task in parallel to compute the coefficients s_p and t_p.
4: Compute the final solution by Eq. (11) in each task.

$$
\begin{bmatrix}
1 & \alpha & & & & & \\
\alpha & 1 & \alpha & & & \mathbf{0} & \\
& \alpha & 1 & \alpha & & & \\
& & \ddots & \ddots & \ddots & & \\
\mathbf{0} & & & \alpha & 1 & \alpha \\
& & & & \alpha & 1
\end{bmatrix}
\begin{bmatrix}
x_{p,1}^U \\
x_{p,2}^U \\
x_{p,3}^U \\
\vdots \\
\vdots \\
x_{p,M}^U
\end{bmatrix}
=
\begin{bmatrix}
0 \\
0 \\
\vdots \\
\vdots \\
0 \\
\alpha
\end{bmatrix}.
\tag{15}
$$

Applying the Gaussian elimination to the system, we can rewrite the system with a coefficient matrix having diagonal and super-diagonal elements, that is

$$
\begin{bmatrix}
\beta_1 & \alpha & & & & \\
& \beta_2 & \alpha & & \mathbf{0} & \\
& & \beta_3 & \alpha & & \\
& & & \ddots & \ddots & \\
\mathbf{0} & & & & \beta_{M-1} & \alpha \\
& & & & & \beta_M
\end{bmatrix}
\begin{bmatrix}
x_{p,1}^U \\
x_{p,2}^U \\
x_{p,3}^U \\
\vdots \\
\vdots \\
x_{p,M}^U
\end{bmatrix}
=
\begin{bmatrix}
0 \\
0 \\
\vdots \\
\vdots \\
0 \\
\alpha
\end{bmatrix},
\tag{16}
$$

where $\beta_1 = 1$, $\beta_j = 1 - \frac{\alpha^2}{\beta_{j-1}}$ $(j = 2, \cdots, M)$. The system can be easily solved by backward substitution and the solution is

$$
x_{p,M}^U = \frac{\alpha}{\beta_M},
$$

$$
x_{p,j}^U = -\frac{\alpha \cdot x_{p,j+1}^U}{\beta_j} = \frac{(-\alpha)^{M-j+1}}{\prod_{k=j}^M \beta_k} \quad (j = M-1, \cdots, 1).
\tag{17}
$$

Since $\lim_{M \to \infty} \beta_M = \frac{1+\sqrt{1-4\alpha^2}}{2}$, $x_{p,M}^U = \frac{1-\sqrt{1-4\alpha^2}}{2\alpha}$ and $x_{p,1}^U = 0$ $(p = 1, \cdots, P)$ as M reaches infinity. Since $\alpha = 3/8$, $x_{p,M}^U$ and $x_{p,1}^U$ converge rapidly as shown in Table 1.

Table 1. Valuses of $x_{p,M}^U$ and $x_{p,1}^U$ when increasing M

M	$x_{p,M}^U$	$x_{p,1}^U$
8	0.451415160966647	$-1.372943122571951 \times 10^{-3}$
16	0.451416229641959	$-2.367389055633383 \times 10^{-6}$
32	0.451416229645136	$-7.038892637296893 \times 10^{-12}$
64	0.451416229645136	$-6.222626756534802 \times 10^{-23}$
128	0.451416229645136	$-4.863092990426517 \times 10^{-45}$

If the number of grid points M in the x-direction of the subregion is taken appropriately, at least $M \geq 64$, the reduced system (14) can be approximately expressed with the asymptotic values as follows;

$$\begin{bmatrix} d & 0 & & & \cdots & & 1 \\ 0 & d & 1 & & & & \\ & 1 & d & 0 & & & \\ & & 0 & d & & & \\ \vdots & & & & \ddots & \ddots & \ddots \\ & & & & 1 & d & 0 \\ 1 & \cdots \cdots & & & & 0 & d \end{bmatrix} \begin{bmatrix} s_1 \\ t_1 \\ s_2 \\ t_2 \\ \vdots \\ s_P \\ t_P \end{bmatrix} = \begin{bmatrix} x_{1,1}^R \\ x_{1,M}^R \\ x_{2,1}^R \\ x_{2,M}^R \\ \vdots \\ x_{P,1}^R \\ x_{P,M}^R \end{bmatrix}, \tag{18}$$

where $d = x_{p,M}^U = \frac{1-\sqrt{1-4\alpha^2}}{2\alpha}$.

The coefficients s_p and t_p $(p = 1, \cdots, P)$ can be computed in each task independently of the others, using Eqs. (19) by transferring $x_{p+1,1}^R$ and $x_{p-1,M}^R$ from adjacent tasks. The final solution is computed by Eq. (11). Hereafter, the procedure is referred to as an asymptotic solver and is shown in **Algorithm 2**.

$$\begin{cases} t_p & = (d \cdot x_{p,M}^R - x_{p+1,1}^R)/(d^2 - 1) \\ s_p & = (x_{p-1,M}^R - d \cdot x_{p,1}^R)/(d^2 - 1) \end{cases} \tag{19}$$

Algorithm 2. (Asymptotic Solver)

Initialization

1: Compute the Cholesky decomposition of the matrix A_M in all tasks.
2: Solve the equation $A_M x_p^U = [0, \cdots, 0, \alpha]^T$ in all tasks.

In time advancement loop

1: Solve the equation $A_M x_p^R = b_p$ in each task after computing the vector b_p.
2: Send the values $x_{p,1}^R$ and $x_{p,M}^R$ to the $(p-1)$ and $(p+1)$ task, respectively, by using **MPI_Sendrecv** function.
3: Compute the coefficients t_p and s_p by the equations (19).
4: Compute the final solution by Eq. (11) in each task.

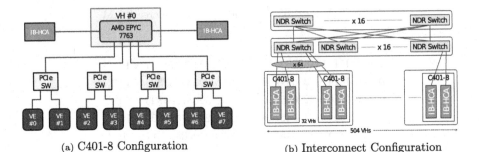

(a) C401-8 Configuration   (b) Interconnect Configuration

Fig. 2. AOBA-S Configuration

4   Performance Evaluation of the Code on SX-Aurora TSUBASA

We have first built a parallel DNS code for compressible isothermal turbulence by the cuboid decomposition, as shown in Fig. 1, without considering optimization. Both the normal and asymptotic solvers are implemented in the code. When the first derivatives in a direction of the Cartesian coordinates are computed, the normal solver is used if the number of grid points in its direction is less than 64. Otherwise, the asymptotic solver is chosen. We confirmed that the computational results by the two solvers is the same as the results by the code used in [8].

In this section, an optimization of the first-created code on SX-Aurora TSUB-ASA is stated. Performance comparison of the two solvers and weak scaling characteristics of the code are also described.

4.1   Measurement Environment

A vector-type supercomputer AOBA-S installed at Cyberscience Center, Tohoku University is used for optimization and performance evaluation of the code [4,9].

AOBA-S consists of 504 compute nodes that are connected by a two-layer fat tree non-blocking network configured by the 32 Infiniband NDR switches. Each node, SX-Aurora TSUBASA C401-8, is configured by a vector host (VH) and eight vector engines (VEs) that are connected by four Peripheral Component Interconnect express (PCIe) Gen4 switches (Fig. 2a). The VH is a standard x86 server with an AMD EPCY 7736 processor running a standard Linux operating system, and has two InfiniBand NDR200 Host Channel Adapters connecting to the fat tree network (Fig. 2b). A VE is a vector accelerator implemented as a PCIe card, on which a vector engine Type 30A (VE30) and six extended high-bandwidth memory (HBM2E) modules, 96GB in total, are mounted. The VE30 processor integrates 16 vector cores, a 64MB shared last level cache (LLC), and a VE direct memory access (DMA) engine that are interconnected through a two-dimensional network (called Network on Chip) with a total bandwidth of 3.0 TB/s. Peak performance of the core is 307.2 Gflop/s in double precision floating-point operations, and therefore 4.92 Tflop/s in total for a VE30.

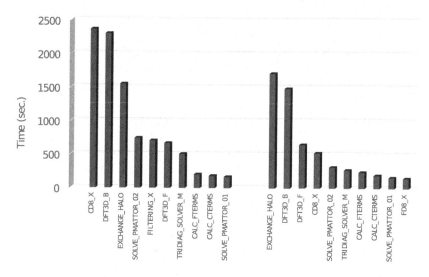

Fig. 3. Top 10 procedures of the code analyzed by Ftrace

Operating system functions have been run on the VH to the greatest extent possible so that the VEs are able to effectively use their computational power by concentrating on program executions with vector instructions. The GNU C library (*glibc*) is ported to the VEs and applications can call *glibc* functions as normal I/O functions [10]. Vectorization should also be applied to achieve high computation performance.

4.2   Cost Distribution and Optimization

Computational cost distribution of the code is analyzed by using a performance analyzer **Ftrace** equipped on the SX-Aurora TSUBASA. The left bar graph in Fig. 3 shows the first 10 high-cost parts of the code, in the case that a DNS with the grid points $N^3 = 2048^3$ was performed by 512 MPI processes with an $8 \times 8 \times 8$ layout on 32 VEs (512 cores). Since the number of grid points in each direction assigned to a process is 256, the asymptotic parallel solver is used.

The first part CD8_X is the first-order derivative calculation in the x-direction. In the right-hand-side calculation of Eq. (6), since the coefficient matrix is the same for all gird points (j,k) of the y-z cross-section of an assigned subregion to an MPI process, all right-hand-side vectors corresponding to the grid points (j,k) are computed as follows:

```
for all grid points (j,k) of the y-z cross-section
  for i = is, ie
    b(jk,i) = a1*( f(i+1,j,k) - f(i-1,j,k) )
            + b1*( f(i+2,j,k) - f(i-2,j,k) )
            + c1*( f(i+3,j,k) - f(i-3,j,k) )
```

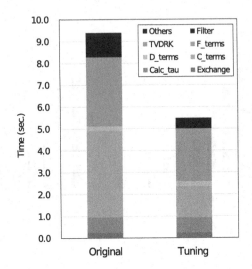

Fig. 4. Comparison of Calculation time between original and tuning codes

where a1, b1, and c1 correspond to the coefficients in Eq. (4) and is and ie are the start and end grid numbers of the x-direction in each subregion. It is clear that the right-hand-side array "f" accesses memory addresses at equal intervals.

The second part DFT3D_B is the backward 3D-DFT used to calculate the Fourier coefficients of the low wavenumbers side applying external force. The third part EXCHANGE_HALO is a halo region exchange procedure by the MPI_Sendrecv function. The fourth part SOLVE_PMATTOR_02 is the asymptotic solver. The fifth part FILTERING_X is the low-pass filter procedure and has almost the similar memory access pattern to the CD8_X.

The loop order of indexes is changed to obtain the highest performance on the SX-Aurora TSUBASA, and it is found that the order of k, i, and j from the outer loop is the best. The right bar chart in Fig. 3 is the top 10 cost-consuming parts after the loop order exchanges, in that the CD8_X is in the fourth and 4.5 times faster than the original.

Figure 4 shows computational time for both the original and tuning codes. We confirmed that the results of the two DNS codes are identical. The times of the parts C_terms and Filter of the tuning code, in which the CD8_X or similar is used, are found to be shorten compared to the original one, resulting in 1.7 times faster.

Table 2. Computation time for a sigle solver call (in seconds)

Solvers	Calculation	Communication	Synchronization
Normal solver	0.540	0.419	0.041
Asymptotic solver	0.181	0.040	0.031

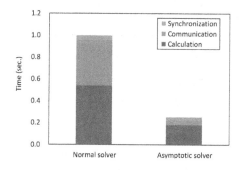 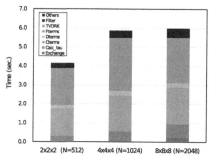

Fig. 5. Comparison of parallel solvers

Fig. 6. Weak scaling characteristics of the code

4.3 Comparison of the Two Parallel Solvers

Here, the two parallel solvers in Algorithms 1 and 2 are compared for the DNS with the grid points $N^3 = 2048^3$. The calculation with the optimized DNS code is performed on 512 MPI processes with an $8 \times 8 \times 8$ layout on 32 VEs (512 cores). Since the number of grid points in any direction in a process is 256, the asymptotic solver is usually selected for the calculation of the first-order derivatives. The normal solver, however, is used for comparison.

Table 2 and Fig. 5 show the computation times of both solvers for a single solver call. These solvers are called multiple times in one time step. Calculation and communication times are greatly decreased in the asymptotic solver, and approximately the 4-fold performance of the asymptotic solver is achieved. The normal solver uses the MPI function MPI_Allgather to construct the right-hand-side in the system (14) in order to calculate the coefficients s_p and t_p, and has to solve the same linear system constructed in duplicate in all MPI processes simultaneously. On the other hand, a function MPI_Sendrecv is sufficient to get the values $x^R_{p+1,1}$ and $x^R_{p-1,M}$ from the adjacent MPI processes and the values are calculate by Eq. (19) in all processes concurrently in Algorithm 2.

In addition, the number of compute nodes used increases to solve problems with a large number of grid points. The communication paths of the MPI_Allgather function among all compute nodes are more complicated than the MPI_Sendrecv neighbor communications, and a larger communication time is required. Therefore, the asymptotic solver can significantly reduce computation time not only for large time-steps simulations, but also for simulations with a larger number of grid points.

4.4 Weak Scaling Characteristics

Here, the same number of grid points is assigned to all MPI processes to check for weak scaling characteristics. Three cases of DNSs with

(1) the number of grid points 512^3 partitioning into 8 MPI processes with a $2 \times 2 \times 2$ layout on a VE,
(2) the number of gird point 1024^3 partitioning into 64 MPI processes with a $4 \times 4 \times 4$ layout on 4 VEs, and
(3) the number of gird point 2048^3 partitioning into 512 MPI processes with a $8 \times 8 \times 8$ layout on 32 VEs

are run on SX-Aurora TSUBASA. The number of grid points computed in one MPI process is 256^3 in all cases. Computation time for the three cases are shown in Fig. 6.

Computation times for Cases 2 and 3 are longer than that for Case 1. This is because Case 1 can be carried out within a VE and therefore MPI communications in the x direction is processed within the VE via interconnection between cores, whereas data by MPI communications for Cases 2 and 3 are transferred via PCIe switches in the same node and among four nodes, respectively. The communication time is slightly increased for Cases 2 and 3. It is found that good weak scalability of the code is achieved with several VEs.

5 Conclusions

Turbulence plays an important role in many flow-related phenomena that occur in various fields of science and technology. It is important to understand and elucidate their universal nature. The DNS of turbulent flows is a widely used method for clarifying the properties of turbulence because it solves the governing equations directly without modeling and resolves eddy motion on the smallest scales.

In this study, at first, the asymptotic parallel solver for the reduced linear system obtained from the parallelization of the linear system for the compact finite difference scheme is proposed. A parallel DNS code incorporating the solver as well the Mattor's solver has been developed for isothermal compressible turbulent flows in a cube using a finite difference method. The convective terms of the governing equations are calculated using an eighth-order compact difference scheme, while the viscous term is calculated using an eighth-order central difference scheme.

The computational cost of the asymptotic solver is lower than that of the normal solver in the condition that the number of grid points along a Cartesian coordinate in a task is greater than or equal to 64. In addition, the code was been optimized on the supercomputer SX-Aurora TSUBASA. As a result, a 1.7-fold speedup and excellent weak scalability were performed. A further code tuning for scalar processors like the supercomputer Fugaku is underway. Larger DNSs with over 8192^3 grid points are possible to compute.

Acknowledgements. This work was partially conducted and funded at Joint-Research Division of High-Performance Computing (NEC) of Cyberscience Center at Tohoku University. This study was partially supported by JSPS KAKENHI Grant number JP23K11124 and the HPCI System Research project (Project ID: hp230143).

References

1. Gottlieb, S., Shu, C.W.: Total variation diminishing Runge-Kutta schemes. Math. Comput. **67**, 73–85 (1998). https://doi.org/10.1090/S0025-5718-98-00913-2
2. Ishihara, T., Gotoh, T., Kaneda, Y.: Study of high-Reynolds number isotropic turbulence by direct numerical simulation. Annu. Rev. of Fluid Mech. **41**(1), 165–180 (2009). https://doi.org/10.1146/annurev.fluid.010908.165203
3. Ishihara, T., Morishita, K., Yokokawa, M., Uno, A., Kaneda, Y.: Energy spectrum in high-resolution direct numerical simulations of turbulence. Phys. Rev. Fluids **1**, 082403 (2016). https://doi.org/10.1103/PhysRevFluids.1.082403
4. Komatsu, K., et al.: Performance evaluation of a vector supercomputer SX-Aurora TSUBASA. In: SC'18: International Conference for High Performance Computing, Networking, Storage and Analysis, pp. 685–696. Dallas, TX, U.S.A. (2018). https://doi.org/10.1109/SC.2018.00057
5. Lele, S.K.: Compact finite difference schemes with spectral-like resolution. J. Comput. Phys. **103**(1), 16–42 (1992). https://doi.org/10.1016/0021-9991(92)90324-R
6. Mattor, N., Williams, T.J., Hewett, D.W.: Algorithm for solving tridiagonal matrix problems in parallel. Parallel Comput. **21**(11), 1769–1782 (1995). https://doi.org/10.1016/0167-8191(95)00033-0
7. Petersen, M.R., Livescu, D.: Forcing for statistically stationary compressible isotropic turbulence. Phys. Fluids **22**(11), 116101 (2010). https://doi.org/10.1063/1.3488793
8. Sakurai, Y., Ishihara, T.: Direct numerical simulations of compressible isothermal turbulence in a periodic box: Reynolds number and resolution-level dependence. Phys. Rev. Fluids **8**, 084606 (2023). https://doi.org/10.1103/PhysRevFluids.8.084606
9. Takahashi, K., et al.: Performance evaluation of a next-generation SX-Aurora TSUBASA vector supercomputer. In: Bhatele, A., Hammond, J., Baboulin, M., Kruse, C. (eds.) High Performance Computing - Proc. 38th Int. Conf. High Performance 2023, pp. 359–378. Springer Science and Business Media Deutschland GmbH, Germany (2023). https://doi.org/10.1007/978-3-031-32041-5_19
10. Yokokawa, M., et al.: I/O performance of the SX-Aurora TSUBASA. In: 2020 IEEE International Parallel and Distributed Processing Symposium Workshops (IPDPSW), pp. 27–35 (2020). https://doi.org/10.1109/IPDPSW50202.2020.00014

Author Index

© The Editor(s) (if applicable) and The Author(s), under exclusive license
to Springer Nature Switzerland AG 2024
L. Franco et al. (Eds.): ICCS 2024, LNCS 14833, pp. 399–400, 2024.
https://doi.org/10.1007/978-3-031-63751-3

Printed in the United States
by Baker & Taylor Publisher Services